AF703404

Advances in Asphalt Materials and Technological Applications

Advances in Asphalt Materials and Technological Applications

Edited by
Philip Webster

Willford Press
www.willfordpress.com

Published by Willford Press,
118-35 Queens Blvd., Suite 400,
Forest Hills, NY 11375, USA

Copyright © 2023 Willford Press

This book contains information obtained from authentic and highly regarded sources. Copyright for all individual chapters remain with the respective authors as indicated. All chapters are published with permission under the Creative Commons Attribution License or equivalent. A wide variety of references are listed. Permission and sources are indicated; for detailed attributions, please refer to the permissions page and list of contributors. Reasonable efforts have been made to publish reliable data and information, but the authors, editors and publisher cannot assume any responsibility for the validity of all materials or the consequences of their use.

Trademark Notice: Registered trademark of products or corporate names are used only for explanation and identification without intent to infringe.

ISBN: 978-1-64728-508-1

Cataloging-in-Publication Data

Advances in asphalt materials and technological applications / edited by Philip Webster.
p. cm.
Includes bibliographical references and index.
ISBN 978-1-64728-508-1
1. Asphalt. 2. Bituminous materials. 3. Materials science. 4. Asphalt--Technological innovations. I. Webster, Philip.

TN853 .A76 2023
553.27--dc23

For information on all Willford Press publications
visit our website at www.willfordpress.com

Willford Press

Contents

Preface

Asphalt refers to a semi solid or viscous liquid which is primarily used for road construction. It comprises hydrogen and carbon compounds along with minor proportion of sulfur, oxygen and nitrogen. It is mixed with aggregate particles, such as stone, gravel and sand, for making asphalt concrete. Due to its high viscosity, asphalt needs to be heated before mixing with the aggregates. New technologies that could help in producing a consistently high-quality pavement are thermal imaging and intelligent compaction. Thermal imaging helps in measuring the temperature of the mat behind the paver. These new technologies may provide information which can be utilized for improving the consistency of compaction and paving operations. This book elucidates the prospective developments with respect to asphalt materials and paving technologies. It covers in detail some innovative concepts revolving around this domain. This book will serve as a reference to a broad spectrum of readers.

After months of intensive research and writing, this book is the end result of all who devoted their time and efforts in the initiation and progress of this book. It will surely be a source of reference in enhancing the required knowledge of the new developments in the area. During the course of developing this book, certain measures such as accuracy, authenticity and research focused analytical studies were given preference in order to produce a comprehensive book in the area of study.

This book would not have been possible without the efforts of the authors and the publisher. I extend my sincere thanks to them. Secondly, I express my gratitude to my family and well-wishers. And most importantly, I thank my students for constantly expressing their willingness and curiosity in enhancing their knowledge in the field, which encourages me to take up further research projects for the advancement of the area.

Editor

1

The Design and Material Characterization of Reclaimed Asphalt Pavement Enriched Concrete for Construction Purposes

Jaroslav Pokorný [1], Radek Ševčík [1,2,*] and Jiří Šál [1]

[1] Department of Civil Engineering, Faculty of Technology, Institute of Technology and Business, Okružní 517/10, 370 01 České Budějovice, Czech Republic; jaroslav.pokorny@mail.vstecb.cz (J.P.); sal@mail.vstecb.cz (J.Š.)

[2] Institute of Theoretical and Applied Mechanics of the Czech Academy of Sciences, Prosecká 809/76, 190 00 Praha, Czech Republic

* Correspondence: sevcik@itam.cas.cz

Abstract: Reclaimed asphalt pavement (RAP) is a valuable commodity originating during processes of road/highways rehabilitations, resurfacing in the cases of the revelation of underneath-placed layers. Removed material can be successfully recycled and utilized as a supplementing material for new hot asphalt mixes. However, its dosages are limited because of variations in properties of aged bitumen compared to fresh material and, thus, a significant amount of waste material is remaining as waste products. Nonetheless, this commodity may find usage in the construction industry that suffers from a shortage of high-quality and easily available aggregates. This work aimed to investigate the optimal substitution of mined natural aggregate with commonly available RAP in order to produce composites with the comparable mechanical performance of reference ordinary Portland concrete. The aggregate substitutions up to 100% with RAP have been studied with a combination of mechanical and analytical techniques. Obtained experimental data showed changes in the porous structure, mineralogy, and in the amount of formed cement-related hydration products that influenced the mechanical performance of produced composites. Composite with 10% of natural aggregate substitution with RAP has shown the strength class C16/20 after 28 days of water curing, according to the EN 206-1. Such innovative products could be utilized in the construction industry. The usage of waste RAP could contribute to preservation of our environment for future generations.

Keywords: concrete; reclaimed asphalt pavement; construction industry

---

1. Introduction

With regard to the development of society and urbanism, concrete-based composites and all modifications constitute a fundamental and available building material all over the world. The main advantages of concrete structures are considered to be quick construction processes, high compressive strengths (even in term of a few days), durability, etc. [1]. In this context and according to the available data, the world concrete production covers about 25 billion tons annually and many predictions claimed that its consumption will have a growing tendency in the future [2,3]. Commonly used concrete consists of three fundamental constituents: Portland cement, blend of fine and coarse aggregates, and batch water. The aggregate mix fills ~75% of the volume unit of produced concrete. Therefore, the quality of used aggregates and their overall interactions with cement-based matrix are crucial for the quality of formed concrete [4].

Currently, many natural sources of high-quality aggregates are depleted or their stocks are significantly decreasing, which is the cause of their increasing costs on the market [5]. The global aggregate consumption for the construction industry exceeded 48.3 million tons annually [6]. Another factor affecting final aggregate cost is related to energy consumption for quarrying and subsequent manufacturing. In addition, costs related to transportation of raw materials for long distance have to be considered as well [7,8]. Different waste materials were tested to partially and/or full replacement of natural aggregates [9,10]. At the present time, the commonly used coal-derived aggregates start to became a scarce commodity on the market and, thus, the usage of different waste materials is intensively investigated and their advantages/disadvantages were reported e.g., in Reference [11] and references therein.

Reclaimed asphalt pavement (RAP), which is one of the waste materials, is obtained as a removed product of road/highways rehabilitations, resurfacing in the cases of the revelation of underneath-placed layers [12]. The recycling procedure of asphalt pavement is known and applied in more than one century. Nevertheless, the main extend of RAP recycle started during the oil crisis in 1973 [13,14]. Asphalts are usually composed of granular aggregate (often high-quality limestone) and bitumen in the percentage representation of 95% and 5%, respectively [15]. The mentioned material is applied for road structures in the form of Hot Asphalt Mixes (HAM) that can occur after the end of their service life recycled [16,17]. However, obtained RAP in the portion not exceeding 25% for new mix design is recommended only [16,18]. If more than 25% of RAP is used, higher stiffness of aged asphalt leads to structure cracking even at low temperatures and a higher inclination to fatigue compared to fresh raw materials [19]. Additionally, about 85.1 million tons of RAP was stockpiled in the U.S. in 2015 [8,20] and, in comparison, about 50 million tons of RAP is produced annually in Europe [13,14]. Notwithstanding, only limited research papers focused on the investigation of RAP as a potential candidate for natural aggregates replacement can be found [15,21,22]. Singh et al. [21] investigated the usage of adjusted RAP by the Abrasion and Attrition method in concrete mixes with w/b = 0.38, thus natural aggregate was substituted by RAP in the range of 25–100 wt.%. Reported results showed the decrease in strength with increasing RAP content. However, the application of treated RAP with lower asphalt content caused the decrease of mechanical resistivity by only about 20% even if 100% of natural aggregates were replaced with RAP. In addition, it was shown that the presence of RAP enhanced the workability of fresh composites, thus allowing long distance transportation. Al-Mufti and Fried [15] applied RAP as a replacing material of 20-mm gravel in concrete with a water to binder ratio of 0.5 and it was shown that 25% replacement decreased compressive strength by 27% of 28 days water-cured samples. Coppola et al. [22] investigated RAP (in the fraction up to 20 mm) substitution of 20% of coarse silica aggregate with ambiguous results. Obtained results showed a beneficial effect of RAP on workability, but, on the other hand, the mechanical performance of composites with 15% of RAP were found to be two times lower in the comparison with reference samples. Mahmoud et al. [23] applied the RAP aggregate as a substitution of a natural aggregate in the range from 0% up to 50% in self-consolation concrete mixes. The gradual decrease of slump flow values of fresh concrete mixtures together with a decrease of strength performance with increasing RAP incorporation were reported. Dosages up to 25% were determined to be suitable in order to preserve sufficient strength properties. Huang et al. [24] studied the influence of fine RAP (0.0/4.0 mm) and coarse RAP (4.0/20.0 mm) as full replacement of natural aggregate. Results showed systematic reduction of compressive strength reaching up to 27.6% in the comparison with the reference sample.

The main aim of this paper refers to the clarification of the role of RAP on properties of concrete composites and to design RAP-containing concrete composites of class C 16/20 suitable for construction purposes. Composites with different RAP/natural aggregates ratios were produced following standardized procedures and the obtained specimens were tested for their structural properties and mechanical resistance. Furthermore, the influence of RAP on the hydration reaction and morphology of concretes was studied with a combination of instrumental methods, such as X-ray powder diffraction and scanning electron microscope.

2. Experimental Part

2.1. Used Materials

Ordinary CEM II/B-M 32.5 R (CEM II) conforming to the EN 197-1 [25] for experimental investigation was used. The binder comprised an average alite (C_3S) content of 7.13 wt.% and possessed a relatively high specific surface area exceeding 380 $m^2 \cdot kg^{-1}$. Chemical composition and basic material characteristics, reported by the producer (Cemex, Ltd., Prachovice, Czech Republic), are listed in Tables 1 and 2. Two sorts of fine aggregates in the fraction of 0/4 mm were used in this study. Mined Nature Silica Sand (NSS) was delivered by Mane Holding, Plc., České Budějovice, Czech Republic, and meets requirements for concrete manufacturing according to EN 12 620 [26]. This silica aggregate was substituted by Reclaimed Asphalt Pavement (RAP) obtained as a recycled material when resurfacing the road I/3 close to Benešov town, Czech Republic. The surface layer of asphalt pavement was removed by a cold milling machine and stockpiled in the nearby landfills. Collected RAP aggregate comprised a mix of coarse particles up to 20 mm and fine fractions below 4 mm. In order to get a comparable particle size distribution (PSD) with NSS, fine aggregates were separated by sieving with a 4-mm mesh sieve. The comparison of RAP before and after sieving together with NSS is depicted in Figure 1. The comparison of PSD of aggregates used in this study is reported in Figure 2. Lower and upper limits highlighted in PSD curves specify an optimal area in which granulometry of aggregate is suitable for concrete manufacturing according to EN 206-1 [27] and EN 12 620 [26].

Table 1. Chemical composition and basic material properties of used cement.

Oxide Composition (wt.%)	
SiO_2	15.84
Al_2O_3	4.57
Fe_2O_3	2.33
CaO	62.69
MgO	2.64
K_2O	0.63
Na_2O	0.25
SO_3	2.93
Cl^-	0.08

Table 2. Material properties of CEM II.

Characteristic	CEM II	Standard
Powder density ($kg \cdot m^{-3}$)	965	EN 1097-3 [28]
Specific density ($kg \cdot m^{-3}$)	3090	EN 1097-7 [29]
Specific surface area ($m^2 \cdot kg^{-1}$)	344	EN 196-6 [30]
Initial setting time (min)	181	EN 196-3 [31]
Final setting time (min)	273	
Loss on ignition (wt.%)	4.55	EN 196-2 [32]

Figure 1. Samples of silica aggregate and RAP aggregate after separation of coarse particles.

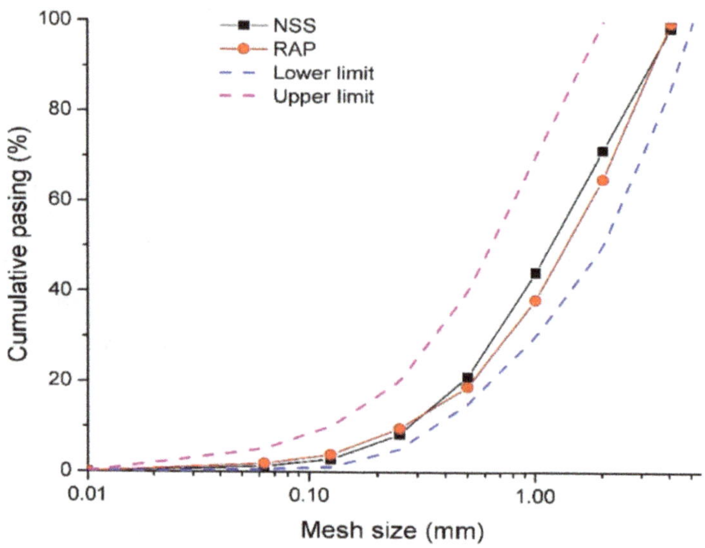

Figure 2. Particle size distribution of silica sand and RAP.

2.2. Mix Design and Preparation of Test Specimens

The reference mix (RC) was designed according to the T.C. Kennedy method [33]. This method takes into consideration the quality of applied aggregate in dependence on the shape of grains, the rock type, and its technological processing. Reference mix was comprised of 362.7 kg of CEM II, 1 760.0 kg of NSS, and 199.5 kg of water for 1 m^3 of produced fresh concrete. Based on the control mix, three types of RAP fine-grained concrete mortars were derived (see in Table 3), whereas natural silica sand (NSS) was substituted with RAP in the amount of 10 (RA-C 10), 50 (RA-C 50), and 100 vol.% (RA-C 100). Both aggregate types were dried before application. The amount of batch water was kept constant for all prepared mixtures with a w/c ratio of 0.55.

Table 3. Composition of produced concretes.

Material	Content ($kg{\cdot}m^{-3}$)			
	RC	**RA-C 10**	**RA-C 50**	**RA-C 100**
CEM II	362.7	362.7	362.7	362.7
NSS 0/4 mm	1760.0	1584.0	880.0	-
RAP 0/4 mm	-	148.1	740.7	1481.4
Water	199.5	199.5	199.5	199.5

Input raw materials were blended with the mixer Spar 200 with vertical axes allowing simulate mixing conditions according to the EN 196-3 [31]. First, dry components including Portland cement, silica aggregate, or a blend of silica aggregate and RAP were homogenized for 60 s at 94 rpm. Then, batch water was gradually added for 30 s and mixing continued 30 s further. Then, sediments of dry material at the bottom of mixing vessels were removed and new mixing at 198 rpm for 120 s was applied. Tested specimens according to EN 12390-2 [34] were prepared. Fresh concrete was placed into iron molds, confirming the requirements of the EN 12390-1 [35], in the shape of prisms (100 × 100 × 400 mm), cylinders (diameter of 100 mm, the height of 200 mm), and cubes (edge of 100 mm). Each specimen was cast in one step and subsequently compacted with the help of a vibrating table (Matest, S.P.A., Treviolo, Italy) operating with the frequency 40 Hz (2400 oscillations per minute) for 2 min. Filled molds were covered by plastic foil and left in laboratory conditions at a temperature of 20 ± 1 °C and 45 ± 5% of relative humidity. After 24 h, partially hardened specimens were demolded and placed in a water environment at a temperature of 20 ± 1 °C for 27 days. The photographic gallery of prepared specimens is depicted in Figure S1 in Supplementary Materials.

2.3. *Instrumental Techniques*

At first, the material properties of the input components were analyzed. The specific gravity with the pycmometric method according to EN 1097-7 [29] was measured whereas the expanded combined uncertainty was about 1.6%. The powder density of CEM II and loose bulk density of both aggregates were obtained on the bases of the standard EN 1097-3 [28]. Compacted bulk density was determined with the same procedure. Only in the case of aggregate samples, the frequency of vibrating table was set to 40 Hz. In both cases, the expanded combined uncertainty reached 2.3%. Voids percentage with the expanded combined uncertainty of 4.4% was subsequently calculated from known values of loose bulk density and specific gravity of a particular aggregate type. PSD of aggregates was measured with the usage of sieving analysis, according to the standard EN 933-1 [36]. Tested samples were dried in hot air dryer at 105 ± 5 °C until the constant mass was reached. After subsequent cooling, representative aggregate samples were sieved through the set of sieves having a mesh size of 4.0, 2.0, 1.0, 0.5, 0.25, 0.125, and 0.063 mm and shaken by vibratory apparatus Retsch AS 200 (Germany). Water absorption of both NSS and RAP was measured in conformity with EN 1097-6 [37].

The consistency of fresh mixes was controlled with a slump test (EN 12350-2 [38]) performed at the end of mixing. The bulk density values were measured according to EN 12390-7 [39] whereas the expanded combined uncertainty of the applied method was 2.5%. Specific densities of hardened tested specimens were measured using the apparatus AccuPyc II 1340 (Micrometrics Co., Ltd., Norcross, USA). The relative standard deviation of specific density was determined to be lower than 1.1%. Total open porosity was calculated from obtained bulk and specific density values. The expanded combined uncertainty of total porosity was 5%. Flexural and compressive strength tests were performed with the usage of hydraulic press Servo Plus Evolution (Matest, S.P.A., Treviolo, Italy) with a disposing loading capacity of 3000/600 kN. Prisms with dimensions of 100 × 100 × 400 mm in a three-point bending setting were tested. On cubes (100 × 100 × 100 mm) and cylinders (diameter of 100 mm, height of 200 mm) maximum compressive forces were recorded and subsequently converted to compressive strengths. The expanded combined uncertainty of compressive strength was 1.8% and 2.0% in the sense of flexural strength determination.

The Mercury Intrusion Porosimetry (MIP) method was employed to determine the specimen's porosity. The set of Pascal devices, Pascal 140 and Pascal 440, producing external pressure of 140 kPa and 440 MPa, respectively, were used. In general, siliceous samples are usually dried at 105 °C in order to remove excess moisture [40]. However, such high temperatures may cause the leakage of asphalt binder in pores in cement-based matrix and, thereby, the analysis can be adversely affected [41]. It was, therefore, approached to vacuum oven drying at 50 ± 5 °C until the constant sample mass was reached.

The adhesion of used aggregates with cement matrix was observed with a polarizing microscope equipped with a digital camera Olympus E-600. Pieces with area of 15 × 15 mm and thickness of 10 mm were cut from composites specimens. After rough grinding, samples were placed into plastic cylindrical molds and casted by epoxide resin Araldite 2020 under vacuum. After hardening at 21 ± 1 °C and 45 ± 5% of relative humidity in order to achieve a smooth surface, samples were ground with an automatic grinder equipped by SiC papers of a grain size of P 800, P 1200, P 2500, and P 4000. After that, samples were hand-polished with a diamond suspension of particle size of 1, 0.25, and 0.1 μm. Prepared polished sections were washed with 96% ethanol (p.a., Lach-Ner, Neratovice, Czech Republic).

The mineralogical composition of hydrated cement pastes was analyzed with X-ray powder diffraction (XRPD) employing a diffractometer D8 Advance (Bruker, Karlsruhe, Germany) with Bragg-Brantano geometry and equipped with LynxEye detector using CuKα radiation and Ni filter. Data were recorded in the angular range from 15 to 90° (2θ) with a step of 0.01° and counting time 0.4 s. Rietveld refinement method was used for a quantitative phase analysis in Topas 4.2 software (Bruker).

Prepared samples were observed under scanning electron microscope (SEM) Quanta 450 FEG (FEI, Brno, Czech Republic) using a backscatter electron detector. Prior to observations, samples were goal-coated (thickness of 5 nm). The acceleration voltage was set to be 20 kV.

3. Results and Discussion

3.1. Material Properties and Morphology of Used Aggregates

The material properties of NSS and its replacing RAP material are summarized in Table 4. In the case of RAP, lower loose bulk density and specific density by more than 20% and 8%, respectively, were detected. This finding pointed out the lower unit weight of produced composites with RAP aggregate regard to traditionally incorporated silica sand. Such light-weighted composites will reduce transportation costs. On the other hand, approximately two times higher water absorption of RAP in comparison with NSS was observed. Such a significance increase in water absorption is assumed to angular-shaped grains of RAP with the rough surface [42].

Table 4. Physical properties of used aggregate types.

Characteristic	NSS	RAP
Loose bulk density ($kg{\cdot}m^{-3}$)	1592	1340
Compacted bulk density ($kg{\cdot}m^{-3}$)	1810	1610
Specific density ($kg{\cdot}m^{-3}$)	2517	2310
Water absorption (%)	0.78	1.55
Voids (%)	37	42

SEM images of NSS and RAP aggregates are depicted in Figure 3. Micrographs of RAP shows fine aggregate particles fully incorporated within the asphalt matrix (Figure 3a,b). Moreover, the presence of micro filler and agglomerated dust particles on the surface of bulky grains was observed (Figure 3b). These adsorbed predominantly flat-shaped particles with apparent highly specific surface area can be, to a certain extent, responsible for increased water absorption of asphalt-covered aggregate as already reported in Reference [43]. In addition, bonding ability of dust and agglomerated particles with coarser RAP grains can be importantly limited and, thus, in a certain way, it may be responsible for the limited strength properties of produced composites [21]. In contrast, silica sand grains represent a much smoother surface with randomly occurring particles of a few micrometers in size. This surface morphology of natural silica sand particles seems to be beneficial in term of uniform redistribution of batch water that is subsequently available for hydration of cement grains.

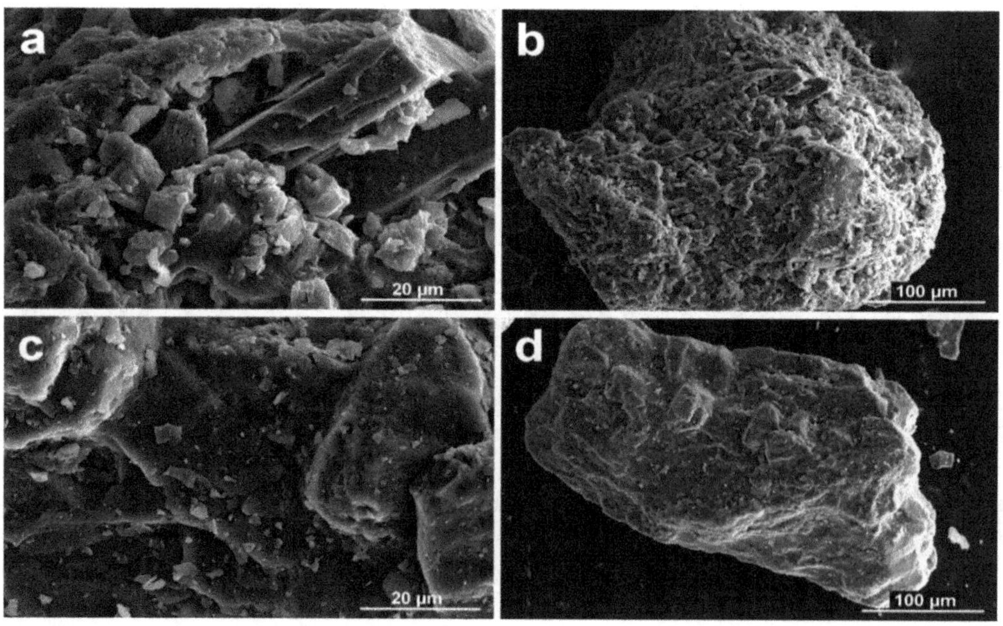

Figure 3. SEM image of RAP (**a**,**b**) and NSS (**c**,**d**) sample.

3.2. *Characterization of Fresh Mixtures*

Physical properties determined for fresh concrete mixes are given in Table 5. The addition of waste aggregate resulted in a visible workability deterioration of developed mixes. In this context, performed slump tests revealed the decrease of the initial slump with respect to the control mix of about 29.4% and 70.6% for the concrete mix with 50% and 100% contents of RAP, respectively. The control mix with a measured slump value of 170 mm met the class S4, according to the EN 206-1 [27]. This phenomenon can be connected, on the one hand, to the angular shape of asphalt-covered aggregate and, on the other hand, to its considerable water absorption (two times higher as discussed above) in comparison with NSS. Similar findings were reported in the study performed by Singh et al. [42]. Accordingly, with higher RAP incorporations into the concrete mix, the amount of water needed to preserve required workability should be increased or a plasticizer should be added. In addition, an adverse effect of bleeding caused by high dosages of RAP exceeding 50 vol.% was detected after several minutes once the specimens were casted. To avoid the bleeding issue, Coppola et al. [22] recommended to limit the amount of reclaimed asphalt to 15 wt.%. However, replacing of NSS with RAP helped to mitigate the unit weight of fresh concretes up to 12 wt.% in comparison with the reference mix. Although lower dosages of RAP (below 10%) did not reduce the weight of fresh mixes with respect to RC, the slump mitigation recorded was only 6% [22].

Table 5. Properties of fresh concrete mixes.

Property	RAP Content (Vol.%)			
	0	**10**	**50**	**100**
Initial slump (mm)	170	160	120	50
Density (kg·m^{-3})	2322	2258	2183	2044

3.3. *Structural Properties*

Structural properties, bulk density, specific density, and total porosity of investigated samples are given in Table 6. The substitution of natural sand by artificial aggregate led to a decrease in bulk densities by 1.1% recorded for concrete with 10 vol.% of RAP, 4.4% determined for concrete with 50 vol.% of RAP, and 9.9% in the case of silica sand-free samples. In a similar way, specific density values of modified samples were reduced. Due to the gradual decline of both bulk and specific densities, total porosity values showed a growing trend. The total porosity of control specimens reached 17.6% contrary to 21.6% observed for material without the presence of natural silica sand. On the contrary, the difference in porosity of RC and RA-C 10 samples were found to be negligible. The incorporation of reclaimed asphalt pavement into concrete mixes partially decreased their unit weight, which introduces a beneficial aspect in view of dead load reduction in relation to bearing structures.

Table 6. Structural properties determined on dried hardened samples.

Property	RAP Content (Vol.%)			
	0	**10**	**50**	**100**
Bulk density (kg·m^{-3})	2078	2055	1987	1872
Specific density (kg·m^{-3})	2521	2498	2430	2387
Total porosity (%)	17.6	17.7	18.2	21.6

3.4. *Mechanical Resistance*

Flexural and compressive strength results obtained for all prepared fine-grained concretes, expressed as average values from six independent measurements, are shown in Figures 4–6. Linearly decreasing trends with regression coefficient ~0.999, ~0.999, and 0.997 determined for compressive strengths measured on cylinders, cubes, and for flexural strength measurements, respectively, were plotted. This negative effect on the mechanical performance could be ascribed, in particular, to oil traces that may influence the hydration process of cement [22] and to weak bonds of the interfacial transition zone (ITZ) between cement paste and bitumen covered aggregate [44,45]. The compressive strength obtained on cylinders decreased from the initial value of 17.2 MPa detected for RC to 11.1 MPa (RA-C 50) and 6.9 MPa (RA-C 100). It represents the decrease of 35.5% and 59.9% with respect to reference samples. A similar trend was observed for cubic samples. The compressive strength was found to decrease of 35.4% for RA-C 50 and 54.0% for RA-C 100 in the comparison to the RC sample. The flexural strength was detected to decrease by 20.0% and 38.3% for RA-C 50 and RA-C 100, respectively, in the comparison with reference.

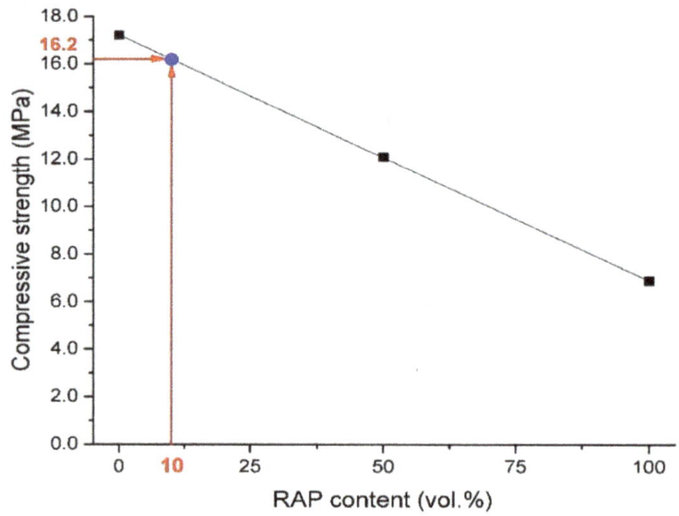

Figure 4. Comparison of compressive strengths obtained on cylindrical samples of concretes with a different content of the RAP aggregate.

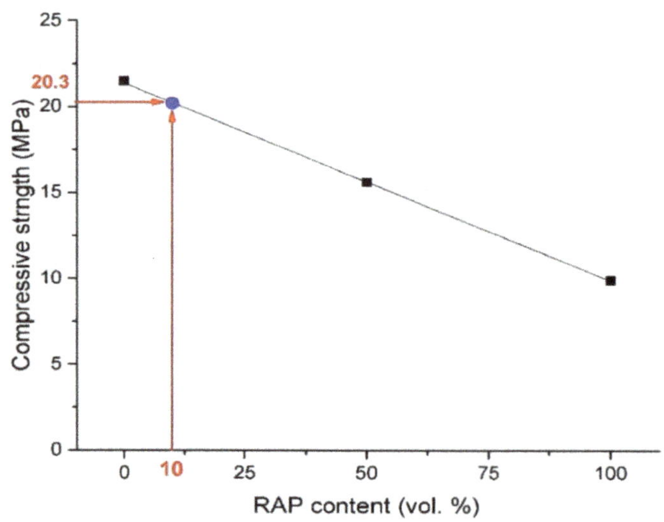

Figure 5. Comparison of compressive strengths obtained on cubic samples of concretes with a different content of the RAP aggregate.

According to trends plotted in Figures 4–6, RAP content of 10 vol.% seems to be optimal to produce composites of the strength class C 16/20. Such a composite should dispose of the average compressive strength of 16.2 MPa measured on cylinders and 20.3 MPa measured on cubes. In the case of flexural performance, RAP content limited to 10 vol.% should result in the strength of 3.3 MPa. Based on this prediction, a composite with only 10 vol.% of RAP (called RA-C 10) was produced by following the same procedure and being tested after 28 days of curing. The experimental values of compressive and flexural strengths were detected to be slightly higher than expected theoretical strengths. The compressive strength was found to be 16.6 MPa and 21.1 MPa measured on cylinders and cubes, respectively. The flexural strength was detected similarly to the theoretical prediction—3.3 MPa.

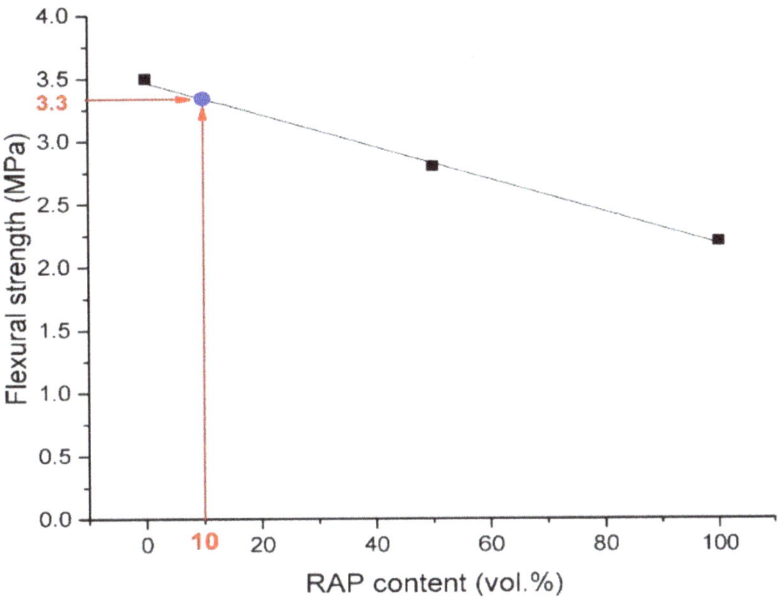

Figure 6. Flexural strength development of concretes with a different content of the RAP aggregate.

3.5. *Mercury Intrusion Porosimetry (MIP)*

Cumulative pore volume cures measured on tested concretes are given in Figure 7. The gradual increase of intruded volume accompanied with increased NSS substitutions with RAP was found to be in agreement with the structural properties mentioned above (and summarized in Table 5) and also with the observed decreasing mechanical resistivity of tested samples. In Figure 8, the dependence of the critical pore size (d_{cr}) on RAP content in studied mixtures is depictured. A critical pore size refers to the steepest slope of the cumulative pore curve and its value is derived from differential distribution curves shown in Figure 9. The procedure for obtaining this parameter is described in detail in Reference [46]. The d_{cr} characterize the interconnection of pores that allow more efficient water transport and, thus, the higher rate of transmissivity for chemical substances through the cement matrix [46]. Plotted values of d_{cr} in the dependence for increasing RAP content in concrete showed a growing linear trend with a high regression coefficient of ~0.999. It indicates a large volume of interconnected pores in RAP samples increasing with the higher substitutions of NSS.

The pores sizes classification, graphically highlighted in Figure 7, may be used as a tool revealing important changes in porous structures of tested concretes caused by RAP additions. Pores could be sorted according to their size into several groups: gel pores and small capillaries (diameter of 0.0025–0.01 μm), medium capillaries (diameter of 0.01–0.05 μm), large capillaries (0.05–10 μm), and entrained air (diameter higher than 10 μm) [47–49]. The main changes were detected in pores with a diameter below 1 μm and above 3 μm. The group of entrained air is composed of spherically shaped pores and helps to decrease the risk of damage by physically and chemically related volumetric changes (frost, salts) [46]. Gel and capillary pores play an essential role on transport processes and permeability in cement-based materials [50]. The RA-C 50 sample showed only a slightly difference in the porous structure, particularly in the area of gel pores and capillaries, in comparison with reference material. In view of concrete without silica sand, the porosity increase is more apparent in both the capillary/gel pores region and in the entrained air section. On the contrary, 10 vol.% of NSS substitution with RAP had a minimum impact on the porosity growth, which indicates the cumulative pore volume curve obtained for concrete RA-C 10.

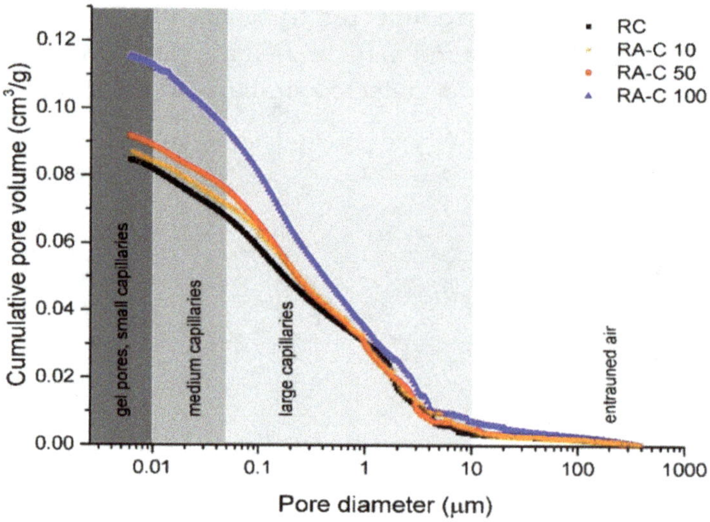

Figure 7. Cumulative pore volume curves obtained for reference (RC) and RAP (RA-C 10, RA-C 50, RA-C 100) specimens.

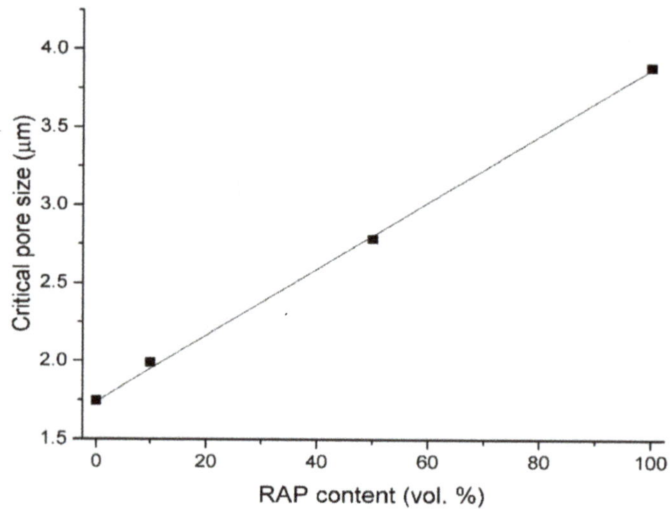

Figure 8. Critical pore diameter (d_{cr}) in dependence on RAP content in concrete mixes.

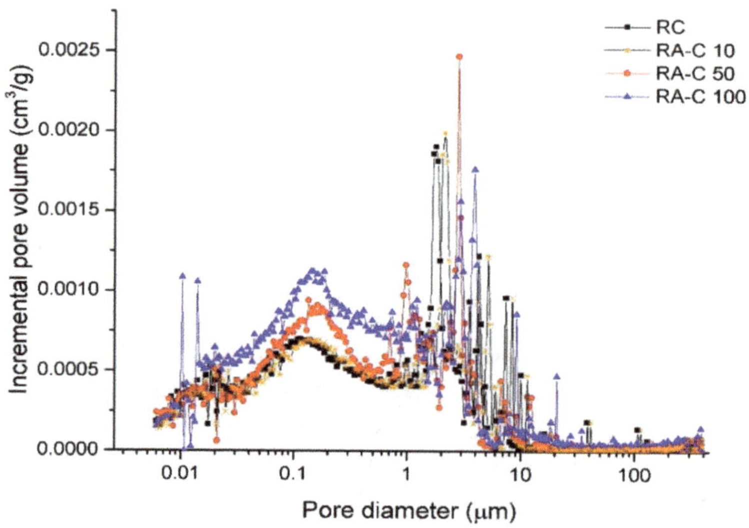

Figure 9. Incremental pore volume curves obtained for reference (RC) and RAP modifies (RA-C 10, RA-C 50, RA-C 100) specimens.

3.6. Mineralogical Composition

Mineralogical compositions of the RAP aggregate as well as developed reference and RAP-enriched concretes are listed in Table 7. RAP, as an input raw material, contained predominantly mineral phases, such as quartz, calcite, albite, and orthoclase. Looking at reference concrete composition (RC), quartz in the amount exceeding 50 wt.% was recorded due to usage of silica aggregate in concrete mix. However, with the growing RAP dosage, quartz content was detected to gradually decrease up to 16.8 wt.% (RA-C 100). The wt.% of minerals, such as muscovite, clinochlore, amphibole, and orthoclase presented in raw RAP increased gradually with the higher content of RAP in produced composites. The same tendency was found for the amount of an amorphous phase. Some changes in Portland cement-related hydration products formation were observed as well, especially the increased formation of ettringite in the case of composites containing RAP. The portlandite formation rate increased around 1 wt.% for both RA-C 50 and RA-C 100 concretes in comparison to a reference sample. Sample RA-C 10 showed only a minor increase (about 0.2 wt.%) in Portlandite content with respect to the reference. It was reported that the formation of Portlandite could be affected by possible leaching of organic compounds from RAP in pore solution [51]. The content of calcite and albite increased gradually with the increasing RAP dosages, which suggest the increment of cementitious products [52].

Table 7. Mix proportions of RC and RAP derived concretes.

Mineral Substance (wt.%)	RAP	RC	RA-C 10	RA-C 50	RA-C 100
Quartz—SiO_2	20.0	50.8	46.3	29.1	16.8
Vaterite—$CaCO_3$	-	4.3	4.2	3.5	4.6
Calcite—$CaCO_3$	19.2	2.0	3.5	8.3	11.8
Alite—$3CaO{\cdot}SiO_2$	-	-	-	-	0.3
Belite—$2CaO{\cdot}SiO_2$	-	-	1.0	1.2	0.5
Portlandite—$Ca(OH)_2$	0.1	1.4	1.6	2.2	2.1
Ettringite—$Ca_6Al_2(SO_4)_3(OH)_{12}{\cdot}26H_2O$	-	0.8	1.8	4.3	3.0
Albite—$NaAlSi_3O_8$	23.1	6.5	8.9	14.4	15.1
Microcline—$KAlSi_3O_8$	0.6	6.7	6.7	7.2	2.8
Muscovite—$KAl_2(AlSi_3O_{10})(F,OH)_2$	6.1	1.2	1.7	3.1	4.4
Clinochlore—$(Mg,Fe)_3(Si,Al)_4O_{10}(OH)_2{\cdot}(Mg,Fe)_3(OH)_6$	6.1	-	1.0	3.3	1.1
Amphibole—$(Mg,Fe)_4Al(Si_7Al)O_{22}(OH,F)_2$	3.7	-	0.2	0.5	4.5
Orthoclase—$KAlSi_3O_8$	12.1	-	0.4	1.5	0.8
Amorphous content	9.0	16.4	17.3	21.4	32.2

3.7. Microscopic Analysis

The quality of Interfacial Transition Zone (ITZ) was initially evaluated on prepared polished sections. The images of reference material (RC) and concrete with 50 vol.% of RAP are depicted in Figures 10 and 11. The sample with NSS contentment showed very compact adhesion and interaction of aggregate grains and cement matrix. There is no evidence of porous layers and damages of the ITZ in the surrounding of coarser particles. In the case of samples with RAP, it is visible with the naked eye that aggregate grains are mostly covered with aged bitumen. Additionally, the preferential adhesion of cement matrix was with asphalt layers. The more detailed observations were performed using SEM.

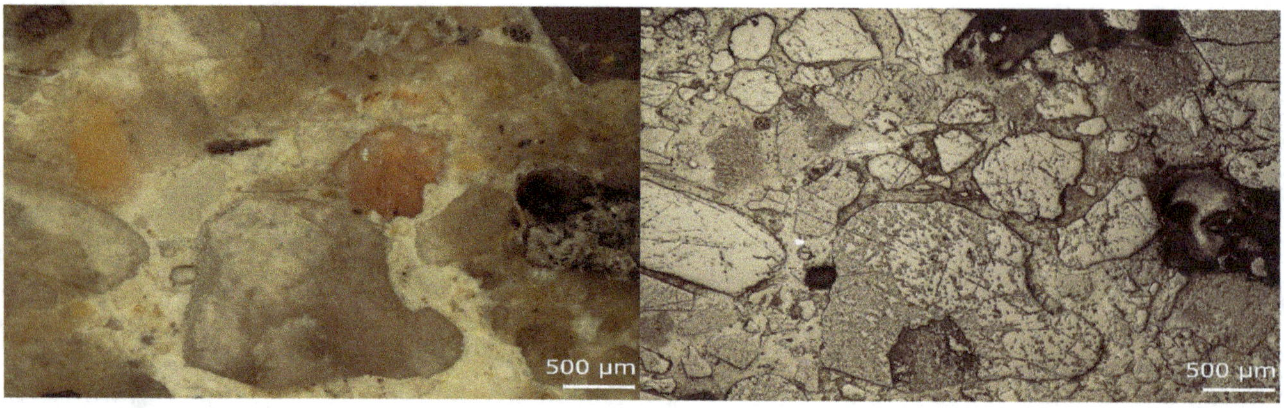

Figure 10. Microscopic images of the RC concrete sample observed under unpolarized (**left**) and polarized (**right**) light.

Figure 11. Microscopic images of concrete sample RA-C 50 observed under unpolarized (**left**) and polarized (**right**) light.

Images of the microstructure of all tested composites including the reference sample after 28 days of water curing are shown in Figure 12. Some phases, based on their morphology, were indicated (Port—Portlandite, CSH—calcium silicate hydrate, Etr—ettringite, Cal—calcite, Alb—albite). The inner structure of hydrated RC consists of well observed, larger crystals of Portlandite (Port) surrounded by hydrated calcium silicate (CSH) and gels creating the main solid skeleton. In addition, smaller irregular-shaped crystals of vaterite supplement the overall composition. Focusing on the RAP modified sample, all the above-described mineral and gel phases were observed as well. The higher abundance of calcite (Cal) and albite (Alb) in samples with RAP confirmed the results of quantitative phase analysis of obtained XRPD patterns. If the samples without and with RAP are compared, it is possible to notice larger voids and a less compact microstructure in composite structures, especially for RA-C 50 (Figure 12c) and RA-C 100 (Figure 12d). Such observations correspond to the measured higher values of open porosity and reduced mechanical resistance as discussed in the previous text. The ITZ between RAP and cement matrix and the bonding of RAP grains with cement are depicted in Figure 12e,f. Only limited adhesion of cement to RAP is visible in both cases. Consequently, the asphalt film act as a separating layer hindering the creation of hard and durable cement-related bonds on the aggregate surface, which was already reported in Reference [53]. These observations were found to be in line with the observed samples after mechanical testing. Crack propagations during compressive strength tests were mainly noticed close to RAP grains rather than at the interfacial of silica aggregate. The cement matrix was recorded.

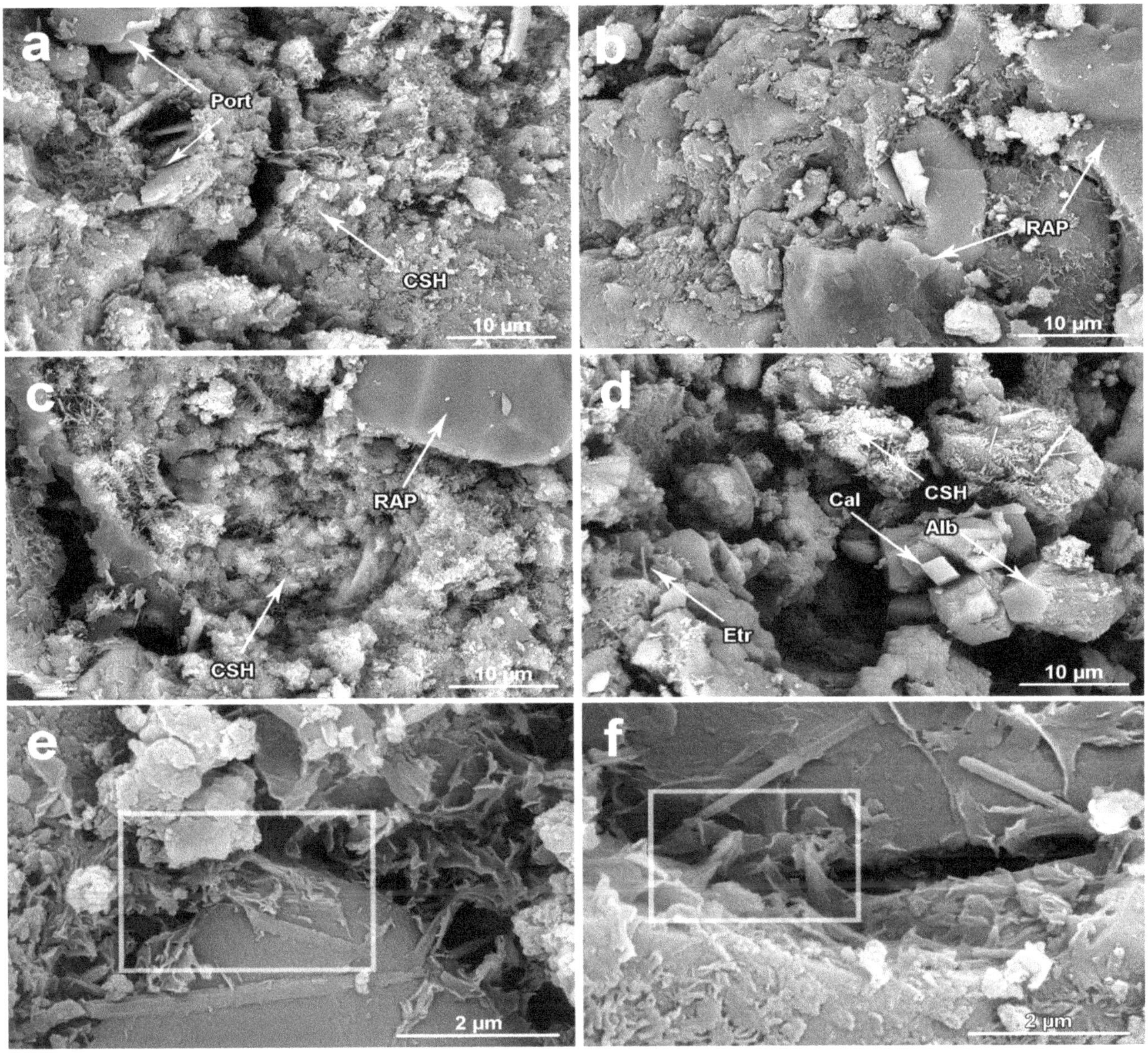

Figure 12. SEM micrographs of RC (**a**), RA-C 10 (**b**), RA-C 50 (**c**), RA-C 100 (**d**) specimens and an examples of ITZ zones (highlighted within inserted rectangles between RAP and cement matrix (**e**) and in the system of RAP-cement-RAP) (**f**). Some phases were identified according to their typical morphology.

With respect to experimental analyses performed, reclaimed asphalt pavement introduces a valuable replacing commodity to natural silica sand in the sense of the preparation of composite materials intended for the building industry. However, RAP dosages in concrete mixes are limited because of its specific material properties that differ from those measured for dense aggregates types commonly used in the concrete industry.

4. Conclusions

The potential usage of reclaimed asphalt pavement as a substitution of mined nature silica aggregate for the design of concrete C 16/20 for construction purposes was studied. Properties of raw input materials as well as developed concretes with various RAP content after 28 days of water curing were examined and completed by a combination of experimental procedures and instrumental techniques. Obtained data showed notable effects of natural silica sand substitutions with waste RAP aggregate on the behavior of the ordinary Portland concrete as follows.

- The behavior of fresh mixes was affected by specific properties of RAP, in particular, importantly increased water absorption, compared with NSS, which resulted in the considerable mitigation of workability. The slump test revealed very good workability of reference mix (slump of 170 mm) meeting the class S4 while the full replacement of NSS with RAP resulted in a very stiff mixture inappropriate for construction purposes. Moreover, high dosages of RAP reaching up 100% led to the bleeding of placed concretes. Reduced RAP dosages up to 10 vol.%, however, mitigated the value of slump in about 6% only, thereby the class S4 was kept.
- Reclaimed asphalt pavement addition helped to decrease the unit weight of produced concretes, thereby the transportation costs can be partially reduced. However, a lowered weight of RAP-dosed concretes is accompanied with the increase in total porosity by 0.6% (50% of RAP content) and 3.6% (100% of RAP content) with regard to the control samples. At lower NSS substitutions with a tested alternative material, not exceeding 10 vol.%, this adverse effect will be negligible. Increased critical pore size induced by waste aggregate brought the growth of interconnected pores and the increase of medium-sized capillaries responsible for transport processes in porous structure. On the contrary, detected higher presence of entrained air, in the form of isolated bubbles, in RAP-enriched concretes can result in improved frost and salt resistivity.
- Incorporation of the waste aggregate resulted in the deterioration of the strength resistance of developed concretes, linearly proportional to its dosage. With the total substitution of NSS with RAP, the average compressive strength measured on cylinders decreased by 60% and by 54% in the case of cubic specimens when compared to the reference material. The maximum flexural strength drop reached approximately 38%. From interpolation of obtained experimental data, the concrete with RAP content of 10 vol.% was found to be optimal in the sense of meeting requirements for the class C 16/20. The experimental mechanical testing of prepared samples with 10 vol.% of reclaimed asphalt (RA-C 10) revealed the average compressive strength of 16.6 MPa and 21.1 MPa for cylindrical and cubic samples, respectively. The flexural strength was determined to be 3.3 MPa. Thus, the design concrete with 10 vol.% substitution of NSS with RAP met requirements for practical usage in civil engineering applications.
- The mineralogical composition obtained by XRPD analysis showed the increased contents of calcite, albite, muscovite, and amorphous phase induced with the increased substitutions of NSS with RAP. On the other hand, silica was detected to gradually decrease. Moreover, the ettringite was detected to be higher in samples containing RAP. SEM observations shown the morphology of hydration products and/or phases are presented in RAP. Microscopic analysis revealed prevailing adhesions occurring between asphalt cover presented on aggregate grains and cement matrix, which cause the decreased mechanical performance of concretes with high RAP content.

From the provided experimental investigation, it can be deduced that reclaimed asphalt pavement can be potentially applied in the concrete C 16/20 production as a partial substitution of nature silica aggregate in limited amount up to 10 vol.% in order to maintain strength properties. The incorporation of unutilized waste aggregate help save depleted nature aggregate recourses and mitigate quarrying-induced impacts on the environment. The findings obtained in the field of the pore structure of produced composites showed the increased potential of RAP-dosed concretes in terms of improved durability that could be the aim of future research.

Author Contributions: Conceptualization, J.P. and R.Š. Methodology, J.Š. Software, R.Š. Validation, J.P. and R.Š. Format analyses, J.Š. Investigation, J.P., R.Š., and J.Š. Resources, J.P. and R.Š. Data curation, R.Š. Writing—original draft preparation, J.P. Writing—reviewing and editing, R.Š. Visualization, J.P. and R.Š. Supervision, R.Š. Project administration, R.Š. and J.Š. Funding acquisition, R.Š. and J.Š. All authors have read and agreed to the published version of the manuscript.

References

1. Meyer, C. The greening of the concrete industry. *Cem. Concr. Compos.* **2009**, *31*, 601–605. [CrossRef]
2. Chan, R.; Liu, X.; Galobardes, I. Parametric study of functionally graded concretes incorporating steel fibres and recycled aggregates. *Constr. Build. Mater.* **2020**, *242*, 118186. [CrossRef]
3. Tiwari, A.; Singh, S.; Nagar, R. Feasibility assessment for partial replacement of fine aggregate to attain cleaner production perspective in concrete: A review. *J. Clean. Prod.* **2016**, *135*, 490–507. [CrossRef]
4. Verian, K.P.; Ashraf, W.; Cao, Y. Properties of recycled concrete aggregate and their influence in new concrete production. *Resour. Conserv. Recycl.* **2018**, *133*, 30–49. [CrossRef]
5. Prem, P.R.; Verma, M.; Ambily, P.S. Sustainable cleaner production of concrete with high volume copper slag. *J. Clean. Prod.* **2018**, *193*, 43–58. [CrossRef]
6. Evangelista, A.C.J.; Tam, V.W.Y.; Santos, J. Recycled ceramic fine aggregate for masonry mortar production. *Proc. Inst. Civ. Eng. Constr. Mater.* **2019**, *172*, 225–234. [CrossRef]
7. Hossain, M.U.; Poon, C.S.; Lo, I.M.C.; Cheng, J.C.P. Comparative environmental evaluation of aggregate production from recycled waste materials and virgin sources by LCA. *Resour. Conserv. Recycl.* **2016**, *109*, 67–77. [CrossRef]
8. Shi, X.; Mukhopadhyay, A.; Zollinger, D. Sustainability assessment for portland cement concrete pavement containing reclaimed asphalt pavement aggregates. *J. Clean. Prod.* **2018**, *192*, 569–581. [CrossRef]
9. Ladomerský, J.; Janotka, I.; Hroncová, E.; Najdená, I. One-year properties of concrete with partial substitution of natural aggregate by cupola foundry slag. *J. Clean. Prod.* **2016**, *131*, 739–746. [CrossRef]
10. Gesoğlu, M.; Güneyisi, E.; Özturan, T.; Öz, H.Ö.; Asaad, D.S. Permeation characteristics of self compacting concrete made with partially substitution of natural aggregates with rounded lightweight aggregates. *Constr. Build. Mater.* **2014**, *59*, 1–9. [CrossRef]
11. Mo, K.H.; Thomas, B.S.; Yap, S.P.; Abutaha, F.; Tan, C.G. Viability of agricultural wastes as substitute of natural aggregate in concrete: A review on the durability-related properties. *J. Clean. Prod.* **2020**, *275*, 123062. [CrossRef]
12. Jahanbakhsh, H.; Karimi, M.M.; Naseri, H.; Nejad, F.M. Sustainable asphalt concrete containing high reclaimed asphalt pavements and recycling agents: Performance assessment, cost analysis, and environmental impact. *J. Clean. Prod.* **2020**, *244*, 118837. [CrossRef]
13. Van den bergh, W.; Kara, P.; Anthonissen, J.; Margaritis, A.; Jacobs, G.; Couscheir, K. Recommendations and strategies for using reclaimed asphalt pavement in the Flemish Region based on a first life cycle assessment research. *IOP Conf. Ser. Mater. Sci. Eng.* **2017**, *236*, 012088. [CrossRef]
14. Larsen, J.K.; Eapa, P. *Asphalt in Figures 2015*; EAPA—European Asphalt Pavement Assotiation: Arlington County, VA, USA, 2015; pp. 1–9.
15. Al-Mufti, R.L.; Fried, A.N. Improving the strength properties of recycled asphalt aggregate concrete. *Constr. Build. Mater.* **2017**, *149*, 45–52. [CrossRef]
16. Shah, A.; McDaniel, R.S.; Huber, G.A.; Gallivan, V.L. Investigation of Properties of Plant-Produced Reclaimed Asphalt Pavement Mixtures. *Transp. Res. Rec. J. Transp. Res. Board* **2007**, *1998*, 103–111. [CrossRef]
17. Morales Fournier, J.; Acosta Álvarez, D.; Alonso Aenlle, A.; Tenza-Abril, A.J.; Ivorra, S. Combining Reclaimed Asphalt Pavement (RAP) and Recycled Concrete Aggregate (RCA) from Cuba to Obtain a Coarse Aggregate Fraction. *Sustainability* **2020**, *12*, 5356. [CrossRef]
18. Willis, J.R.; Turner, P.; De Goes Padula, F.; Tran, N.; Julian, G. Effects of Changing Virgin Binder Grade and Content on High Reclaimed Asphalt Pavement Mixture Properties. *Transp. Res. Rec. J. Transp. Res. Board* **2013**, *2371*, 66–73. [CrossRef]
19. Zaumanis, M.; Mallick, R.B.; Poulikakos, L.; Frank, R. Influence of six rejuvenators on the performance properties of Reclaimed Asphalt Pavement (RAP) binder and 100% recycled asphalt mixtures. *Constr. Build. Mater.* **2014**, *71*, 538–550. [CrossRef]
20. Han, S.; Chung, H.-W.; Subgranon, T.; Tia, M. Development of Mix Designs for Minimally Refined Reclaimed Asphalt Pavement (RAP) Concrete for Florida Concrete Test Road. *Sustainability* **2018**, *10*, 2598. [CrossRef]
21. Singh, S.; Ransinchung, G.D.R.N.; Kumar, P. Feasibility study of RAP aggregates in cement concrete pavements. *Road Mater. Pavement Des.* **2019**, *20*, 151–170. [CrossRef]

22. Coppola, L.; Kara, P.; Lorenzi, S. Concrete manufactured with crushed asphalt as partial replacement of natural aggregates. *Mater. Constr.* **2016**, *66*, 101. [CrossRef]
23. Mahmoud, E.; Ibrahim, A.; El-Chabib, H.; Patibandla, V.C. Self-Consolidating Concrete Incorporating High Volume of Fly Ash, Slag, and Recycled Asphalt Pavement. *Int. J. Concr. Struct. Mater.* **2013**, *7*, 155–163. [CrossRef]
24. Huang, B.; Shu, X.; Li, G. Laboratory investigation of portland cement concrete containing recycled asphalt pavements. *Cem. Concr. Res.* **2005**, *35*, 2008–2013. [CrossRef]
25. *EN 197-1, Cement—Part 1: Composition, Specifications and Conformity Criteria for Common Cements*; European Comitee for Standardization (CEN): Brussels, Belgium, 2011.
26. *EN 12620, Aggregates for Concrete*; European Comitee for Standardization (CEN): Brussels, Belgium, 2004.
27. *EN 206-1, Concrete—Part 1: Specification, performance, production and conformity*; European Comitee for Standardization (CEN): Brussels, Belgium, 2017.
28. *EN 1097-3, Test of Mechanical and Physical Properties of Aggregate—Part 3: Determination of Loose Bulk Density and Voids*; European Comitee for Standardization (CEN): Brussels, Belgium, 1999.
29. *EN 1097-7, Test of Mechanical and Physical Properties of Aggregate—Part 7: Determination of Particle Density of Filler—Pyknometr Method*; European Comitee for Standardization (CEN): Brussels, Belgium, 2008.
30. *EN 196-6, Methods of Testing Cement—Part 6: Determination of Fineness*; European Comitee for Standardization (CEN): Brussels, Belgium, 2010.
31. *EN 196-3, Methods of Testing Cement—Part 3: Determination of Setting Times and Soudness*; European Comitee for Standardization (CEN): Brussels, Belgium, 2017.
32. *EN 196-2, Method of Testing Cement—Part 2: Chemical Analysis of Cement*; European Comitee for Standardization (CEN): Brussels, Belgium, 2013.
33. Kennedy, C.T. The Design of Concrete Mixes. *J. Proc.* **1940**, *36*, 373–400.
34. *EN 12390-2, Testing of Hardened Concrete—Part 2: Making and Curing Specimens for Strength Tests*; European Comitee for Standardization (CEN): Brussels, Belgium, 2019.
35. *EN 12390-1, Testing of Hardened Concrete—Part 1: Shape, Dimensions and Other Requirements for Specimens and Moulds*; European Comitee for Standardization (CEN): Brussels, Belgium, 2013.
36. *EN 933-1, Test of Geometrical Properties of Aggregate—Part 1: Determination of Particle Size Distribution—Sieving Method*; European Comitee for Standardization (CEN): Brussels, Belgium, 2012.
37. *EN 1097-6, Tests for Mechanical and Physical Properties of Aggregates—Part 6: Determination of Particle Density and Water Absorption*; European Comitee for Standardization (CEN): Brussels, Belgium, 2014.
38. *EN 12350-2, Testing Fresh Concrete—Part 2: Slump Test*; European Comitee for Standardization (CEN): Brussels, Belgium, 2000.
39. *EN 12390-7, Testing Hardened Concrete—Part 7: Density of Hardened Concrete*; European Comitee for Standardization (CEN): Brussels, Belgium, 2020.
40. Zhang, M.; Li, H. Pore structure and chloride permeability of concrete containing nano-particles for pavement. *Constr. Build. Mater.* **2011**, *25*, 608–616. [CrossRef]
41. Abraham, S.M.; Ransinchung, G.D. Effects of Reclaimed Asphalt Pavement aggregates and mineral admixtures on pore structure, mechanical and durability properties of cement mortar. *Constr. Build. Mater.* **2019**, *216*, 202–213. [CrossRef]
42. Singh, S.; Ransinchung, G.D.; Kumar, P. Laboratory Investigation of Concrete Pavements Containing Fine RAP Aggregates. *J. Mater. Civ. Eng.* **2018**, *30*, 04017279. [CrossRef]
43. Rahardjo, H.; Satyanaga, A.; Leong, E.-C.; Wang, J.-Y. Unsaturated properties of recycled concrete aggregate and reclaimed asphalt pavement. *Eng. Geol.* **2013**, *161*, 44–54. [CrossRef]
44. Brand, A.S.; Roesler, J.R. Bonding in cementitious materials with asphalt-coated particles: Part I—The interfacial transition zone. *Constr. Build. Mater.* **2017**, *130*, 171–181. [CrossRef]
45. Brand, A.S.; Roesler, J.R. Bonding in cementitious materials with asphalt-coated particles: Part II—Cement-asphalt chemical interactions. *Constr. Build. Mater.* **2017**, *130*, 182–192. [CrossRef]
46. Pipilikaki, P.; Beazi-Katsioti, M. The assessment of porosity and pore size distribution of limestone Portland cement pastes. *Constr. Build. Mater.* **2009**, *23*, 1966–1970. [CrossRef]
47. Winslow, D.; Liu, D. The pore structure of paste in concrete. *Cem. Concr. Res.* **1990**, *20*, 227–235. [CrossRef]
48. Zhao, H.; Xiao, Q.; Huang, D.; Zhang, S. Influence of Pore Structure on Compressive Strength of Cement Mortar. Available online: https://www.hindawi.com/journals/tswj/2014/247058/ (accessed on 29 April 2020).

49. Dong, H.; Gao, P.; Ye, G. Characterization and comparison of capillary pore structures of digital cement pastes. *Mater. Struct.* **2017**, *50*, 154. [CrossRef]
50. Diamond, S. Aspects of concrete porosity revisited. *Cem. Concr. Res.* **1999**, *29*, 1181–1188. [CrossRef]
51. Nestle, N.; Zimmermann, C.; Dakkouri, M.; Niessner, R. Action and Distribution of Organic Solvent Contaminations in Hydrating Cement: Time-Resolved Insights into Solidification of Organic Waste. *Environ. Sci. Technol.* **2001**, *35*, 4953–4956. [CrossRef]
52. Hoy, M.; Horpibulsuk, S.; Arulrajah, A.; Mohajerani, A. Strength and Microstructural Study of Recycled Asphalt Pavement: Slag Geopolymer as a Pavement Base Material. *J. Mater. Civ. Eng.* **2018**, *30*, 04018177. [CrossRef]
53. Singh, S.; Ransinchung, G.D.; Kumar, P. An economical processing technique to improve RAP inclusive concrete properties. *Constr. Build. Mater.* **2017**, *148*, 734–747. [CrossRef]

2

Study of the Stiffness of the Bitumen Emulsion Based Cold Recycling Mixes for Road Base Courses

Katarzyna Konieczna [1,*], Piotr Pokorski [1], Wojciech Sorociak [2,3], Piotr Radziszewski [1], Dawid Żymełka [3] and Jan Bolesław Król [1]

1 Faculty of Civil Engineering, Warsaw University of Technology, Al. Armii Ludowej 16, 00-637 Warsaw, Poland; p.pokorski@il.pw.edu.pl (P.P.); p.radziszewski@il.pw.edu.pl (P.R.); j.krol@il.pw.edu.pl (J.B.K.)

2 Faculty of Civil Engineering, Silesian University of Technology, ul. Akademicka 5, 44-100 Gliwice, Poland; wojciech.sorociak@eurovia.pl

3 Eurovia Polska S.A., Bielany Wrocławskie, ul. Szwedzka 5, 55-040 Kobierzyce, Poland; dawid.zymelka@eurovia.pl

* Correspondence: k.konieczna@il.pw.edu.pl

Abstract: The benefits of the use of cold recycling mixtures (CRMs) in pavement rehabilitation are associated with both the reduction of natural resource consumption by replacing them with recycled materials and the reduction of energy consumption during their production and paving. The evolution of the stiffness of CRMs in road construction and the fatigue life of pavements with CRM base layers are still being investigated. In this paper, CRMs with 1% cement content, called bitumen-stabilized materials with bitumen emulsion (BSM-Es), were examined. Mixtures that were differentiated in terms of Reclaimed Asphalt Pavement (RAP) content, as well as the amount and type of bitumen emulsions, were subjected to indirect tensile stiffness modulus (ITSM) tests at 5 °C, 13 °C, and 20 °C. The thermal sensitivities of the BSM-E mixtures were analyzed. BSM-E mixture stiffness modulus levels at various temperatures were determined using a statistical approach. On the basis of the results obtained, a discussion on the mechanistic-empirical design of flexible pavements with BSM-E base layers is presented. The potential benefits of using BSM-E materials in road construction in certain aspects of pavement life are indicated.

Keywords: cold recycling mixtures; bitumen-stabilized materials; ITSM; mechanistic-empirical pavement design; bitumen emulsion

---

1. Introduction

In the era of limited availability of natural resources and growing awareness of environmental issues, cold recycling mixtures (CRMs) are gaining popularity in terms of pavement base layer construction. The main advantages of this technology are the reduction of costs, energy consumption, and greenhouse gas emissions in comparison with conventional hot mix asphalt (HMA) production [1,2].

The properties of CRMs depend on the qualitative and quantitative selection of their components, which are Reclaimed Asphalt Pavement (RAP), virgin aggregates, bitumen emulsion or foamed bitumen, water, and active filler. Depending on the applied combination of binding agents, various types of cold recycling mixtures can be distinguished, such as cement-treated materials (CTMs) with cement as the only binding agent, cement–bitumen-treated materials (CBTMs) containing bitumen emulsion or foamed bitumen with a ratio of residual bitumen (B) to cement (C) less than or equal to one ($B/C \leq 1$), and bitumen-stabilized materials (BSMs) with a residual bitumen content of approximately 3% and a maximum amount of active filler (e.g., cement) equal to 1% ($B/C \geq 1$) [3,4].

While CTMs are considered to be bound materials with high stiffness, rigidity, and brittleness [5–7], bitumen-stabilized materials (BSMs), due to a relatively high bitumen content and reduced amount of cement, are characterized by flexibility, lower stiffness, and a lower tendency to experience shrinkage cracking. This is due to the fact that the binder in BSMs is selectively dispersed among the fine aggregate particles, creating local bonds between the coarse aggregate skeleton [8,9]. Due to this phenomenon, mixtures are characterized by the combination of the mechanical properties of unbound (stress-dependent behavior) and viscoelastic materials (time and temperature-dependent behavior) [10–12]. The role of an active filler in BSMs with bitumen emulsion (BSM-Es) is mostly related to the acceleration of the bitumen emulsion breaking process and decreasing the moisture susceptibility of the material [8,9,13]. The mode of failure for BSMs is considered to be permanent deformation [3,8,9].

The quantitative and qualitative selection of ingredients plays an extremely important role in developing the mechanical and durability properties of CBTM and BSM mixtures, as reported by many researchers [14–17]. According to the extensive research of Dołzycki and Jaskula [18], pavement sections used in recent decades in Poland which were made with base layers with CBTM mixtures (referred to in the paper as mineral-cement-emulsion (MCE) mixtures) developed transverse reflective cracking. The main reasons for this phenomenon were considered to be excessively high cement contents (above 2.5%) in the CBTM (MCE) mixtures and, simultaneously, insufficient bitumen emulsion contents (often below 3% of the mixture mass), which resulted in an overly stiff mixture prone to cracking. This can lead to the conclusion that CBTM base layers with high cement contents (2.5% and more) may exhibit a two-phase behavior: a pre-cracked phase with higher initial stiffness and a post-cracked phase with a significantly lower stiffness of the material.

The aspects of fatigue behavior, as well as the cracking resistance of cold recycling mixtures, are considered complex and have not been fully investigated [19–22]. As BSMs cannot be characterized and modelled as purely viscoelastic nor as granular materials [11,12,23], the behavior of the BSM base layers in the pavement structure is still under discussion. Two leading hypotheses that present different approaches regarding the evolution of the material's stiffness in construction have been developed so far. The first hypothesis, developed by Liebenberg and Visser [24,25], is based on the structural design of semi-rigid pavements. It assumes two phases of BSM performance: an effective fatigue phase, in which a high initial stiffness gradually diminishes due to fatigue, and an equivalent granular phase when the BSM layer is cracked and behaves similarly to an unbound granular material, although with a slightly greater stiffness modulus than the parent material (terminal stiffness).

According to the second hypothesis, presented by Ebels [13], the service life of BSMs consists of a curing phase, with an increase in initial stiffness resulting from moisture reduction and densification of the BSM layer, and a stiffness reduction phase, when the BSMs demonstrate the quasi-viscoelastic behavior with temperature, time, and load frequency dependence characteristics for hot mix asphalt (HMA). The second concept of stiffness development is likely to be more appropriate to the service life of BSMs, according to indirect tensile tests (ITTs) and falling weight deflectometer (FWD) measurement results in the framework of internationally conducted pilot projects [26].

The research presented in this paper consists of laboratory testing and discussion on the mechanistic-empirical design of flexible pavements with base layers consisting of bitumen-stabilized materials with bitumen emulsion (BSM-E). The experimental program focuses on the determination of the stiffness modulus (ITSM) of BSM-E mixtures at 5 °C, 13 °C, and 20 °C. Based on the obtained results, the thermal sensitivity and stiffness levels for BSM-E mixtures at various temperatures were characterized. In the second part of the paper, the fatigue life of flexible pavement with BSM-E base layers was determined according to the AASHTO 2004 pavement design procedure. An assumption was made that the BSM-E mixture retained its homogeneity in its entire service life (i.e., no brittle fracture occurred). The calculations performed for the pavement structures with BSM-E base layers were compared to those performed for reference structures with CBTM base layers used on site in Poland.

2. Materials and Methods

2.1. Materials and BSM-E Mix Design

Eight types of BSM-E mixtures were designed. Differentiation of the mixtures was made in terms of the amount of RAP (70% and 50% of the mineral mixture mass), as well as the amount and type of bitumen emulsion used. Additionally, one reference CBTM mixture with 50% RAP, 3.0% cement and 3.0% bitumen emulsion addition was designed.

Grading curves of the BSM-E mixtures were designed using 50% and 70% RAP 0/31.5 mm, derived from milling and wearing and base asphalt courses originally containing dolomite 0/11 and basalt 11/16 aggregates. The average bitumen content in the RAP was determined to be 5.0% by its aggregate weight (PN-EN 12697-1). In order to meet the grading criteria presented in the international standards for BSMs [8], virgin aggregates, such as continuously graded basalt aggregate 0/31.5 mm and fine basalt aggregate 0/2 mm, were added to the mixture. The mixing proportions of the RAP and virgin aggregates were set to obtain a similar shape for the grading curves of the BSM-E and CBTM mixtures, as shown in Figure 1.

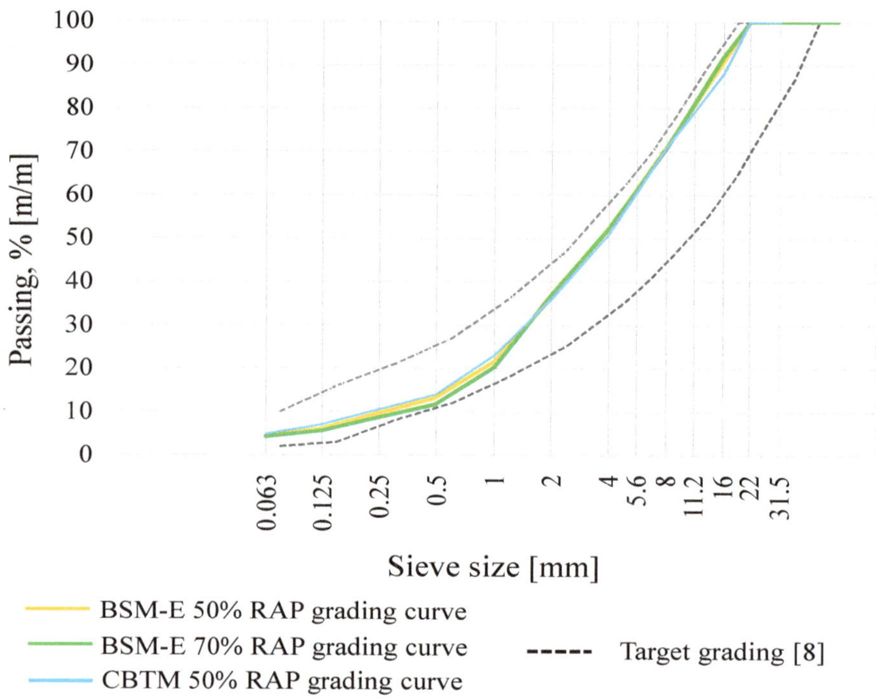

Figure 1. Bitumen-stabilized materials with bitumen emulsion (BSM-E) and cement–bitumen-treated material (CBTM) grading curves with target grading [8].

Two types of slow-setting cationic bitumen emulsions with 60% residual bitumen content (type 1: E1, with a bitumen emulsion produced on the base of 50/70 penetration grade bitumen, and type 2: E2, with a bitumen emulsion of C60B10 ZM/R with 70/100 penetration grade bitumen) and the cement type CEM II/B-S 32.5 R (EN 197-1) were used to prepare the BSM-E mixtures. The amount of cementitious binder was set to a constant 1% of the mixture mass. The optimum moisture content (OMC) for mixtures was determined in accordance with the modified Proctor moisture–density relationship test procedure (PN–EN 13286-2). The OMC was set to 6.8% for BSM-E mixtures containing 70% RAP and 7.6% for BSM-E mixtures containing 50% RAP. The OMC of the CBTM mixture was set to 5.0%. The mixture compositions of the BSM-E mixtures and CBTM mixture are listed in Table 1.

Table 1. BSM-E and CBTM mixture compositions.

Mixture Type	Mixture Designation	Ingredient Content [%]					
		RAP	Basalt Aggregate 0/31.5 mm	Basalt Aggregate 0/2 mm	Bitumen Emulsion	Bitumen Type	Cement
BSM-E	70%_3.0%E1 70%_3.0%E2 70%_4.5%E1 70%_4.5%E2	70	–	30	3.0 4.5	50/70 70/100 50/70 70/100	1.0
BSM-E	50%_3.0%E1 50%_3.0%E2 50%_5.2%E1 50%_5.2%E2	50	37.5	12.5	3.0 5.2	50/70 70/100 50/70 70/100	1.0
CBTM	CBTM 3% Cement	50	50	–	3.0	70/100	3.0

2.2. *Specimen Preparation and Curing*

The BSM-E and CBTM mixtures were mixed and compacted at ambient temperatures (25 °C). The test specimens, with a height of 63.5 mm ± 3.5 mm and a diameter of 101.6 mm, were compacted using a Marshall hammer by applying 75 blows per side, according to national requirements [27]. Perforated molds with 24 holes 2 mm in diameter were used in order to facilitate the drainage of water during the compaction process. Once compacted, specimens were left in moulds for 24 h in order to develop sufficient strength before demoulding. After that, for the BSM-E specimens, the accelerated curing protocol was adopted [8]. The specimens were placed in a forced-air oven at 40 °C and cured for 72 h. After the curing process, specimens were cooled at ambient temperature (25 °C). For the CBTM specimens, due to the higher content of cement, the curing conditions presented in national requirements were applied (28 days, 25 °C, and air humidity 40–70%) [27].

Bulk densities of the BSM-E specimens ranged from 2.092 Mg/m^3 to 2.138 Mg/m^3 for mixtures containing 70% RAP in the mineral mixture and from 2.144 Mg/m^3 to 2.160 Mg/m^3 for mixtures containing 50% RAP. The average bulk density of the CBTM mixture specimens was 2.207 Mg/m^3.

2.3. *Indirect Tensile Stiffness Modulus (ITSM) Testing*

A UTM-25 universal testing machine was used to carry out stiffness modulus tests (ITSM) at 5 °C, 13 °C, and 20 °C in an indirect tensile configuration (IT-CY) according to PN-EN 12697-26. A target deformation value of 5 μm was selected in accordance with national instructions [27] in order to avoid premature brittle failure by the specimens during the early stages of the test. Measurements were conducted on at least 3 specimens (on average, 6 samples for each mixture) along two perpendicular diameters for each specimen, and the average ITSM values for each mixture type and each testing temperature were calculated.

2.4. *Methodology of Flexible Pavement Design with a Cold Recycling Mix (CRM) Base Layer*

Design of the flexible pavement with a BSM-E base layer was carried out according to fixed pavement structures presented in the national design and construction guide *The Catalogue of Typical Flexible and Semi-rigid Pavements* [28]. The mechanistic-empirical method was used based on the following criteria: bottom-up fatigue cracking of the bituminous layers and permanent deformation (AASHTO 2004 pavement design procedure). In order to calculate the stress and strain state in the pavement structures, the theory of elastic layered half space was applied. Calculations were conducted using BISAR 3.0 software under the assumption that the single axis traffic load was represented by a circular contact area with P = 50 kN load and q = 850 kPa contact pressure.

The comparison of the fatigue lives was carried out between flexible pavement structures with base layers constructed with BSM-E and CBTM mixtures. Both the BSM-E and CBTM layers were considered as flexible. In accordance with the national design guide recommendations [28], the thickness (h_i), Poisson's ratio (ν_i), and modulus of elasticity (stiffness) (E_i) at the equivalent temperature of 13 °C were specified for each pavement layer. The stiffness modulus of the traditionally used CBTM mixtures was set to 1500 MPa, and the Poisson's ratio was set to 0.30. The value of 1500 MPa was the stiffness modulus value, which is recommended for calculations in the Polish guidelines for CBTM mixtures [28]. There is no recommended stiffness modulus value separately established for BSM-E mixtures. Due to the variety of stiffness modulus values obtained both in the laboratory and during in situ testing for the CRMs, the assumption of a single stiffness modulus value of 1500 MPa for a wide range of cold recycling mixtures is still being discussed. As stated in the paper by Dolzycki and Jaskula [18], although the 1500 MPa value was lower than the values obtained during laboratory and in situ testing of the CBTM mixtures (referred to in the paper as mineral-cement-emulsion (MCE) mixtures), it was supposed to represent the inhomogeneous material after a certain period of performance, namely when the first distress and micro-cracks began to occur. The BSM-E stiffness modulus for pavement design purposes was calculated using a statistical approach on the basis of the results obtained in this research.

3. Results and Discussion

3.1. Indirect Tensile Stiffness Modulus (ITSM)

The results of the ITSM tests at 5 °C, 13 °C, and 20 °C, presented in Figure 2a–i, show that the ITSM values of the BSM-E and CBTM mixtures decreased with the increase of the testing temperature. The average stiffness moduli of all BSM-E mixtures ranged from 4382 MPa to 5215 MPa at a temperature of 5 °C, from 3037 MPa to 3721 MPa at 13 °C, and from 1146 MPa to 2913 MPa at 20 °C. While comparing these values with the values obtained for the CBTM mixture containing 3% cement, it could be stated that the average stiffness of the BSM-E mixtures was approximately 50–60% lower.

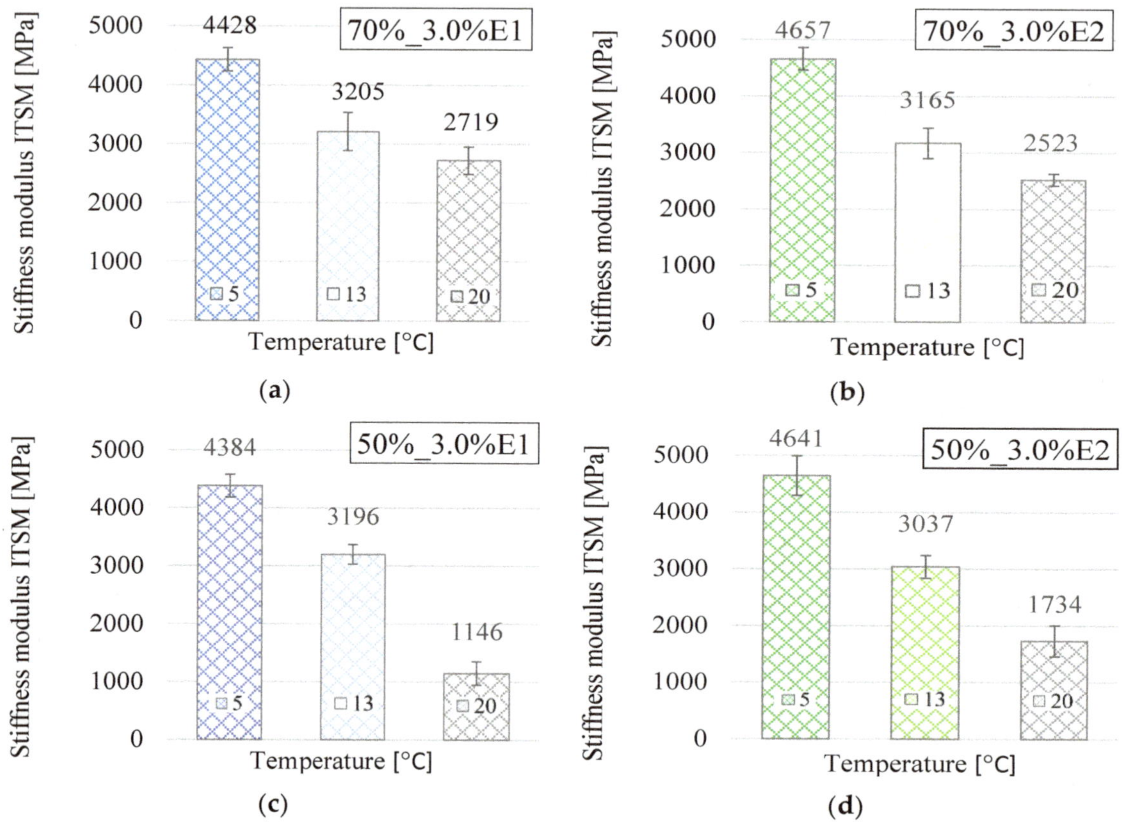

Figure 2. *Cont.*

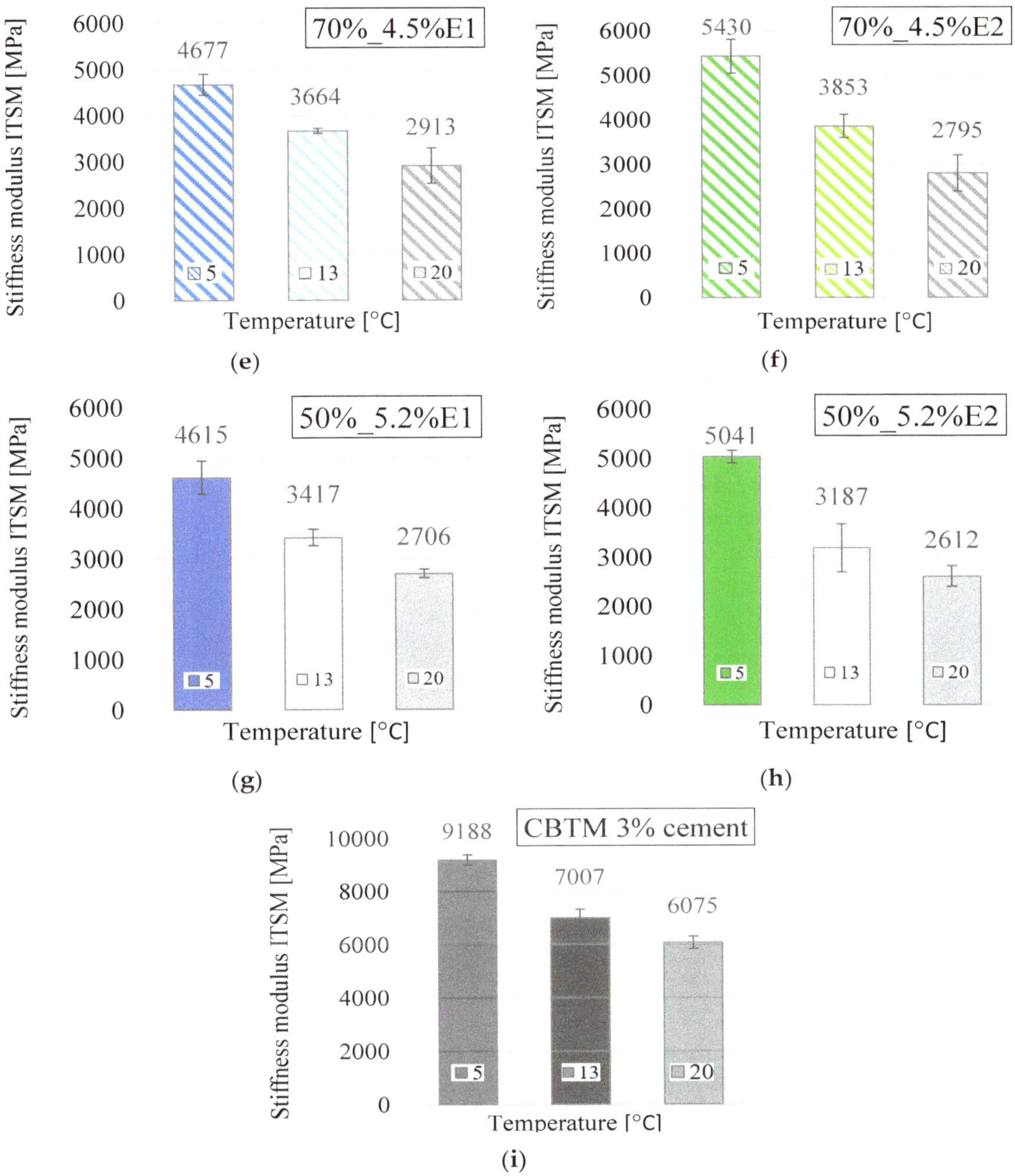

Figure 2. ITSM test results for BSM-E and CBTM mixtures at 5 °C, 13 °C, and 20 °C. (**a**) BSM-E 70%_3.0%E1, (**b**) BSM-E 70%_3.0%E2, (**c**) BSM-E 50%_3.0%E1, (**d**) BSM-E 50%_3.0%E2, (**e**) BSM-E 70%_4.5%E1, (**f**) BSM-E 70%_4.5%E2, (**g**) BSM-E 50%_5.2%E1, (**h**) BSM-E 50%_5.2%E2, and (**i**) CBTM with 3% cement content.

To evaluate statistically significant differences in the BSM-E stiffness modulus values, one-way analysis of variance (ANOVA) using Statgraphics Centurion 18.1.06 software was performed. By means of the F-test, it was concluded that there was a statistically significant difference between the ITSM mean values at the 5% significance level for each testing temperature. Based on the stiffness modulus values presented in Figure 2a–i, it can be observed that the BSM-E mixtures containing 50% RAP were charaterized by lower ITSM values than the corresponding BSM-E mixtures with a higher RAP content. However, the multiple range tests did not show statistically significant differences between corresponding mixtures with different RAP percentages.

In the case of the BSM-E mixtures with 50% RAP and 3.0% bitumen emulsion content, a significant decrease in the value of the ITSM stiffness modulus at 20 °C was observed. Such a tendency was not observed for the BSM-E mixtures with 70% RAP content.

It can be assumed that this reduction of the stiffness is related to the loss of cohesion of the mixtures with low contents of bitumen emulsion (1.8% residual binder) and relatively low contents of RAP. Although RAP has so far been mainly treated as a black rock in cold recycling mixtures, it is not without significance that the old binder surrounding the RAP particles could still take part in binding the aggregates together during compaction [29]. Ebels [13] reported that the RAP particles present in cold recycling mixtures were characterized by a lower absorption rate of fresh bitumen compared with virgin aggregates, which resulted in a greater affinity between the fresh bitumen and the RAP particles.

3.2. Temperature Dependency

As mentioned in Section 3.1, the obtained ITSM test results confirmed the temperature dependency of the BSM-E mixture, displaying a decrease in ITSM values with the increase in the testing temperature (Figure 3). The stiffness and temperature relationships were based on the average values at three different temperatures. The error bars are not shown in Figure 3 (given in Figure 2), taking into account the clarity of the presented relationships.

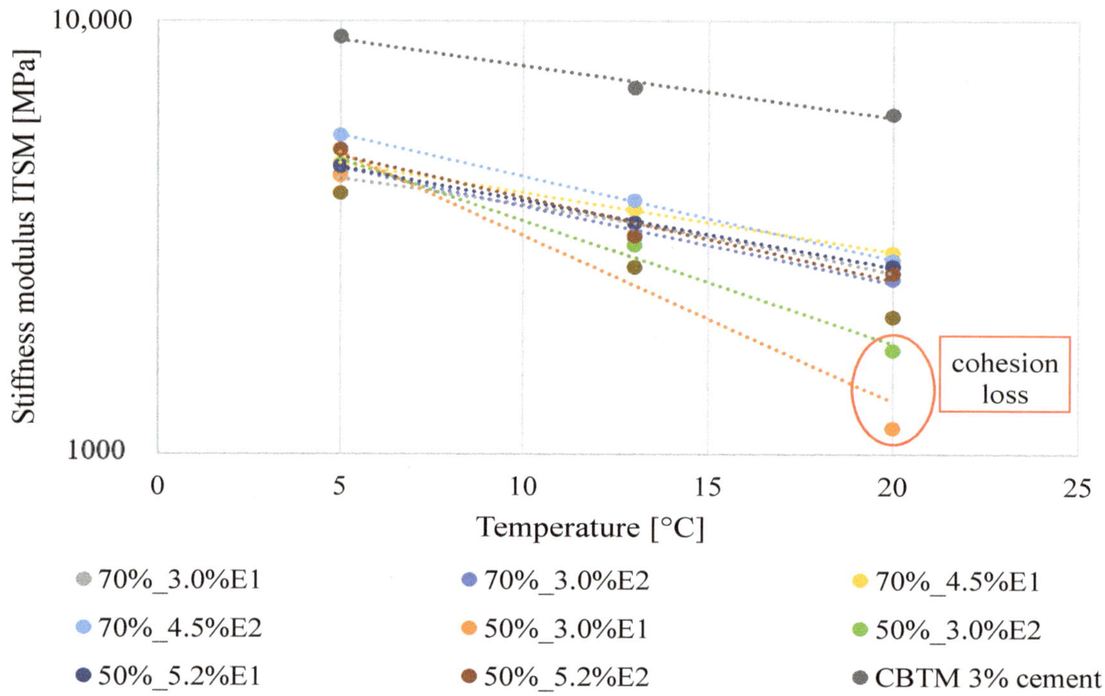

Figure 3. ITSM values as a function of the testing temperature.

In order to quantify the temperature dependency of the BSM-E mixtures, the following equation was proposed:

$$\log S = -A_{ITSM} \cdot t + B_{ITSM} \quad (1)$$

where S is the ITSM stiffness modulus at temperature t and A_{ITSM} and B_{ITSM} are parameters depending on the BSM-E mixture characteristics. The fitting parameters of the regression curve for the BSM-E mixtures are presented in Table 2.

Table 2. Fitting parameters for the ITSM temperature dependency, calculated in Equation (1).

ITSM	CBTM 3% Cement	70%_3.0% E1	70%_3.0% E2	50%_3.0% E1	50%_3.0% E2	70%_4.5% E1	70%_4.5% E2	50%_5.2% E1	50%_5.2% E2
A_{ITSM} (logMPa/°C)	0.0120	0.0142	0.0178	0.0383	0.0284	0.0137	0.0193	0.0155	0.0192
B_{ITSM} (logMPa)	4.0166	3.7086	3.7492	3.8872	3.8221	3.7395	3.8279	3.7395	3.7838
R^2	0.9809	0.9783	0.9878	0.8918	0.9861	0.9996	0.9983	0.9988	0.9659

Considering the R^2 values presented in Table 2, it can be stated that the applied analytical model allowed a good fitting with the ITSM results (R^2 = 0.9659 – 0.9983). The 50%_3.0%E1 mixture was the exception, with an R^2 value of 0.8918. Regression curves for the BSM-E mixtures with 50% RAP and 3.0% bitumen emulsion signifficantly differed from those obtained for other mixtures as a result of a significant drop in the stiffness modulus value at a temperature of 20 °C (cohesion loss). The regression curve slope values (represented by A_{ITSM}) described the gradient of stiffness modulus change as a function of the testing temperature; the higher the slope (A_{ITSM}) value, the more temperature-sensitive the BSM-E mixture was. Based on the calculated A_{ITSM} values, it can be stated that all the BSM-E mixtures containing type E2 bitumen emulsion with a 70/100 binder (except for 50%_3,0%E2) were characterized by a higher temperature dependency than corresponding mixtures containing type E1 bitumen emulsion with a 50/70 penetration grade binder. In comparison with the CBTM mixture, the BSM-E mixtures presented higher A_{ITSM} values, which stood for a higher sensitivity to changes in the test temperature.

3.3. Statistical Analysis of the BSM-E Stiffness Level

Statistical analysis of the results was performed to evaluate the level of BSM stiffness at the three test temperatures (5 °C, 13 °C, and 20 °C). The ITSM test results, divided into three groups, were evaluated using statistical data analysis tools (Statgraphics Centurion 18.1.06 software). Probability distribution fitting was conducted, and box plots were created to illustrate the dispersion of the obtained ITSM values in each group. Statistical parameters of the tested statistical samples are presented in Table 3.

Table 3. Statistical parameters of tested ITSM values samples.

Temperature (°C)	Statistical Sample Size	Average Stiffness Modulus Value (MPa)	St. Dev. (MPa)	*p*-Value (Shapiro–Wilk Test)
5	40 specimens	4678	406	0.056872
13	50 specimens	3289	355	0.777681
20	40 specimens	2619	329	0.370573

Based on the application of the Shapiro–Wilk test, it was concluded that at the adopted level of significance $\alpha = 0.05$, there were no grounds for rejecting the hypothesis that the analyzed ITSM results in all three samples came from the normal distribution (Shapiro–Wilk test p-value $> \alpha = 0.05$). The general stiffness levels of the BSM-E mixtures were defined as stiffness modulus values S_{temp}, for which the probability of getting a lower ITSM value was approximately equal to 5% ($P(X < S_{temp}) \cong 0.05$).

For the test temperature of 5 °C, the average ITSM value was calculated as 4678 MPa, while the standard deviation of the results was equal to 406 MPa. As shown in Figure 4, the BSM-E mixture 70%_4.5%E2 was characterized by the highest average stiffness modulus at 5 °C. Individual values for the other mixtures were in the range of 3990–5159 MPa.

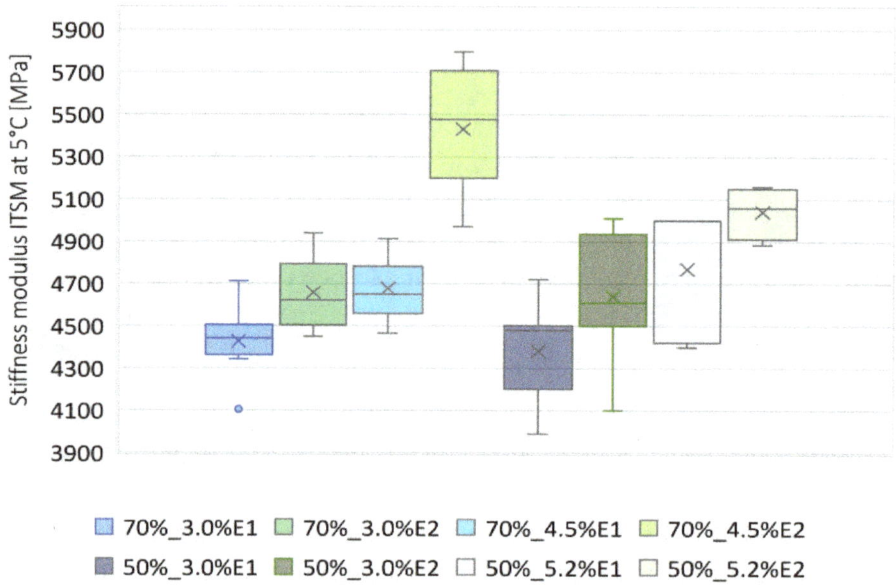

Figure 4. Box plot of ITSM values at 5 °C.

On the basis of the conclusions made, the probability density function of a normal distribution with a mean value μ and a standard deviation σ: N ($\mu = 4678$, $\sigma = 406$) was created (Figure 5).

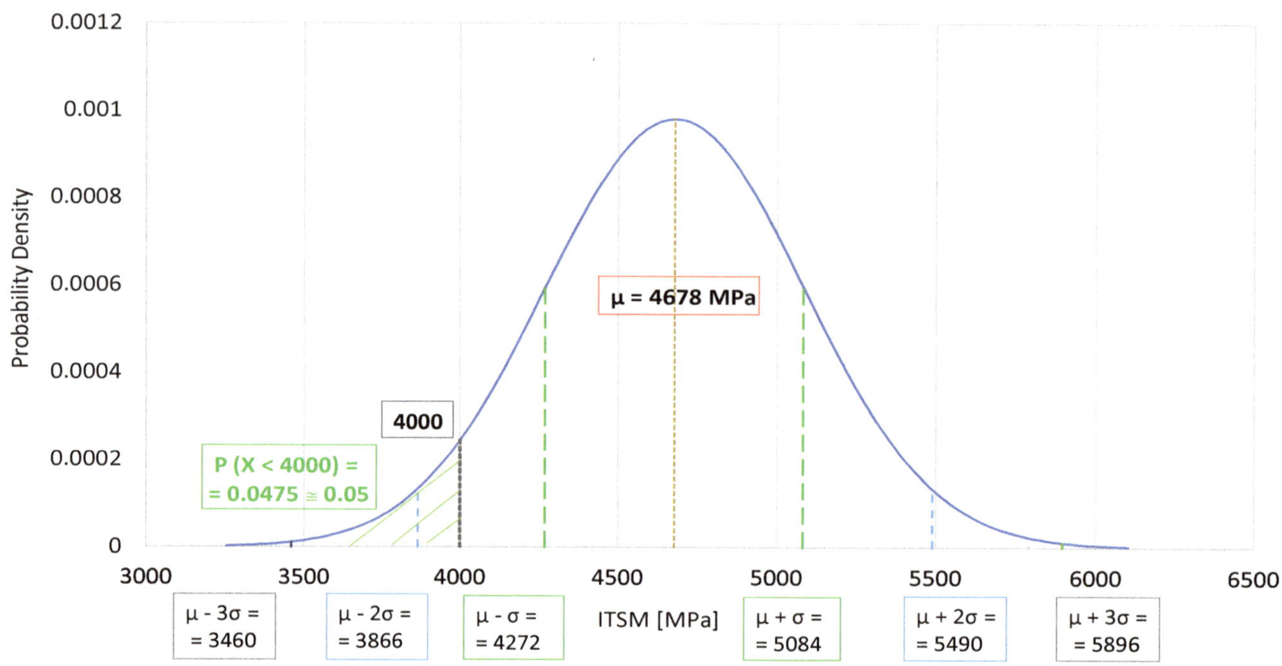

Figure 5. Probability density function of a normal distribution N (4678; 406) for ITSM values at 5 °C.

Based on the areas limited by subsequent standard deviation values that had been marked in the Gaussian curve, the value of 4000 MPa was defined as the BSM-E stiffness level at 5 °C.

For the test temperature of 13 °C, the average ITSM value was calculated to be 3289 MPa, while the standard deviation of the results was equal to 355 MPa. Individual ITSM values for the BSM-E mixtures were in the range of 2695–4089 MPa, with the 70%_4.5%E2 mixture characterized by the highest average stiffness modulus (Figure 6). The probability density function of a normal distribution N (3289; 355) for the ITSM values is presented in Figure 7. The value of 2700 MPa was defined as the BSM-E stiffness level at 13 °C. According to national requirements [28], the stiffness modulus value at 13 °C is used as a parameter for flexible pavement design.

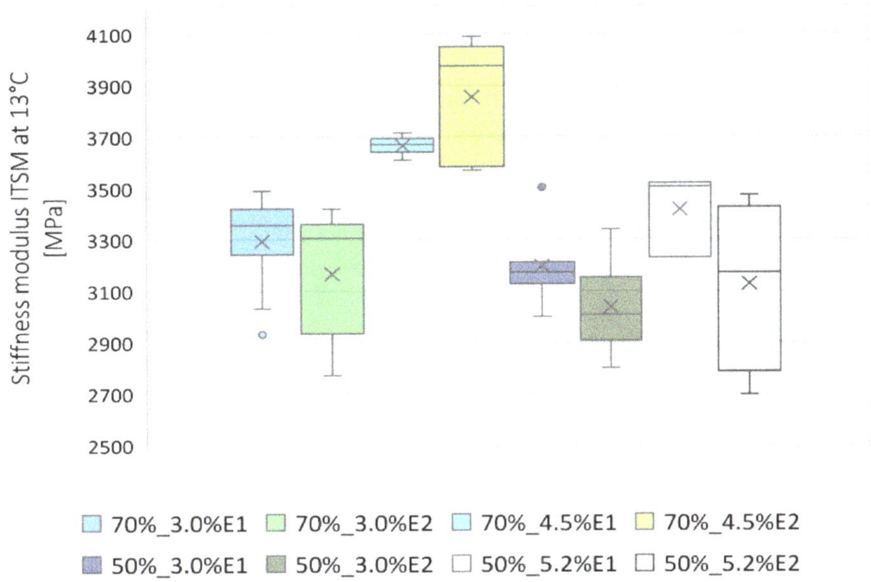

Figure 6. Box plot of ITSM values at 13 °C.

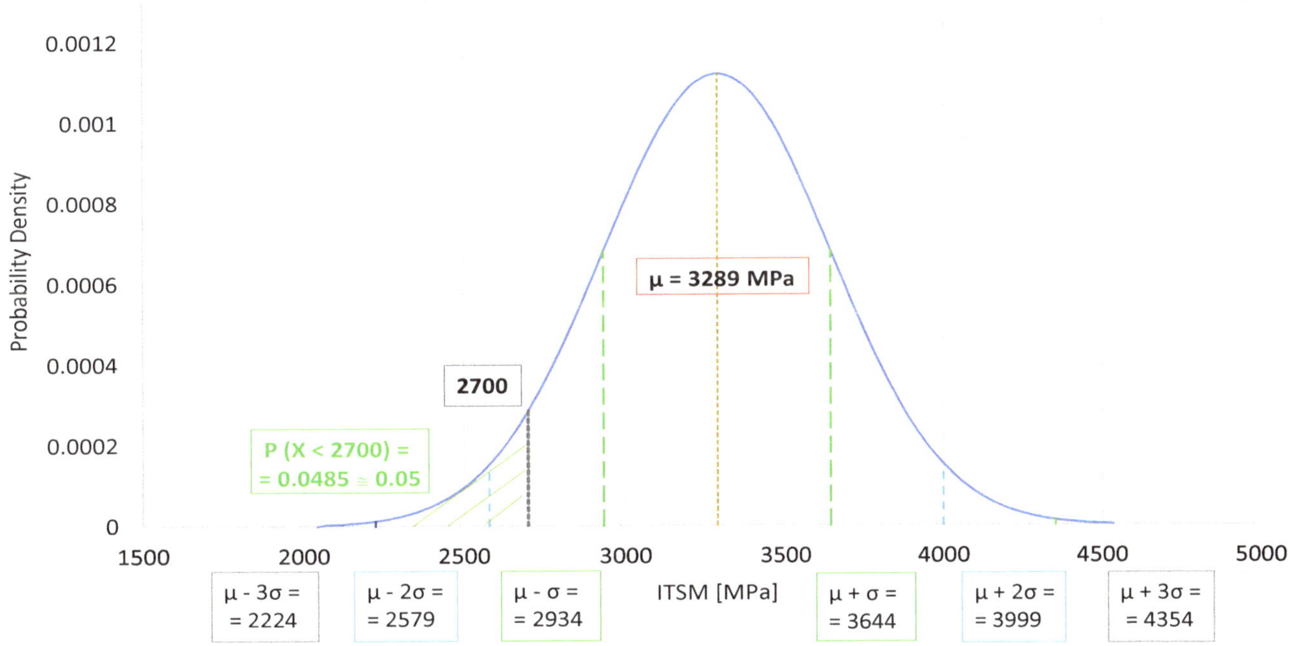

Figure 7. Probability density function of a normal distribution N (3289; 355) for ITSM values at 13 °C.

As presented in Figure 8, at a temperature of 20 °C, the ITSM values of the BSM-E mixtures with 50% RAP and 3.0% bitumen emulsion were significantly lower than the values obtained for other mixtures. Therefore, the stiffness moduli of those mixtures were considered as outliers and were not taken into account in the statistical analysis. The average ITSM value, based on the rest of the results, was calculated as 2619 MPa, while the standard deviation of the results was equal to 329 MPa. The probability density function of a normal distribution N (2619; 329) for the ITSM values at 20 °C is presented in Figure 9. The value of 2070 MPa was defined as the stiffness level of the BSM-E mixtures at the temperature of 20 °C.

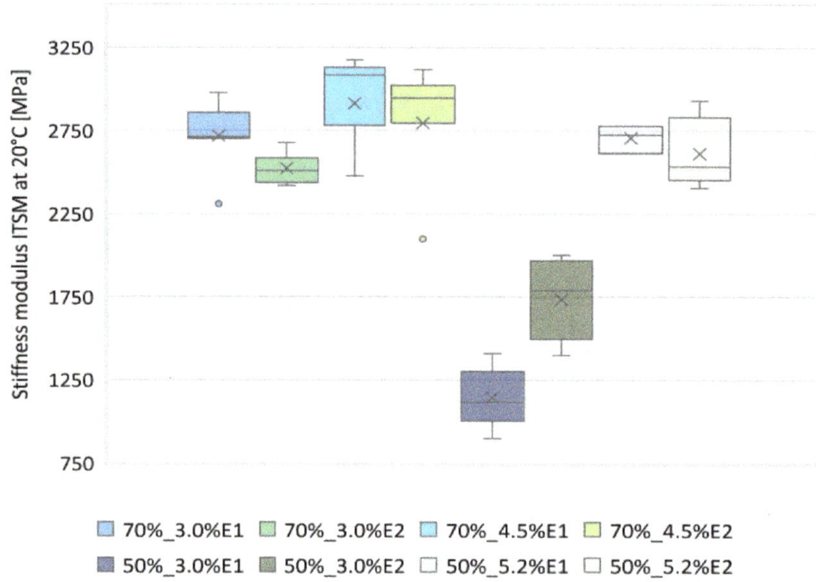

Figure 8. Box plot of ITSM values at 20 °C.

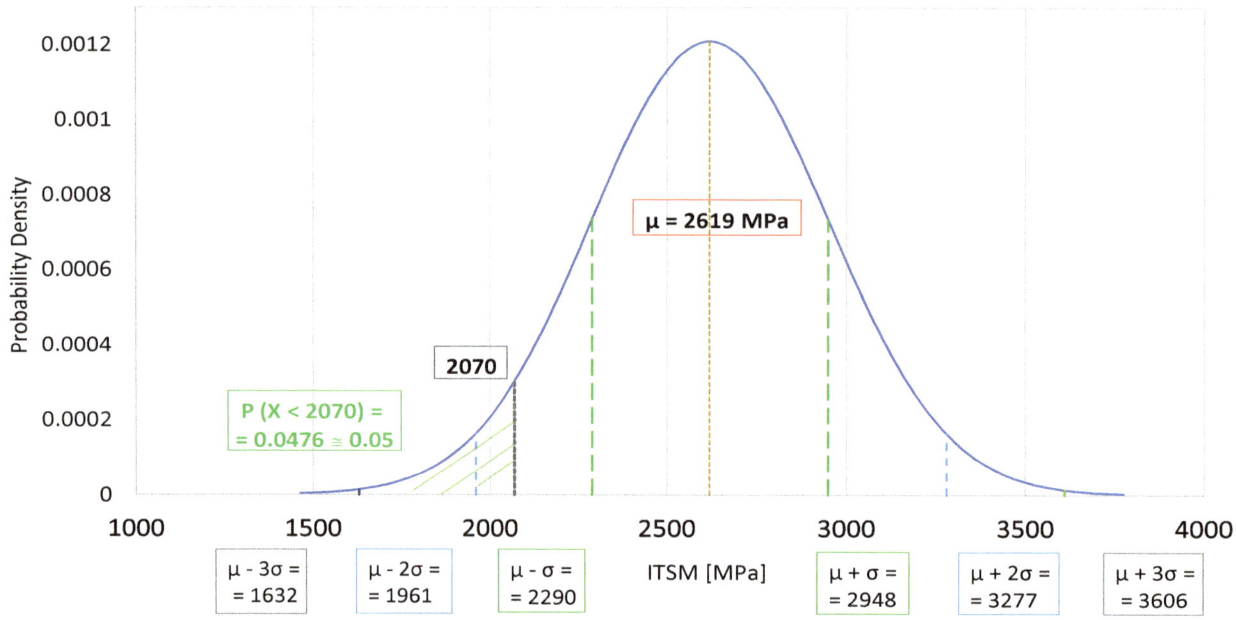

Figure 9. Probability density function of a normal distribution N (2619; 329) for ITSM values at 20 °C.

Figure 10 presents the regression curve for the calculated BSM-E stiffness modulus levels (S_{temp}) and single points that represent the average ITSM values for the designed BSM-E mixtures at temperatures of 5 °C, 13 °C, and 20 °C. It can be seen that the specified S_{temp} values were lower than the laboratory-tested average ITSM values for the BSM-E mixtures (except for the results obtained for the BSM-E mixtures with 50% RAP and 3.0% bitumen emulsion at 20 °C).

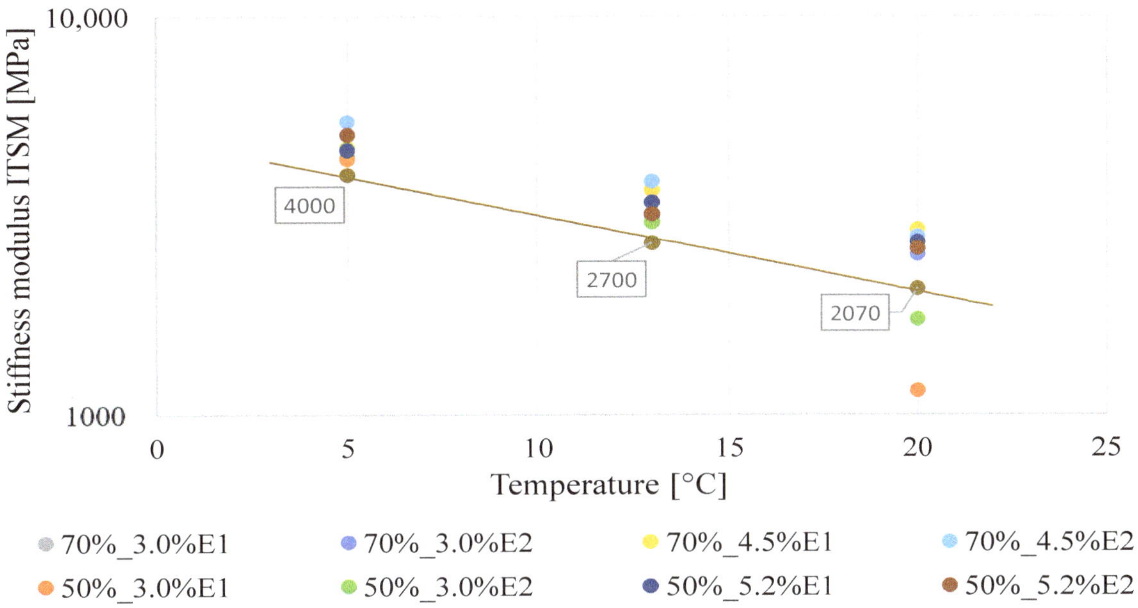

Figure 10. Regression curve for the specified BSM-E stiffness values at 5 °C, 13 °C, and 20 °C.

4. Analysis of Flexible Pavement Structures with BSM-E Base Layers

Table 4 presents the results of the pavement life analysis conducted for the reference structure, consisting of flexible pavement with a base layer made of a CBTM mixture traditionally used in Poland and a pavement structure with a BSM-E base layer. The Polish traffic categories KR1–KR2 are considered to be light traffic, whereas categories KR3–KR4 stand for medium traffic.

Table 4. Calculated pavement life of flexible pavement structures with CBTM and BSM-E base layers.

Base Layer		CBTM Mixture; E = 1500 MPa				BSM-E Mixture; E = 2700 MPa			
Traffic Category		KR1	KR2	KR3	KR4	KR1	KR2	KR3	KR4
Asphalt/CBTM or BSM-E Layers Thickness (cm)		8/15	12/15	12/20	16/20	8/15	12/15	12/20	16/20
Pavement Life (No. of Equivalent Single Axle Loads)	N1 [1]	0.1	0.8	2.8	10.5	0.5	2.1	8.7	28.0
	N2 [2]	17.2	9.6	12.6	17.8	252.4	60.1	87.5	86.8
	Min (N1, N2)	0.1	0.8	2.8	10.5	0.5	2.1	8.7	28.0
Required Pavement Life acc. to Traffic Category (No. Equivalent Single Axle Loads) [28]		0.03–0.09	0.09–0.50	0.5–2.5	2.5–7.4	0.03–0.09	0.09–0.50	0.5–2.5	2.5–7.4
Reserve of Pavement Life (%)		+11%	+60%	+12%	+42%	+456%	+320%	+248%	+278%

[1] N1 = permanent deformation criterion. [2] N2 = fatigue cracking of bituminous layers criterion (FC = 10%).

It can be observed that for both variants of the pavement structure, pavement life was determined by the permanent (structural) deformation criterion. The results of calculations of the flexible pavement structure with a base layer of a BSM-E mixture indicated a significant reserve of the required pavement life (over 200%) for all traffic categories. The obtained numbers of equivalent single axle loads (100 kN) to failure for pavements with a BSM-E layer were approximately three to five times greater than those calculated for pavement structures with a CBTM mixture characterized by a stiffness modulus equal to 1500 MPa. Such a significant overdesign of the pavement structures with BSM-E encouraged authors to propose alternative design solutions. For technological reasons, in the case of traffic categories KR1 and KR2, the thickness of the asphalt layer packet had not been changed. The analysis of the fatigue life of category KR1–KR2 pavement structures led to a conclusion that the use of BSM-E base layers could bring significant benefits in terms of the extended service life and reduction of possible maintenance and rehabilitation cost of local roads with light traffic loading. Alternative design solutions for flexible pavements with BSM-E base layers in medium traffic categories (KR3–KR4) are shown in Table 5.

Table 5. Alternative design solutions for flexible pavements with BSM-E base layers.

Traffic Category		**KR3**		**KR4**	
Pavement Design Solution		Basic Solution CBTM	Alternative Solution BSM-E	Basic Solution CBTM	Alternative Solution BSM-E
Asphalt/CBTM or BSM-E Layers Thickness (cm)		12·(4 + 8) [i]/20	10·(4 + 6) [i]/20	16·(4 + 5 + 7) [ii]/20	14·(4 + 4 + 6) [ii]/20
Tensile Strain: Bottom of Asphalt Layers (μm/m)		82.50	48.58	70.45	55.32
Compressive Strain: Top of Subgrade (μm/m)		381.8	342.9	285.5	259.50
Pavement Life (No. of Equivalent Single Axle Loads)	N1 [1]	2.8	4.6	10.5	16.1
	N2 [2]	12.6	118.9	17.8	48.1
	min (N1, N2)	2.8	4.6	10.5	16.1
Required Pavement Life Acc. to Traffic Category (No. of Equivalent Single Axle Loads) [28]		0.5–2.5		2.5–7.4	
Basic/Alternative Solution Pavement Life Ratio		1	1.6	1	1.5

[i] Surface course + binder course. [ii] Surface course + binder course + base course. [1] N1 = permanent deformation criterion. [2] N2 = fatigue cracking of bituminous layers criterion (FC = 10%).

The main goal of creating pavement structure alternatives was to obtain similar values of pavement life for both basic and alternative design solutions while ensuring construction criteria was followed. In that case, the incorporation of the BSM-E layer allowed authors to propose the reduction of the suggested thicknesses of asphalt layer packets by 2 cm for KR3 as well as for the KR4 traffic category. It can be observed that the values of the tensile stresses at the bottom of the asphalt layers decreased by approximately 40% for the KR3 traffic category and 20% for the KR4 traffic category, in comparison with the basic pavement solutions with CBTM mixtures. The potential possibility of reducing the thickness of asphalt layer packets in the case of medium traffic category roads without negatively affecting pavement life has unquestionable benefits in terms of construction material savings, as well as construction, maintenance, and rehabilitation cost reduction.

In order to introduce alternative solutions for high traffic categories (KR5–KR7) that are not included in the design guide [28], the results of laboratory testing and material behavior assumptions should be confirmed beforehand and complemented by a real scale analysis. For instance, falling weight deflectometer (FWD) tests are advised to be carried out in certain exploitation time intervals in order to analyze the responses of pavements subjected to high traffic loads in terms of surface deflections, strain, and stress measurements.

5. Conclusions

The following conclusions can be drawn, based on laboratory test results and pavement design analysis:

- Due to the reduced amount of cement, BSM-E mixtures exhibit 50–60% lower stiffness modulus values than CBTM mixtures in which a 3% cement addition is applied;
- The thermal sensitivities of the BSM-E mixtures depend on the type of bitumen emulsion used. Thermal sensitivities for the BSM-E mixtures are higher in comparison with CBTM mixtures, which may indicate that base layers made of CBTM do not exhibit the properties of flexible layers, but rather manifest behavior similar to rigid materials;

- Based on statistical analysis, assuming a 5% probability of obtaining lower stiffness modulus results, the following stiffness levels of BSM-E mixtures S_{temp} were determined at three testing temperatures: S_{temp5} = 4000 MPa at 5 °C, S_{temp13} = 2700 MPa at 13 °C, and S_{temp20} = 2070 MPa at 20 °C;
- Assuming the validity of the BSM-E behavior hypothesis that a layer's stiffness modulus increases during the initial period of pavement exploitation, and the material does not exhibit brittle cracking and a loss of initial stiffness, the constant BSM-E stiffness modulus value of 2700 MPa can be adopted for pavement design calculations;
- The adaptation of a higher stiffness modulus for BSM-E layers (2700 MPa) in comparison with CBTM layers (1500 MPa) is possible due to the expected difference of material characteristics from rigid and more prone to shrinkage and fatigue cracking for CBTMs to flexible and more crack-resistant BSM-E layers;
- The mechanistic-empirical pavement design calculations showed that, in the case of light traffic categories KR1–KR2, the application of BSM-E mixtures in base layers may bring potential benefits in terms of extended pavement life and the durability of local roads;
- For medium traffic categories (KR3–KR4), the thicknesses of asphalt layers presented in the national requirements can be reduced without negatively affecting the pavement life, which can contribute to the reduction of construction material use.

The laboratory testing and pavement design results should be confirmed and complemented by a real scale analysis. Falling weight deflectometer (FWD) tests are recommended to be conducted in order to analyze pavement response and determine the back-calculated stiffness modulus of the BSM-E layer in certain time intervals. Such tests are crucial to verify the assumption of increased cracking resistance and flexibility of BSM-E base layers.

Author Contributions: Conceptualization, K.K. and J.B.K.; methodology, K.K., J.B.K., P.R., and P.P.; software calculations, K.K.; validation, J.B.K., P.R., W.S., D.Ż., and P.P.; formal analysis, P.R. and J.B.K.; resources, J.B.K., D.Ż., and W.S.; data curation, K.K.; writing—original draft preparation, K.K.; writing—review and editing, J.B.K. and W.S.; visualization, K.K.; supervision, J.B.K.; project administration, J.B.K.; funding acquisition, D.Ż. and J.B.K. All authors have read and agreed to the published version of the manuscript.

Funding: This publication was financially supported by an internal research grant in 2020 for the discipline of Civil Engineering and Transport from the Warsaw University of Technology. The research results were obtained during the research project, supported within the European Union under the Regional Operational Program of the Lower Silesian Voivodeship 2014–2020, Priority Axis 1 "Enterprises and innovations", Measure and Sub-measure 1.2 "Innovative enterprises", Scheme 1.2.A "Support for enterprises wishing to start or develop R&D activities", agreement No. RPDS.01.02.01.-02. 0103/17 00.

References

1. Fang, X.; Garcia, A.; Lura, P. Overview on cold cement bitumen emulsion asphalt. *Rilem Tech. Lett.* **2016**, *1*, 116–121. [CrossRef]
2. Thom, N.; Dawson, A.R. Sustainable Road Design: Promoting Recycling and Non-Conventional Materials. *Sustainability* **2019**, *11*, 6106. [CrossRef]
3. Grilli, A.; Graziani, A.; Bocci, M. Compactability and thermal sensitivity of cement–bitumen-treated materials. *Road Mater. Pavement Des.* **2012**, *13*, 599–617. [CrossRef]
4. Cardone, F.; Grilli, A.; Bocci, M.; Graziani, A. Curing and temperature sensitivity of cement-bitumen treated materials. *Int. J. Pavement Eng.* **2015**, *16*, 868–880. [CrossRef]
5. Xuan, D.; Houben, L.; Molenaar, A.A.A.; Shui, Z. Mechanical properties of cement-treated aggregate material—A review. *Mater. Des.* **2012**, *33*, 496–502. [CrossRef]
6. Fedrigo, W.; Peres Núñez, W.; Radünz Kleinert, T.; Ferreira Matuella, M.; Pereira Ceratti, J.A. Strength, shrinkage, erodibility and capillary flow characteristics of cement-treated recycled pavement materials. *Int. J. Pavement Res. Technol.* **2017**, *10*, 393–402. [CrossRef]
7. Nusit, K.; Jitsangiam, P. Damage Behavior of Cement-Treated Base Material. *Procedia Eng.* **2016**, *143*, 161–169. [CrossRef]

8. Wirtgen Group. *Wirtgen Cold Recycling Technology*, 1st ed.; Wirtgen GmbH: Windhagen, Germany, 2012; pp. 18–19.
9. Southern African Bitumen Association (Sabita). *A Guideline for the Design and Construction of Bitumen Emulsion and Foamed Bitumen Stabilised Materials; Technical Guideline: Bitumen Stabilised Materials*, 3rd ed.; Southern African Bitumen Association (Sabita): Cape Town, South Africa, 2020; pp. 1–2, 7–8, 36–37.
10. Perez, I.; Medina, L.; del Val, M.A. Mechanical properties and behaviour of in situ materials which are stabilised with bitumen emulsion. *Road Mater. Pavement Des.* **2013**, *14*, 221–238. [CrossRef]
11. Nivedya, M.; Veeraragavan, A.; Krishnan, J.M.; Erkens, S.; Liu, X.; Anupam, K.; Yiqiu, T. How to characterize the mechanical response of Bitumen Stabilized Material? In Proceedings of the 4th Chinese-European Workshop on Functional Pavement Design, Delft, The Netherlands, 29 June–1 July 2016; pp. 355–364.
12. Jenkins, K. Cracking Behaviour of Bitumen Stabilised Materials (BSMs): Is There Such a Thing? *Strain-Hardening Cem. Based Compos.* **2012**, *4*, 1007–1015. [CrossRef]
13. Ebels, L.J. Characterisation of Material Properties and Behaviour of Cold Bituminous Mixtures for Road Pavements. Ph.D. Thesis, Stellenbosch University, Stellenbosch, South Africa, 2008.
14. Fan, L.; Lyu, Y.; Fan, L.; Zhang, Y. Effects of Cement and Emulsified Asphalt on Properties of Mastics and 100% Cold Recycled Asphalt Mixtures. *Materials* **2019**, *12*, 754. [CrossRef]
15. Iwanski, M.; Chomicz-Kowalska, A. Moisture and Frost Resistance of the Recycled Base Rehabilitated with the Foamed Bitumen Technology/Odpornosc Na Oddziaływanie Wody I Mrozu Podbudowy W Technologii Recyklingu Z Asfaltem Spienionym. *Arch. Civ. Eng.* **2012**, *58*, 185–198. [CrossRef]
16. Iwański, M.; Chomicz-Kowalska, A. Laboratory Study on Mechanical Parameters of Foamed Bitumen Mixtures in the Cold Recycling Technology. *Procedia Eng.* **2013**, *57*, 433–442. [CrossRef]
17. Sainton, A.; Bourdrel, D.Procédé de double traitement à froid des matériaux au ciment et à l'émulsion de bitumen. *Rev. Générale Routes Aérodromes* **1994**, *719*, 34–41.
18. Dołżycki, B.; Jaskula, P. Review and evaluation of cold recycling with bitumen emulsion and cement for rehabilitation of old pavements. *J. Traffic Transp. Eng.* **2019**, *6*, 311–323. [CrossRef]
19. Engbers, G.; Smallegange, J.P. The development of Emulscement®: A base material with a future. In *Reflective Cracking in Pavements: Design and Performance of Overlay Systems*, 1st ed.; Francken, L., Beuving, E., Molenaar, A.A.A., Eds.; CRC Press: London, UK, 1996; pp. 18–29, ISBN 9780429077630.
20. Kukiełka, J.; Bańkowski, W. The experimental study of mineral-cement-emulsion mixtures with rubber powder addition. *Constr. Build. Mater.* **2019**, *226*, 759–766. [CrossRef]
21. Buczyński, P.; Iwański, M.; Mazurek, G.; Krasowski, J.; Krasowski, M. Effects of Portland Cement and Polymer Powder on the Properties of Cement-Bound Road Base Mixtures. *Materials* **2020**, *13*, 4253. [CrossRef]
22. Dolzycki, B.; Szydlowski, C.; Jaczewski, M. The influence of combination of binding agents on fatigue properties of deep cold in-place recycled mixtures in Indirect Tensile Fatigue Test (ITFT). *Constr. Build. Mater.* **2020**, *239*, 117825. [CrossRef]
23. Jenkins, K.J.; Long, F.M.; Ebels, L.J. Foamed bitumen mixes = shear performance? *Int. J. Pavement Eng.* **2007**, *8*, 85–98. [CrossRef]
24. Liebenberg, J.J.E.; Visser, A.T. Stabilization and Structural Design of Marginal Materials for Use in Low-Volume Roads. *Transp. Res. Rec. J. Transp. Res. Board* **2003**, *1819*, 166–172. [CrossRef]
25. Liebenberg, J.J.E.; Visser, A.T. Towards a mechanistic structural design procedure for emulsion- treated base layers. *J. S. Afr. Inst. Civ. Eng.* **2004**, *46*, 2–8.
26. Ebels, L.J.; Jenkins, K.; Collings, D. Cold mix (bitumen stabilization) technology in Southern Africa into the 21st century. In Proceedings of the International Symposium on Pavement Recycling, São Paulo, Brazil, 14–16 March 2005.
27. Dolzycki, B. *Guidelines for Design and Paving of Mineral-Cement-Emulsion Mixes (MCE)*; General Directorate for National Roads and Motorways: Gdansk, Poland, 2014; pp. 17–19. (In Polish)
28. Judycki, J.; Jaskuła, P.; Pszczoła, M. *The Catalogue of Typical Flexible and Semi-Rigid Pavements*; General Directorate for National Roads and Motorways: Gdansk, Poland, 2014; p. 79. (In Polish)
29. Preti, F.; Gouveia, B.C.S.; Rahmanbeiki, A.; Romeo, E.; Carter, A.; Tebaldi, G. Application and validation of the cohesion test to characterise reclaimed asphalt pavement. *Road Mater. Pavement Des.* **2019**, *20*, S434–S445. [CrossRef]

3

Evaluation of Mechanical Properties of Recycled Material for Utilization in Asphalt Mixtures

Farzaneh Tahmoorian [1,*], Bijan Samali [1], Vivian W.Y. Tam [2,3] and John Yeaman [4]

[1] Centre for Infrastructure Engineering, Western Sydney University, Kingswood, NSW 2751, Australia; B.Samali@westernsydney.edu.au
[2] School of Computing, Engineering, and Mathematics, Western Sydney University, Kingswood, NSW 2751, Australia; V.Tam@westernsydney.edu.au
[3] College of Civil Engineering, Shenzhen University, Shenzhen 518060, China
[4] Faculty of Science, Health, Education and Engineering, University of Sunshine Coast, Sippy Downs, Queensland 4556, Australia; jyeaman@usc.edu.au
* Correspondence: F.Tahmoorian@westernsydney.edu.au

Abstract: With an expanding world, the demand for extensive road networks is increasing. As natural resources become scarce, the necessity of finding alternative resources has led to the idea of applying recycled material to pavement construction including asphalt pavements. Amongst all asphalt components, aggregate constitutes the largest part of asphalt mixtures. Therefore, the utilization of recycled material for aggregate will represent an important opportunity to save virgin material and divert material away from landfills. Because of the large amount of construction waste generation around the world, using recycled construction aggregate (RCA) in asphalt mixtures appears to be an effective utilization of RCA. However, as aggregate plays an important role in the final performance of the asphalt mixture, an understanding of their properties is essential in designing an asphalt mixture. Therefore, in this research, the properties of RCA have been evaluated through laboratory investigations. Based on the test results, it is required that combination of RCA with some other targeted waste materials be considered in asphalt mixture. This paper presents the results of an experimental study to evaluate the RCA properties as an alternative for virgin aggregate in asphalt mixture under different percentages and combination with other aggregates, such as reclaimed asphalt pavement (RAP) and basalt.

Keywords: asphalt; basalt; reclaimed asphalt pavement; recycled construction aggregate; recycled aggregate; pavement

---

1. Introduction

The increasing amount of waste all over the world has shown that effective measures have to be implemented to reduce their negative environmental impact. Landfilling of waste is not a solution, due to danger of leaching and soil impregnation with potential subsequent contamination of underground water.

On the other hand, there are important sustainability benefits associated with the use of recycled material in pavement industry. Recycling helps the environment by reducing resource extraction and the use of virgin material, thereby reducing energy and water use, reducing harmful gas emissions and helping reduce waste to landfills. Buying recycled products also in some cases can reduce cost.

Therefore, the idea of using recycled material in pavement construction has attracted the attention of many researchers (e.g., Karlsson and Isacsson, 2006; Huang et al., 2007; Widyatmoko, 2008; Ahmedzade and Sengoz, 2009; Rafi et al., 2011; Khan and Gundaliya, 2012; Menaria and Sankhla, 2015;

Chandh and Akhila, 2016) in the pavement industry to investigate feasibility of the application of some of waste materials as alternative material in pavement construction [1–8].

Among different layers of flexible pavements, asphalt surface layer plays a fundamental role in flexible pavement structure systems as it should withstand varying traffic loads and constantly changing environmental conditions. Moreover, the asphalt surface layer is critical for safe and comfortable driving. Due to the composition nature of asphalt surface layer, application of solid waste in asphalt layer reduces not only environmental issues associated with waste disposal but also the demand for virgin aggregate which will subsequently result in cost savings and economic advantages, representing a value add application for waste material.

However, the selection of waste material to be used for pavement construction, particularly asphalt surface layer, is of high importance as the application of waste should not adversely influence the structural and functional aspects of the pavements [9–12].

Among different asphalt components, coarse aggregate properties are identified by the researchers (e.g., Vavrik, 2009; Zaniewski and Srinivasan, 2004; Husain, 2014; Al-Mosawe et al., 2015) as the second most important parameter after gradation for the performance of hot-mix asphalt (HMA) because coarse aggregate often forms the skeleton of the asphalt structure and transfers traffic and environmental loads to the underlying base, subbase, and subgrade layers [13–16]. Therefore, the behaviour and performance of asphalt mixture and eventually the asphalt surface layer are directly affected by the material properties and composition of this aggregate skeleton. In fact, the low stiffness of the asphalt mixtures and the excessive rutting in hot-mix asphalt (HMA) pavement surfaces are often attributed to the poor asphalt mixture designs which is primarily controlled by the asphalt binder and aggregate properties [17]. Therefore, except for the fine mixes, the selection of coarse aggregate greatly influences the asphalt layer behaviour.

In addition, since the aggregate represents the major portion of the asphalt mix, from the viewpoint of environmental preservation and effective use of resources, a comprehensive understanding of the engineering properties of the recycled aggregate can provide enormous benefits. Recognizing this fact, the reported studies and research on the utilization of recycled aggregate such as reclaimed asphalt pavement (RAP), recycled construction aggregate (RCA), recycled glass, etc. have increased all over the world over the past two decades [18–29]. Among the recycled aggregates that can be utilized in asphalt mixture, RCA obtained from construction and demolition waste constitute a major part of generated solid waste as a result of renovation and construction projects. Referring to literature survey (e.g., Arulrajah, 2012; Bennert et al., 2000; Blankenagel, 2005; Conceicao et al., 2011; Jayakody et al., 2014; Jimenez et al., 2012; Papp et al., 1998; Nataatmadja and Tan, 2001), although RCA has been used effectively as a base course and subbase course material [30–37], but, few research studies (e.g., Celaura et al., 2010; Hossain et al., 1993; Pereira et al., 2004; Rebbechi and Green, 2005; Berthelot et al., 2010; Wu et al., 2013) have reported the use of RCA in hot-mix asphalt [38–42]. Accordingly, in this research, the properties of RCA have been thoroughly evaluated through the laboratory investigation and tests. The results of these tests have showed that RCA has some shortcomings in satisfying design requirements as asphalt mixtures aggregate, in terms of some properties such as absorption and wet/dry strength variation. Therefore, utilization of RCA in asphalt mixture on its own, can result in less efficient asphalt mixtures and it is, hence, required that combination of RCA with some other targeted and acceptable waste materials and aggregates in certain percentages be considered in designing the asphalt mixture. Accordingly, RCA, RAP, and basalt have been considered as coarse aggregate in this research, and various tests have been conducted on each individual component and in combination. The paper will demonstrate the results of the conducted tests leading to the selection of most acceptable combination of aggregates for designing asphalt mixtures.

It should be noted that because of the diversity in quality and composition of the recycled construction aggregates, this research has been performed on aggregate samples which are collected from a recycling unit in Sydney over a period of one year.

2. Aggregate Properties and Their Relationship to Asphalt Performance

The high proportion of aggregate materials in volumetric design of asphalt mixes inherently links aggregate properties to the strength, stiffness, and generally the performance of the asphalt surface layer.

Because of the important impact of aggregate on the properties of asphalt mixture, a better understanding of the aggregates characteristics is essential in selecting the appropriate materials to optimize the asphalt mixture for strength and durability, and subsequently design a pavement with enough resistance to permanent deformation and cracking.

The most important physical and mechanical characteristics of aggregates include size and gradation, shape and angularity, surface texture, absorption, particle density, durability, toughness and hardness, resistance to polishing, soundness, cleanliness and the deleterious materials contained. Many research studies (e.g., Dahir, 1979; Brown et al., 1989; Brown and Bassett, 1990; Button et al., 1990; Elliot et al., 1991; Krutz and Sebaaly, 1993; Oduroh et al., 2000; Chen and Liao, 2002; Sengoz et al., 2014; Masad et al., 2009; Wu and King, 2011) have been conducted to link the properties of the aggregates to the performance of asphalt concrete pavement [43–53].

The results of these studies have shown that the physical and mechanical properties of the aggregates significantly affect the performance of the asphalt pavements.

Referring to the literature and the research conducted to relate aggregate properties and HMA performance, Figure 1 is generated to illustrate a generalized pattern and a summary of the effects of aggregate properties on the asphalt performance. The figure is the result of extensive literature review during the course of this research study and could be used by the practicing engineers as well as researchers to further improve their understanding of the effects of aggregate constituents on asphalt system performance. The reported relations and correlations shown in Figure 1 exemplify the complexities of mix design issues and considerations involved. This is certainly not unexpected considering the heterogeneity of the asphalt mixes. For example, as shown in this figure, different aggregate properties affect different aspects of asphalt mixture performance, which consequently define pavement service life. Accordingly, in order to design asphalt mixtures with longer service lives and lower production and maintenance costs, the aggregate must have appropriate characteristics.

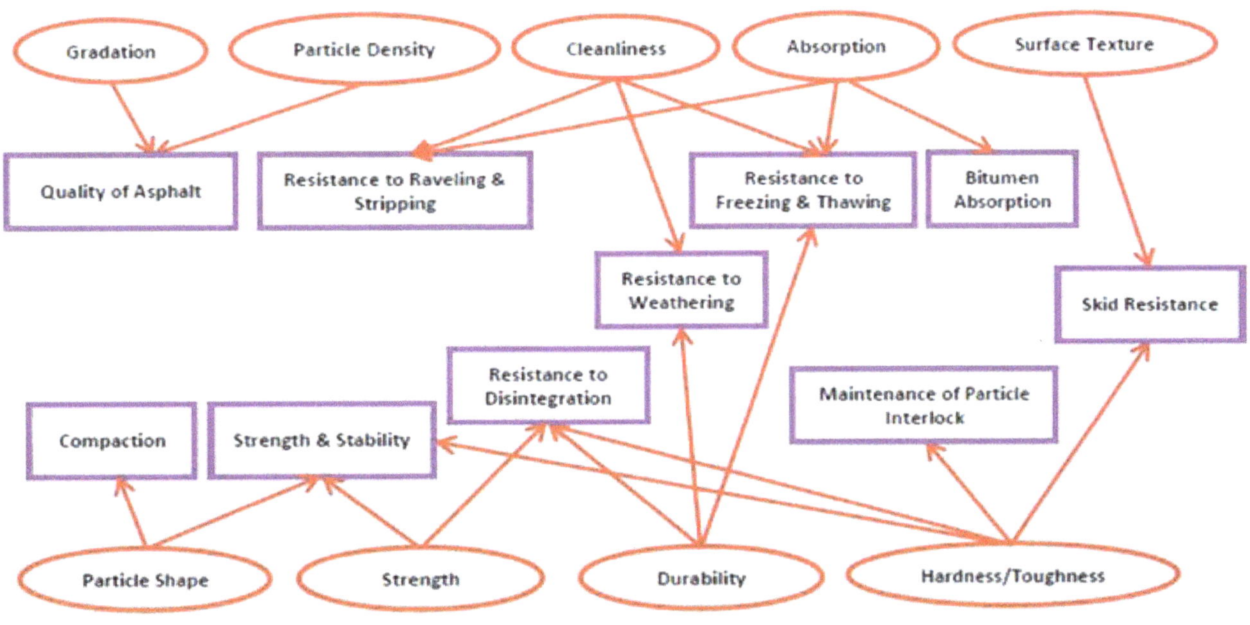

Figure 1. A Summary of the Effects of Aggregate Properties on the Asphalt Performance.

Therefore, the following section describes the experimental work carried out on selected coarse aggregates in order to evaluate the feasibility of using RCA as a part of coarse aggregates in asphalt mixture, and to produce an economical and sustainable asphalt mixture with adequate strength and good workability.

3. Experimental Work

3.1. Materials

In the present study, RCA, RAP, and basalt passing through 20 mm and retained on 4.75 mm I.S sieve have been used throughout the experiments. RAP material used in this research was stockpiled RAP collected from Boral Asphalt Plant (Prospect, NSW, Australia) which is generated from milling and being used in their asphalt projects. It was plant-screened material retained on 19 mm sieve size. The crushed virgin basalt aggregate was obtained from a local supplier. These virgin aggregates were transported from a local quarry (Nepean Quarries) in the vicinity of Sydney. In addition, RCA was collected from a local recycling centre called Revesby Recycling Centre (Revesby, NSW, Australia), a licensed waste facility and transfer station which accepts all construction and demolition waste from both the residential and commercial waste streams. In this centre, RCA is produced through the first sorting process for removing of contaminants such as wood, plastic, metal and glass, then crushing of construction wastes, and finally screening for removal of contaminants such as reinforcement, wood, plastics and gypsum.

3.2. Laboratory Tests

This section reports the laboratory investigation on RCA, RAP and basalt, in order to obtain comprehensive information and data of their properties and to compare these properties with the requirements specified in the standards as well as with the properties of the virgin aggregate. The key properties investigated in this experimental study are presented in Table 1.

Table 1. The Key Properties Investigated in the Experimental Study.

Property	Test Method	Test Name
Gradation and particle size distribution	AS 1141.11.1	Particle Size Distribution (Sieving Method)
Flakiness index of aggregate	AS 1141.15	Flakiness Index
Proportion of misshapen particles	AS 1141.14	Particle Shape by proportional calliper
Water absorption of aggregate	AS1141.6.1	Particle Density and Water Absorption of coarse aggregate
Variation in strength of aggregate in wet/dry condition	AS 1141.22	Wet/Dry Strength Variation
Particle density of aggregate	AS 1141.6.1	Particle Density and Water Absorption of coarse aggregate
Strength and crushing value of aggregate	AS 1141.21	Aggregate Crushing Value
Percentage of weak particles in coarse aggregate	AS 1141.32	Weak particles (including clay lumps, soft and friable particles) in coarse aggregates

In addition, based on the test results on the individual aggregates, necessary tests were conducted on different combinations of these aggregates. The results of these tests are shown in the following sections. It should be noted that three samples were performed for each test and the average of the three samples was reported as the test result.

3.2.1. Particle Size Distribution Test

The gradation of aggregate to be used in asphalt mixtures are evaluated through Particle Size Distribution Test (Figure 2).

Figure 2. Aggregate Gradation by Particle Size Distribution Test.

This test is conducted in accordance with AS 1141.11.1 (2009) and the gradation curves obtained from this test for different coarse aggregates considered in this research, including RCA, RAP and basalt, are shown in Figure 3.

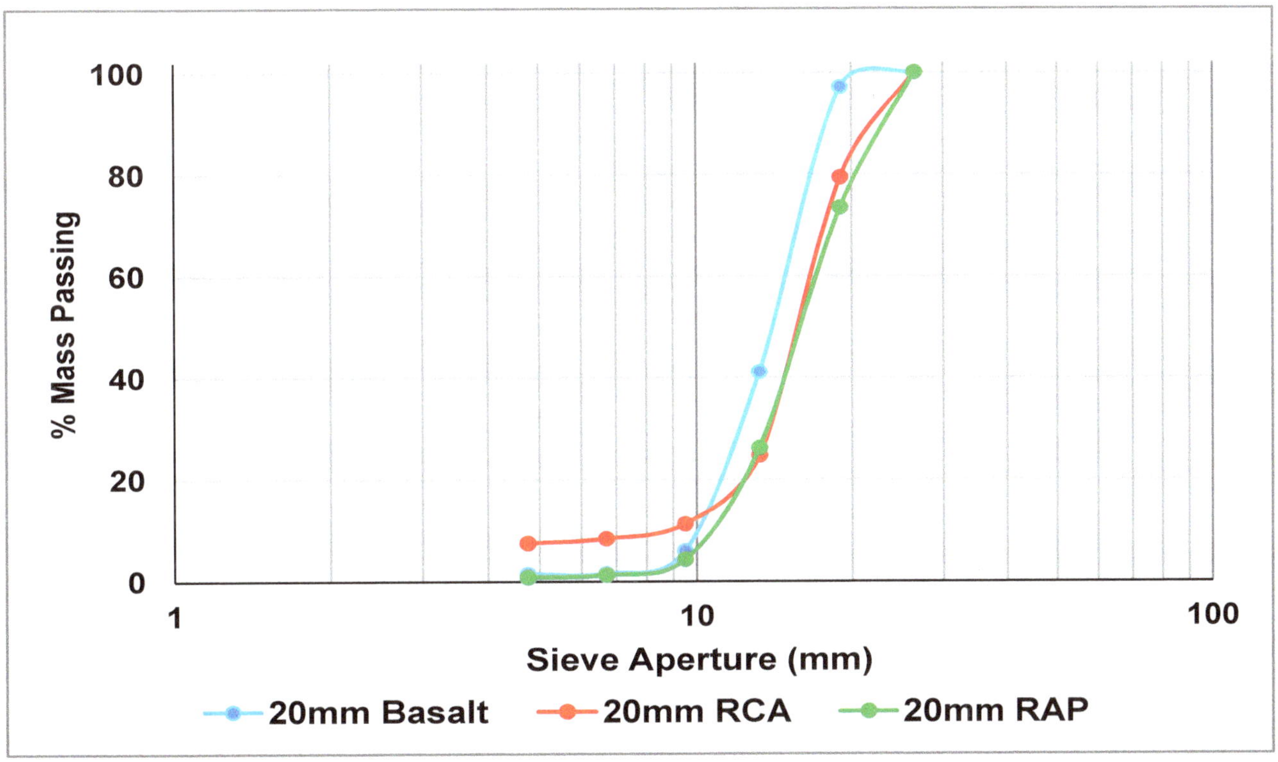

Figure 3. The Results of Particle Size Distribution Test for Coarse Aggregate; RCA: recycled construction aggregate; RAP: reclaimed asphalt pavement.

3.2.2. Particle Shape Test

The results of the studies on aggregate have shown that the aggregate physical shape properties significantly affect both the strength and stability of asphalt mixes [54]. Therefore, in order to design asphalt mixtures with long service lives, the aggregate must have the proper gradation and shape. The particle shape of aggregate substantially influences the mechanical stability of asphalt mix. The presence of excessive flaky and elongated particles is undesirable in asphalt mixtures as they tend to break down during the production and construction, and thus affect the durability of HMAs. Therefore, it is preferable to have rough and angular aggregates rather than smooth and round aggregates.

In this study, the proportion of misshapen aggregates, including the flat particles, elongated particles and, flat and elongated particles found in coarse aggregate is evaluated through the Particle Shape Test (Figure 4). The particle shape test is carried out by proportional caliper, using a 2:1 calliper ratio and based on AS 1141.14 (2007). The results of this test on three samples for each aggregate type (i.e., RCA, RAP and basalt) and the average value are given in Table 2.

Figure 4. Classification of Aggregate Based on Particle Shape Test.

Table 2. The Results of Particle Shape Test for Coarse Aggregates and Misshapen Percentage Limits for Dense Graded Asphalt Based on Australian Standards.

Sample Number	Misshapen Particles (%)			Australian Standards Misshapen Percentage Limits (%)
	RCA	RAP	Basalt	
Sample 1	5.3	12.0	19.4	35% (max) For heavy and very heavy traffic
Sample 2	6.2	7.7	18.3	
Sample 3	7.0	8.7	17.3	
Average Misshapen Particles Percentage (%)	**6.2**	**9.5**	**18.3**	

As presented in Table 2, basalt materials show more of misshapen particles than RAP and RCA while still below the 35% limit of the Australian standard.

3.2.3. Flakiness Index Test

Some aggregates, on account of their shape, would be unsuitable for asphalt mixture as they would have low potential for developing inter-particle interlock. The percentage by mass of this type of aggregates, namely flaky aggregates is determined by the most commonly used test, called Flakiness Index Test (Figure 5). In this test, the flakiness index is determined by direct measurement using a special slotted sieve, from the ratio of the mass of material passing the slotted sieve to the total mass of the size fraction.

Figure 5. Conducting Flakiness Index Test for Coarse Aggregate.

The flakiness index test is performed based on AS 1141.15 (1999) and the results of this test on three samples for each aggregate type (i.e., RCA, RAP and basalt) and the average flakiness index for each aggregate type are given in Table 3.

Table 3. The Results of Flakiness Index Test for Coarse Aggregates and Flakiness Index Limits for Dense Graded Asphalt Based on Australian Standards.

Sample Number	Flakiness Index (%)			Australian Standards Flakiness Index Limits (%)
	RCA	RAP	Basalt	
Sample 1	5.6	12.8	21.3	
Sample 2	8.1	10.1	21.1	25% (max)
Sample 3	7.1	8.4	14.7	For heavy and very heavy traffic
Average Flakiness Index (%)	**6.9**	**10.4**	**19.0**	

The results of flakiness index test shows that RCA has less flakiness index than basalt and RAP which can positively affect the inter-particle interlock in asphalt mixture.

3.2.4. Particle Density and Water Absorption Test

The absorption is an indication of porosity in aggregate which demonstrates the pore structure of the aggregate. In asphalt mixtures, a porous aggregate increases the binder absorption, resulting in a dry and less cohesive asphalt mixture. In addition, the particle density of the aggregate is an essential property of the aggregate which plays an important role in the whole procedure of asphalt mix design. Therefore, in this research, the particle density and water absorption test is conducted on coarse aggregates (i.e., RCA, RAP and coarse basalt) based on the procedure described in AS 1141.6.1 (2000), as presented in Figure 6.

In this test, the amount of water which a dried sample will absorb is measured. This test is performed on three trials and the related test results on RCA, RAP and basalt are given in Table 4, under apparent, dry, and saturated surface dry (SSD) conditions.

Figure 6. Evaluation of Particle Density of Coarse Aggregate.

Table 4. The Results of Particle Density and Water Absorption Test on Coarse Aggregates and Water Absorption Limits for Dense Graded Asphalt Based on Australian Standards.

Sample Name	Apparent Particle Density (g/cm^3)	Particle Density on a Dry Basis (g/cm^3)	Particle Density on a SSD Basis (g/cm^3)	Water Absorption (%)	Australian Standards Limits for Water Absorption (%)
RCA 1	2.375	2.211	2.352	6.39	
RCA 2	2.375	2.211	2.352	6.39	
RCA 3	2.361	2.214	2.349	6.12	
Average Values for RCA	**2.370**	**2.212**	**2.351**	**6.30**	
RAP 1	2.539	2.422	2.468	1.89	2% (max) For heavy and very heavy traffic
RAP 2	2.544	2.437	2.479	1.72	
RAP 3	2.540	2.433	2.475	1.73	
Average Values for RAP	**2.541**	**2.431**	**2.474**	**1.78**	
Basalt 1	2.635	2.528	2.568	1.60	
Basalt 2	2.630	2.521	2.562	1.64	
Basalt 3	2.654	2.542	2.584	1.67	
Average Values for Basalt	**2.640**	**2.530**	**2.571**	**1.64**	

The results of the particle density and water absorption test on different coarse aggregates (i.e., RCA, RAP and basalt) and their average value, as presented in Table 4, indicate the high absorption of RCA in comparison with RAP and basalt. The RCA water absorption exceeds the limit set by the Australian Standard.

As this research aims to investigate the feasibility of the application of RCA as a recycled material for potential partial replacement of coarse virgin aggregate (basalt) in asphalt mixtures, the particle density and water absorption test is also conducted on the mix of coarse aggregates (i.e., RCA, RAP and coarse basalt) considering different percentages of these materials. Such undertaking was needed in order to get a better understanding of an acceptable range of mix proportions in terms of water absorption.

The results of Particle Density and Water Absorption test on six different mixes of RCA, RAP and basalt and the average water absorption and particle density for each mix are given in Table 5. Despite the fact that above mixes (except when there is no RCA) have water absorption of more than 2%, the use of RCA is still a viable option as discussed in Section 3.3.

Table 5. The Results of Particle Density and Water Absorption Test for the Mix of Coarse Aggregates.

Sample Name		Apparent Particle Density (g/cm^3)	Particle Density on a Dry Basis (g/cm^3)	Particle Density on a SSD Basis (g/cm^3)	Water Absorption (%)
75% Basalt and 25% RCA	1	2.579	2.394	2.466	2.98
	2	2.589	2.406	2.476	2.94
	3	2.601	2.420	2.489	2.88
Average Values for 75% Basalt and 25% RCA		**2.590**	**2.407**	**2.477**	**2.93**
50% Basalt and 50% RCA	1	2.520	2.296	2.385	3.86
	2	2.535	2.322	2.406	3.62
	3	2.527	2.313	2.397	3.65
Average Values for 50% Basalt and 50% RCA		**2.527**	**2.310**	**2.396**	**3.71**
25% Basalt and 75% RCA	1	2.477	2.224	2.326	4.57
	2	2.471	2.207	2.313	4.84
	3	2.480	2.234	2.333	4.44
Average Values for 25% Basalt and 75% RCA		**2.476**	**2.222**	**2.324**	**4.62**
80% Basalt and 20% RAP	1	2.727	2.607	2.651	1.68
	2	2.719	2.600	2.644	1.68
	3	2.724	2.596	2.643	1.80
Average Values for 80% Basalt and 20% RAP		**2.723**	**2.601**	**2.646**	**1.72**
25% Basalt and 25% RCA and 50% RAP	1	2.579	2.411	2.476	2.70
	2	2.581	2.401	2.471	2.89
	3	2.585	2.412	2.479	2.78
Average Values for 25% Basalt and 25% RCA and 50% RAP		**2.582**	**2.408**	**2.475**	**2.79**
25% Basalt and 50% RCA and 25% RAP	1	2.606	2.380	2.466	3.63
	2	2.592	2.356	2.447	3.85
	3	2.596	2.355	2.447	3.94
Average Values for 25% Basalt and 50% RCA and 25% RAP		**2.598**	**2.364**	**2.453**	**3.81**

3.2.5. Crushing Value Test

Aggregates used in road construction should be strong enough to resist crushing under traffic wheel loads [55]. The strength of the coarse aggregates can be evaluated by the Aggregate Crushing Value Test. In this test, the aggregate were crushed by a compression testing machine with a load rate of 40 kN/min to reach the peak load of 400 kN. The percentage of particles produced when the aggregate is crushed under this load and which pass a 2.36 mm sieve is called Aggregate Crushing Value.

The aggregate crushing value provides a relative measure of resistance to crushing under a gradually applied compressive load. To achieve a high quality pavement, it is preferred to utilize the aggregate possessing low crushing value.

In this research, the crushing value of RCA, RAP and basalt is assessed through the Aggregate Crushing Value Test in accordance with AS 1141.21 (1997), as presented in Figure 7. This test was performed in two trials, as required in the standard, and the related test results on RCA, RAP and basalt and the average crushing values for each aggregate type are given in Table 6.

Table 6. The Results of Aggregate Crushing Value Test for Coarse Aggregates and Crushing Value Limits for Dense Graded Asphalt Based on Australian Standards.

Sample Number	Crushing Value (%)			Australian Standards Aggregate Crushing Value Limits (%)
	RCA	RAP	Basalt	
Sample 1	29.53	7.04	9.32	35% (max) For heavy and very heavy traffic
Sample 2	28.88	7.76	8.51	
Average Crushing Value (%)	**29.21**	**7.40**	**8.91**	

Figure 7. Crushing Value Test for Coarse Aggregates. (**a**) apparatus to crush aggregate; (**b**) compression machine ; (**c**) aggregates after crushing and sieving.

3.2.6. Weak Particle Test

The aggregate cleanliness refers to the presence of foreign or deleterious substances such as soft particles, weak and weathered materials, friable particles, clay lumps, and organic matters. The presence of these materials in the used aggregate can lead to stripping and ravelling in HMAs, as these materials adversely affect the bond between the aggregate and asphalt, and subsequently the stability of the pavement structure. Moreover, these substances disintegrate under traffic loading and wetting and drying cycles.

The cleanliness of aggregate can be evaluated based on the Weak Particles Test. In this test, the percentage of weak particles in coarse aggregate is determined. These particles will deform under finger pressures when wet. In this study, the percentage of weak particles in RCA, RAP and basalt are determined through the Weak Particle Test in accordance with AS 1141.32 (2008).

The weak particle test is conducted on two samples, as specified in the related standard, and the results of this test on RCA, RAP and basalt and the average weak particle percentage for each type of aggregate are presented in Table 7. The test results show that RCA and basalt have higher percentage of weak particles. However, all aggregates still meet the Standard's requirements.

Table 7. The Results of Weak Particle Test for Coarse Aggregates and Weak Particles Percentage Limits for Dense Graded Asphalt Based on Australian Standards.

Sample Number	Weak Particles (%)			Australian Standards Aggregate Crushing Value Limits (%)
	RCA	RAP	Basalt	
Sample 1	0.21	0.05	0.29	1% (max) For heavy and very heavy traffic
Sample 2	0.25	0.03	0.16	
Average Weak Particles Percentage (%)	**0.23**	**0.04**	**0.23**	

3.2.7. Wet/Dry Strength Variation Test

Strength is an important aggregate property which is related to the satisfactory resistance to crushing under the roller during construction, and adequate resistance to surface abrasion under traffic [56]. Therefore, aggregates used in pavement construction should be strong enough to resist crushing during mixing, laying process, compaction, consolidation and during its service life period when they are subjected to various loads applied by traffic [57].

In this research, the variation in strength of aggregate is evaluated by conducting the Wet-Dry Strength Variation Test on RCA, RAP and basalt in accordance with AS 1141.22 (2008), as shown in Figure 8.

(a) (b)

Figure 8. Conducting the Wet/Dry Strength Variation Test on Coarse Aggregate. (**a**) apparatus to crush aggregate; (**b**) compression machine.

This test determines the variation in strength of the aggregates tested after drying in an oven and then saturated yet with a dry surface. Based on the available standards, the wet/dry strength variation of less than 35% indicate a durable material but values as high as 60% could be used in undemanding circumstances.

In this research, the wet/dry strength variation test was conducted on the RCA, RAP and basalt fraction passed through 13.2 mm and retained on 9.5 mm I.S sieve. Different loading was used in order to adjust the applied load for providing the fines within the range of 7.5% and 12.5%. The results of these tests for coarse aggregates are illustrated in Figures 9 and 10 under dry condition and saturated surface dry condition (SSD), respectively. The wet and dry strengths can be inferred from the test results shown in these figures. Based on the obtained data, the wet/dry strength variation was calculated as follows:

$$\text{Wet/dry strength variation} = \frac{D - W}{D} \times 100 \quad (1)$$

where D is the dry strength in kilonewtons, and W is the wet strength in kilonewtons.

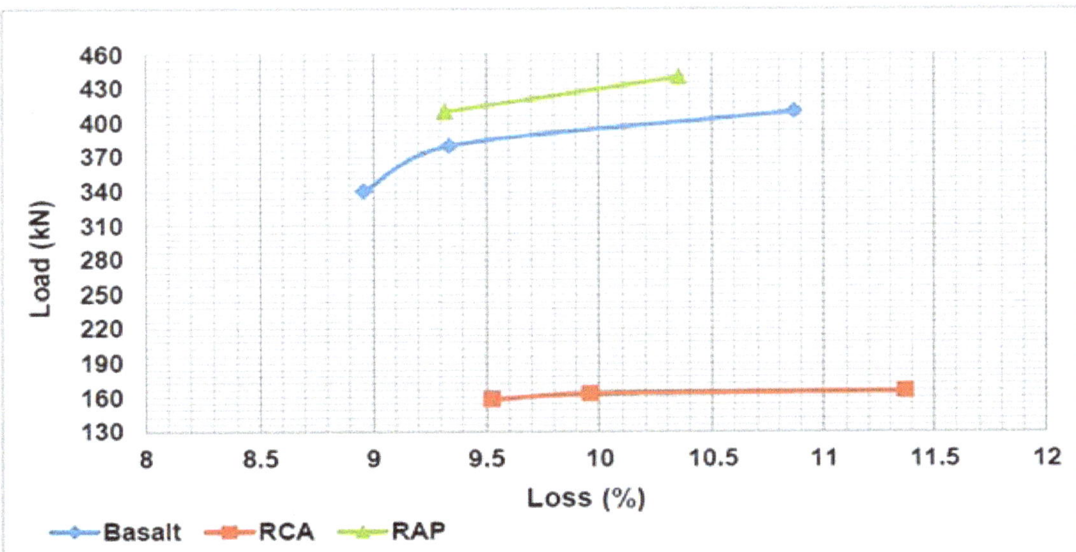

Figure 9. Results of Wet/Dry Strength Test for Coarse Aggregate (Dry Strength).

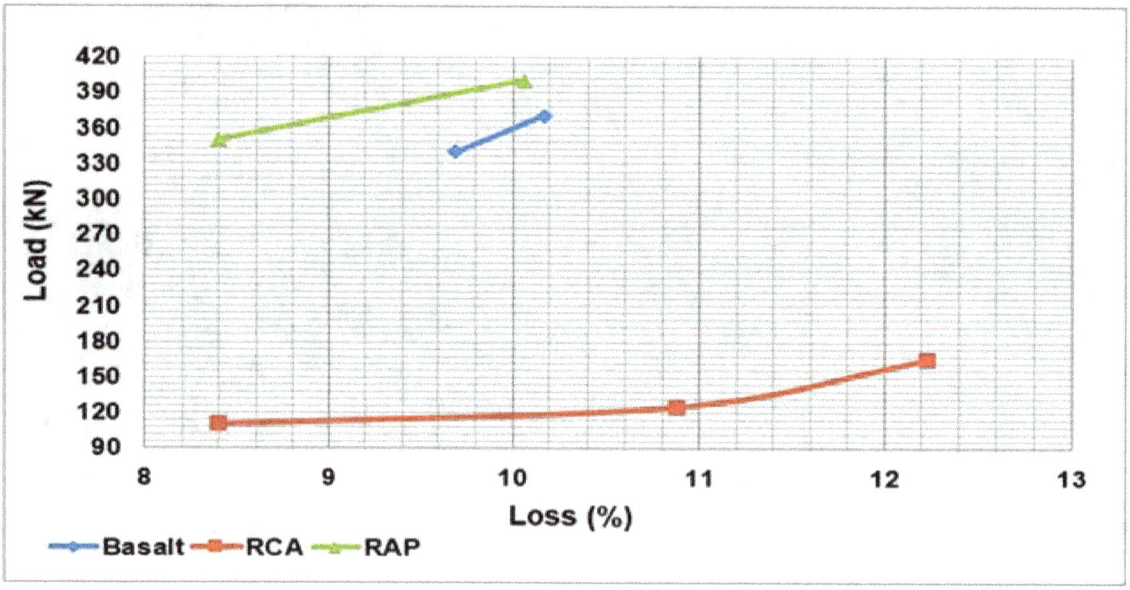

Figure 10. Results of Wet/Dry Strength Test for Coarse Aggregate (Wet Strength).

The results of the calculations for wet strength, dry strength, and wet/dry strength variation for basalt, RAP and RCA are presented in Table 8.

Table 8. The Results of Wet/Dry Strength Variation Test for Aggregates and Strength Limits for Dense Graded Asphalt Based on Australian Standards.

Material	Dry Strength, D (kN)	Wet Strength, W (kN)	Wet/Dry Strength Variation (%)	Australian Standards Wet/Dry Strength Limits (%)
RCA	163.1	119.7	26.6	35% (max) For heavy and very heavy traffic
RAP	429.8	398.2	7.4	
Basalt	392.9	359.4	8.5	

As the results of wet/dry strength test shows, the wet/dry strength variation of RCA is substantially more than the corresponding values for RAP and basalt. Therefore, as mentioned previously, it appears plausible to further investigate the feasibility of the application of RCA for the replacement of part of basalt in asphalt mixtures.

Accordingly, the wet/dry strength variation test was also conducted on different mix of coarse aggregates (i.e., RCA, RAP and coarse basalt) considering different percentages of these materials. Figures 11 and 12 illustrate the results of the wet/dry strength test for several mixes of RCA, RAP and basalt in dry condition and saturated surface dry condition respectively.

Based on the obtained results from these graphs, the wet strength (W) and the dry strength (D) can be determined and subsequently the wet/dry strength variation can be calculated as shown previously. The results of the calculations for wet strength, dry strength, and wet/dry strength variation on different mix of RAP, basalt and RCA are presented in Table 9. The results indicate that all mixes satisfy the maximum 35% limit set by the Australian Standards.

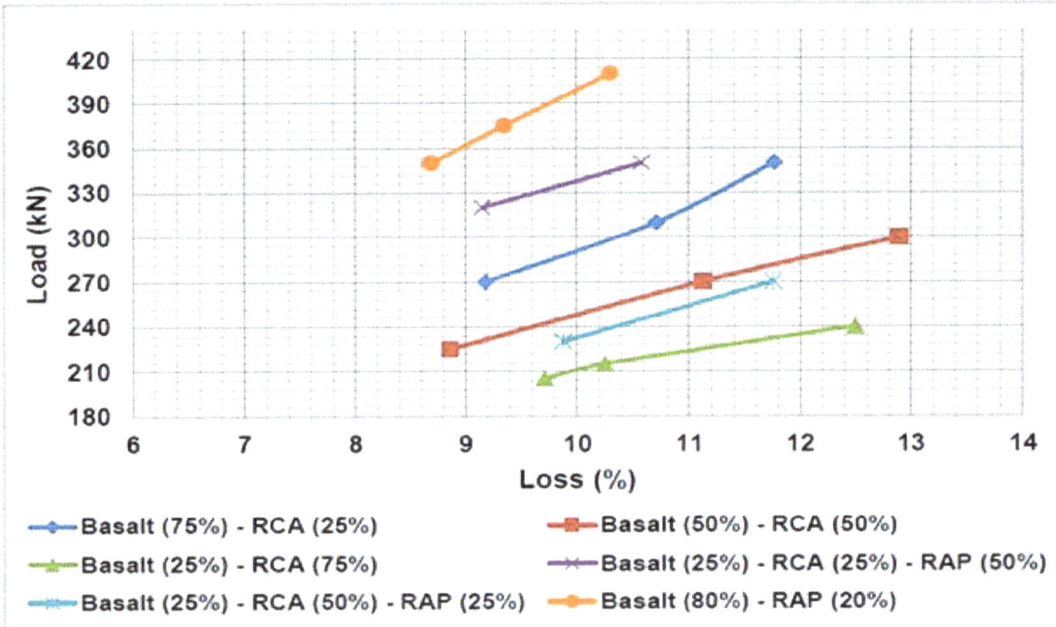

Figure 11. Results of Wet/Dry Strength Test for Mix of Coarse Aggregates (Dry Strength).

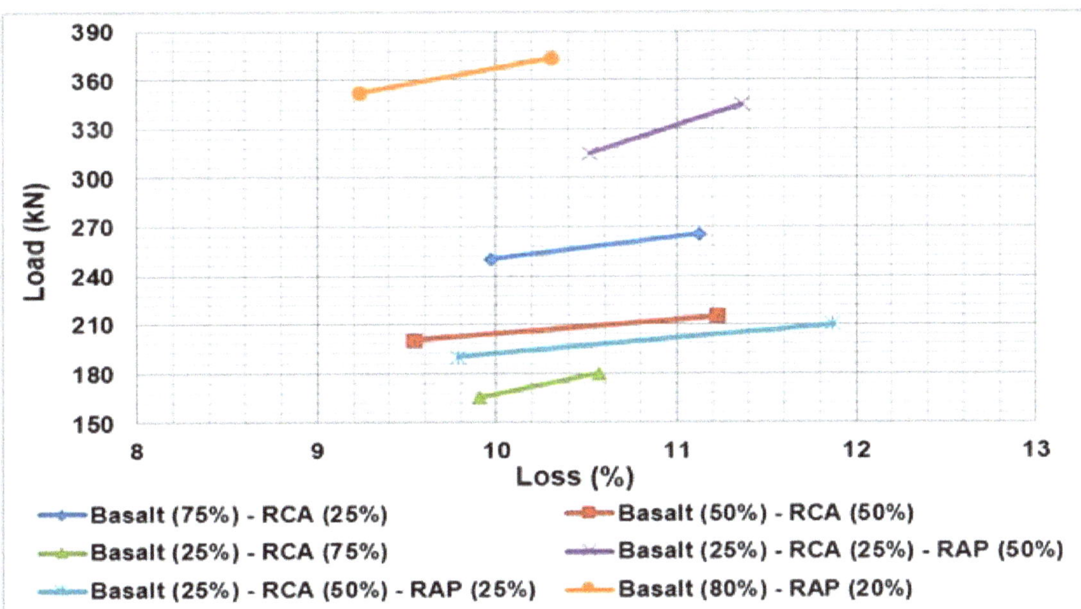

Figure 12. Results of Wet/Dry Strength Test for Mix of Coarse Aggregates (Wet Strength).

Table 9. The Results of Wet/Dry Strength Variation Test for Different Mix of Coarse Aggregates.

Material	Dry Strength, D (kN)	Wet Strength, W (kN)	Wet/Dry Strength Variation (%)	Australian Standards Wet/Dry Strength Limits (%)
Basalt (75%) and RCA (25%)	291.4	250.4	14.1	35% (max) For heavy and very heavy traffic
Basalt (50%) and RCA (50%)	247.6	204.1	17.6	
Basalt (25%) and RCA (75%)	210.4	167.2	20.5	
Basalt (80%) and RAP (20%)	398.9	366.9	8.0	
Basalt (25%), RCA (25%) and RAP (50%)	337.8	296.4	12.3	
Basalt (25%), RCA (50%) and RAP (25%)	232.6	192	17.5	

3.3. Results and Discussions

As presented in the previous sections, in this research, the properties of RCA, RAP, basalt and mix of these aggregates were evaluated by conducting a series of tests. The test results are summarized in Table 10.

Table 10. Summary of the Test Results for the Evaluation of Coarse Aggregate Properties.

Test	Test Method	Aggregate			Typical Limit Based on Australian Standards
		RCA	RAP	Basalt	
Particle Distribution Test	AS 1141.11.1	As presented in relevant Figures and Tables			-
Flakiness Index Test	AS 1141.15	6.91	10.42	19.03	25% (max)
Particle Shape Test	AS 1141.14	6.16	9.47	18.34	35% (max)
Water Absorption	AS 1141.6.1	**6.30**	1.78	1.64	2% (max)
Particle Density	AS 1141.6.1	2.370	2.541	2.640	-
Particle Density on Dry Basis	AS 1141.6.1	2.212	2.431	2.530	-
Particle Density on SSD Basis	AS 1141.6.1	2.351	2.474	2.571	-
Aggregate Crushing Value	AS 1141.21	29.21	7.40	8.91	35% (max)
Weak Particles	AS 1141.32	0.23	0.04	0.23	1% (max)
Wet/Dry Strength Test	AS 1141.22	26.6	7.4	8.5	35% (max)
Wet Strength	AS 1141.22	**119.7**	398.2	359.2	150 kN (min)
Dry Strength	AS 1141.22	163.1	429.8	392.9	-

In addition, to have better comparisons between the aggregate properties, the test results on different aggregates as well as the standard limits are also illustrated in Figure 13.

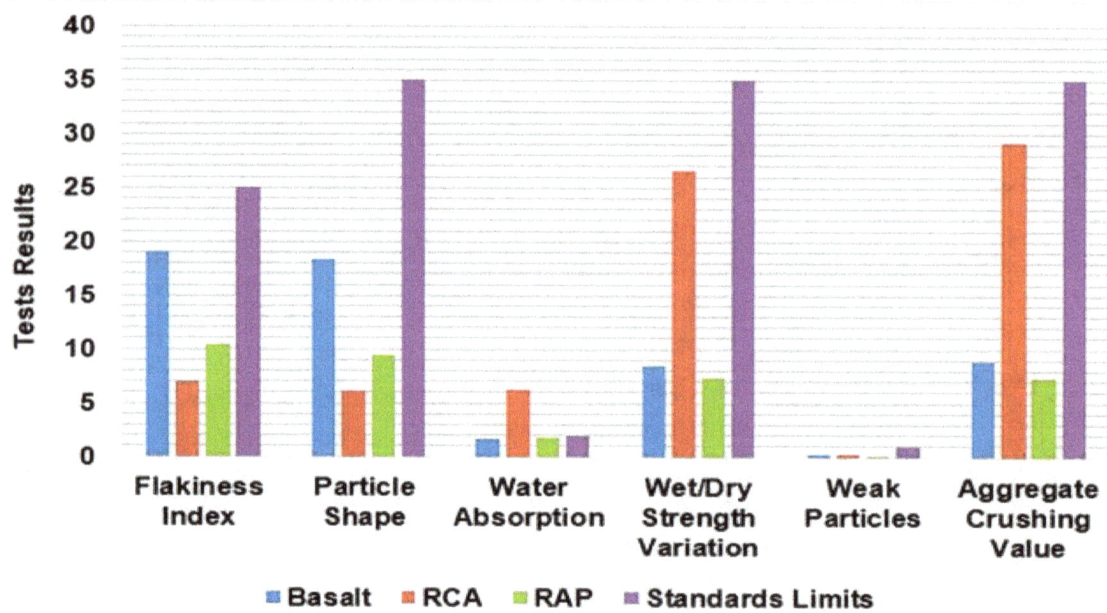

Figure 13. Comparison of Different Aggregate Properties with Standard Limits.

As shown in Figure 13, the results of preliminary tests on coarse aggregates indicate that all properties of RCA, except for water absorption and wet strength (which are shown in bold in Table 10), are within the limits specified by relevant Australian Standards and hence deemed appropriate for use as aggregate in the asphalt mixture. However, for some parameters such as Flakiness Index and Particle Shape which are two dominant characteristics having significant impact on asphalt mixture strength and stability; RCA displays smaller value in comparison with basalt and RAP. This can be one of the strong points of RCA as flakiness index and particle shape are the two important properties for proper compaction, deformation resistance, and workability of asphalt mixture [58].

In addition, as can be observed in Table 10 and Figure 13, the water absorption of RCA is higher than the corresponding value of RAP and basalt and the Australian Standards limit, because it is well known that water absorption requires linked and open cracks in the structure of aggregate and RCA contains cracks due to the crushing processes. Moreover, the great amounts of impurities in RCA can increase the water absorption of RCA. The high water absorption of RCA may result in high bitumen absorption in asphalt mixtures, and hence plays an important role in asphalt mixture design.

Accordingly, since this research aims to investigate the feasibility of the application of RCA for the partial replacement of coarse virgin aggregate (basalt) and in combination with other recycled aggregate (RAP) in asphalt mixtures, the particle density and water absorption tests were conducted on different mix of coarse aggregates while considering different percentages of these materials. The results of these tests are presented in Figure 14.

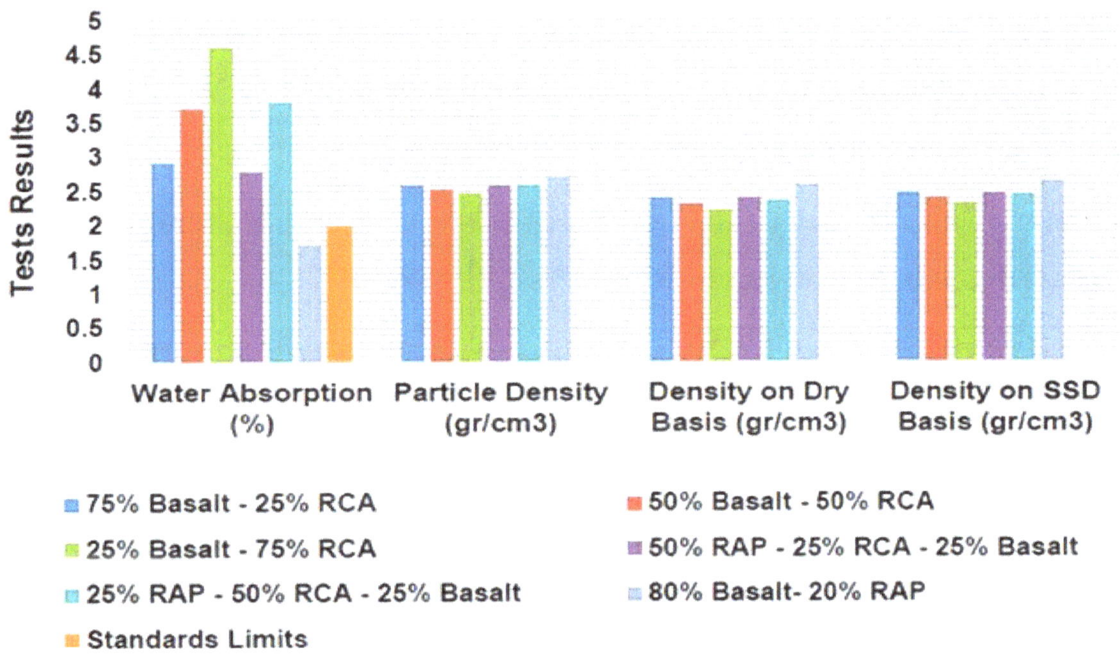

Figure 14. Comparison of Water Absorption and Particle Density of Different Mixes of Coarse Aggregates.

As can be observed in Figure 14, increasing RCA in the mix does not make any substantial change in mix density in comparison with water absorption. In other words, by increasing RCA from 0% to 100% in the mix, the density decreases by 7%, whereas water absorption increases by 74%.

In addition, although wet/dry strength variation of RCA meets the requirements of Australian standards, the test results show that this value is higher than the corresponding value of RAP and basalt. As the wet/dry strength variation is related to the principal mechanical properties which are required for asphalt aggregate, it is of high importance in asphalt mixture design. Therefore, wet/dry strength variation test was also conducted on different mixes of coarse aggregate. Figure 15 shows the comparison of wet strength, dry strength, and wet/dry variation in different mixes of RCA, RAP and basalt.

As illustrated in this figure, the wet/dry strength variation of mix of RCA/basalt increases by increase of the percentage of RCA in the mix, so that the increase of RCA from 0% to 100% will result in 20% increase in wet/dry strength variation. The results of these two tests (i.e., water absorption and particle density test, and wet/dry strength variation test) on mix of coarse aggregates are summarized in Table 11.

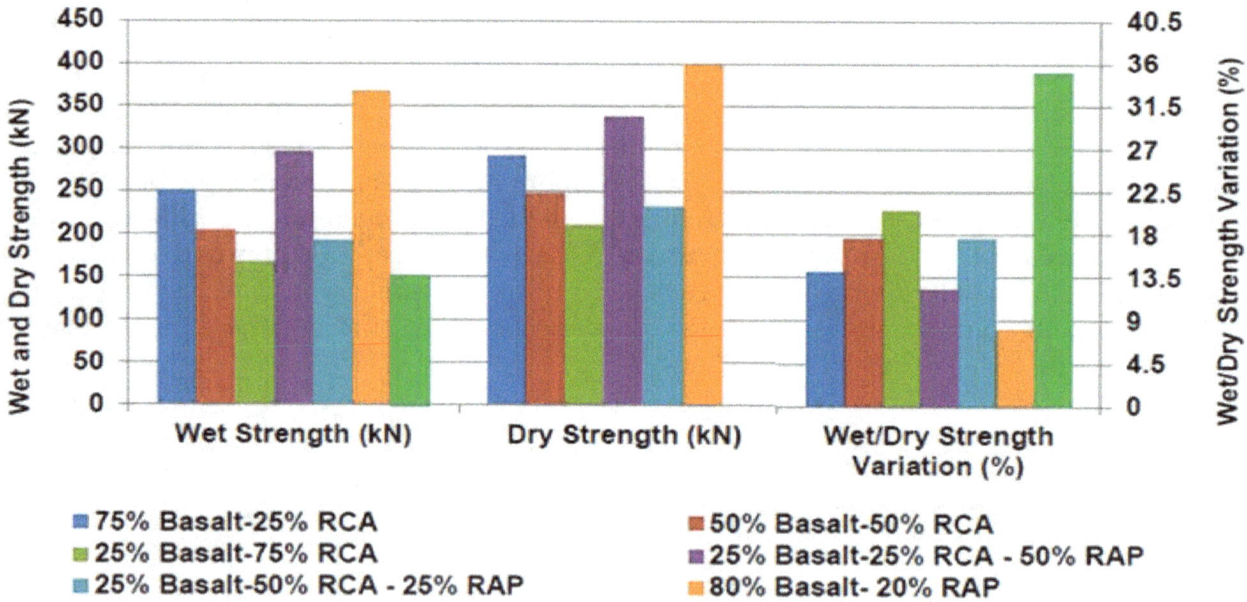

Figure 15. Comparison of the Wet Strength and Dry Strength of Different Mixes of Coarse Aggregates.

Table 11. Summary of Tests Results for Evaluation of Mix of Coarse Aggregates Properties.

Mix	Test Method	Water Absorption	Particle Density	Wet/Dry Strength Variation
Basalt (75%), RCA (25%)	AS 1141.15	2.93	2.590	14.1
Basalt (50%), RCA (50%)	AS 1141.14	3.71	2.527	17.6
Basalt (25%), RCA (75%)	AS 1141.6.1	4.62	2.476	20.5
Basalt (80%), RAP (20%)	AS 1141.6.1	1.72	2.723	8.0
Basalt (25%), RCA (25%), RAP (50%)	AS 1141.21	2.79	2.582	12.3
Basalt (25%), RCA (50%), RAP (25%)	AS 1141.32	3.81	2.598	17.5

The results of tests on mix of coarse aggregate showed that in all cases of RCA ratios, RCA increase causes a decrease in wet and dry strength and an increase in water absorption. This will necessitate the proper selection and optimum combination of RCA and other aggregates.

The coefficient of variation is used as an indication to measure the heterogeneity of test results. The results of calculation of standard deviation (SD) and coefficient of variation (CV) for each set of aggregate mixes are presented in Table 12.

Table 12. Coefficient of Variation and Standard Deviation for Mix of Coarse Aggregates.

Mix	Coefficient of Variation	Standard Deviation
Basalt (100%) + RCA (0%)	2.14	0.035
Basalt (75%) + RCA (25%)	1.71	0.050
Basalt (50%) + RCA (50%)	3.52	0.131
Basalt (25%) + RCA (75%)	4.42	0.204
Basalt (0%) + RCA (100%)	2.47	0.156

As can be observed in Table 12, the coefficient of variation for each data set reveals that the test results dispersion is low and the tests are conducted consistently.

Furthermore, regression analysis is typically applied to the water absorption test results for different combination of RCA and basalt, to show the typical amount of RCA and basalt in a blend to give 2% water absorption which is the standard limit of water absorption based on Australian standards (Figures 16 and 17).

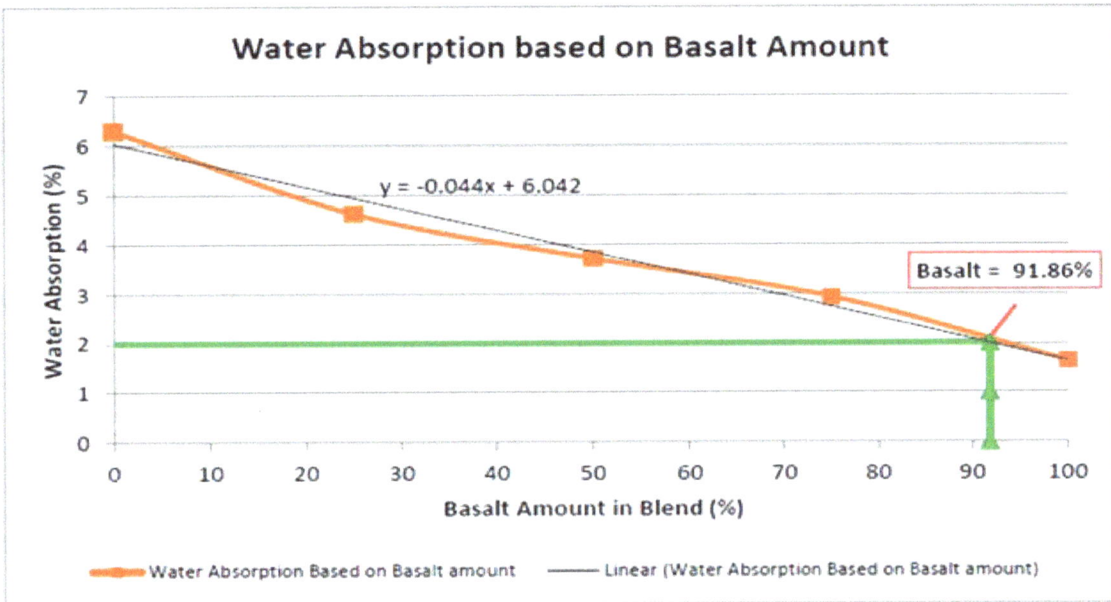

Figure 16. Regression Analysis for Determination of Optimum Basalt Amount.

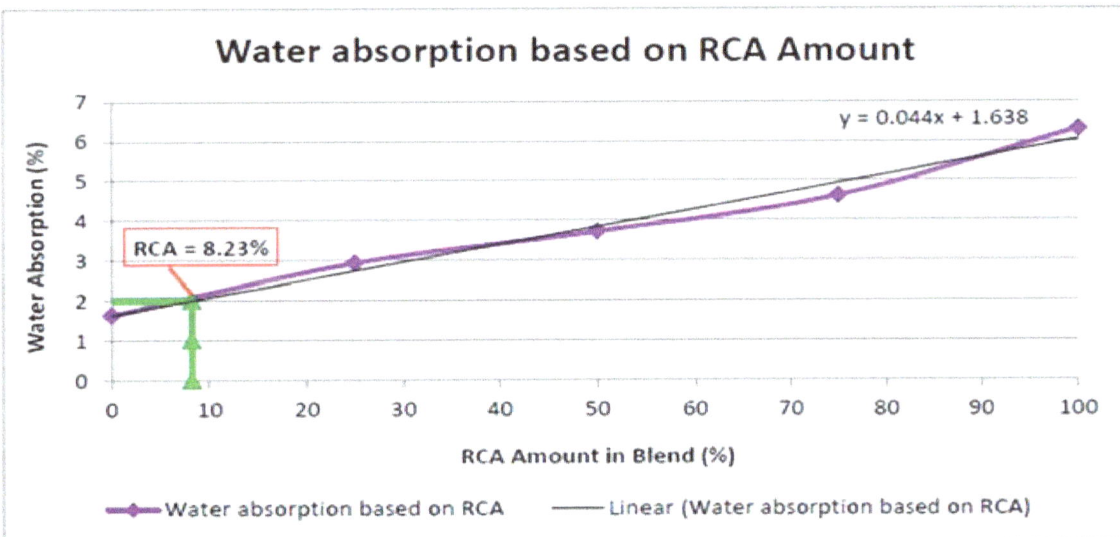

Figure 17. Regression Analysis for Determination of Optimum recycled construction aggregate (RCA) Amount.

As illustrated in Figures 16 and 17, the standard water absorption limit of 2% can be achieved by mixing of almost 8% and 92% of RCA and basalt, respectively. However, based on the available references [59], typically, the amount of binder absorbed by aggregate is 0.3 to 0.7 times the water absorption of the aggregate. In addition, according to this standard, if the sample absorbs between 2% and 4% of its mass, it should be carefully examined by other tests. If the sample absorbs in excess of 4% of its mass, it will rarely prove to be an adequate aggregate for asphalt production. Based on the water absorption results, it can be observed that the combination of 25% RCA and 75% basalt would provide water absorption of 2.93%, and also water absorption of the combination of 50% RCA and 50% basalt would be 3.71%, which are still in the range of aggregate water absorption that suggest further research.

4. Conclusions

Since the coarse aggregate properties are identified by current research as the second most important parameter after gradation for the performance of HMA [49], therefore, in this research, attempts were made to assess the properties of RCA for use in asphalt mixture as coarse aggregate, and this paper presented the summary results of a comprehensive set of preliminary tests on RCA, RAP and basalt as well as different mixes of these aggregates to evaluate their basic mechanical and physical properties. It was argued that information on these fundamental properties were paramount in designing a durable and sustainable asphalt mixtures. To this end, different aggregate and aggregate mixes containing different percentages of RCA, RAP, and basalt were investigated in this research to assess its suitability as coarse aggregate in asphalt. This paper presented the results of this experimental work conducted as a component of a broader research project for designing an asphalt mixture. Based on this research, it was concluded that:

(1) RCA has lower value of flaky and misshapen particles in comparison with RAP and basalt. This implies that asphalt mixtures containing a certain amount of RCA can have better workability, deformation resistance and compaction.
(2) RCA exhibits comparatively more absorption and wet/dry strength variation than conventional aggregate and RAP, while the results of other tests show that RCA still meets the requirements for aggregate in asphalt mixtures. Cracks and adhering mortar and cement paste can be significant reasons for the high water absorbtion of RCA which needs to be compensated for during mix design.
(3) The results of water absorption and particle density test on different mix of coarse aggregates revealed that RCA increase will increase water absorption of the mixture. Therefore, the selection of optimum combination of RCA and other aggregates is required to satisfy the relevant standards requirements.
(4) Regression analysis applied to the results of water absorption test on different combination of RCA and basalt, as illustrated in Figures 16 and 17, indicates that mixing of almost 8% of RCA with natural aggregates will provide the standard water absorption limit of 2%.
(5) Since, according to Austroads (2014), the aggregates with water absorption of between 2% and 4% of their mass should be carefully examined by other tests [59]. This standard limit will allow further investigation of the application of up to 50% of RCA in mixtures because based on the water absorption results, it can be observed that the combination of 25% RCA and 75% basalt would provide water absorption of 2.93%, and the combination of 50% RCA and 50% basalt would provide water absorption of 3.71%, which are still in the range of aggregates that require further research for their water absorption properties.

Acknowledgments: The authors would like to acknowledge Boral Asphalt Company for providing part of materials required for the research work.

Author Contributions: Farzaneh Tahmoorian is the main author and researcher conducting this research as the main component of her PhD studies, Bijan Samali has directed and coordinated the project in a supervisory capacity, Vivian WY Tam and John Yeaman have been consultants to the project providing valuable insights into behavior of aggregates and asphalt.

References

1. Karlsson, R.; Isacsson, U. Material-Related Aspects of Asphalt Recycling-State-of-the-Art. *J. Mater. Civ. Eng.* **2006**, *18*, 81–92. [CrossRef]
2. Huang, Y.; Bird, R.; Heidrich, O. A Review of the Use of Recycled Solid Waste Materials in Asphalt Pavements. *Resour. Conserv. Recycl.* **2007**, *52*, 58–73. [CrossRef]
3. Widyatmoko, I. Mechanistic-Empirical Mixture Design for Hot Mix Asphalt Pavement Recycling. *Constr. Build. Mater.* **2008**, *22*, 77–87. [CrossRef]

4. Ahmedzadeh, P.; Sengoz, B. Evaluation of steel slag coarse aggregate in hot mix asphalt concrete. *J. Hazard. Mater.* **2009**, *165*, 300–305. [CrossRef] [PubMed]
5. Rafi, M.M.; Qadir, A.; Siddiqui, S.H. Experimental Testing of Hot Mix Asphalt Mixture Made of Recycled Aggregates. *Waste Manag. Res.* **2011**, *29*, 1316–1326. [CrossRef] [PubMed]
6. Khan, I.; Gundaliya, P.J. Utilization of Waste Polyethylene Materials in Bituminous Concrete Mix for Improved Performance of Flexible Pavements. *J. Appl. Res.* **2012**, *1*, 85–86.
7. Menaria, Y.; Sankhla, R. Use of Waste Plastic in Flexible Pavements-Green Roads. *Open J. Civ. Eng.* **2015**, *5*, 299–311. [CrossRef]
8. Chandh, K.A.; Akhila, S. A Laboratory Study on Effect of Plastic on Bitumen. *Int. J. Sci. Res.* **2016**, *5*, 1406–1409.
9. Tabsh, S.W.; Abdelfatah, A.S. Influence of Recycled Concrete Aggregates on Strength Properties of Concrete. *Constr. Build. Mater.* **2009**, *23*, 1163–1167. [CrossRef]
10. Mills-Beale, J.; You, Z. The Mechanical Properties of Asphalt Mixtures with Recycled Concrete Aggregates. *Constr. Build. Mater.* **2010**, *24*, 230–235. [CrossRef]
11. Chen, M.; Lin, J.; Wu, S. Potential of Recycled Fine Aggregates Powder as Filler in Asphalt Mixture. *Constr. Build. Mater.* **2011**, *25*, 3909–3914. [CrossRef]
12. Silva, H.M.; Oliveira, J.R.; Jesus, C.M. Are totally recycled hot mix asphalt a sustainable alternative for road paving? *Resour. Conserv. Recycl.* **2012**, *60*, 38–48. [CrossRef]
13. Vavrik, W.R.; Harrell, M.J.; Gillen, S.L. Achieving Perpetual Pavement through Staged Construction. In Proceedings of the International Conference of Perpetual Pavements, Ohio University, Athens, OH, USA, 30 September 2009.
14. Zaniewski, J.P.; Srinivasan, G. Evaluation of Indirect Tensile Strength to Identify Asphalt Concrete Rutting Potential. In *Asphalt Technology Program*; West Virginia University: West Virginia, WV, USA, 2004; Volume 3, pp. 154–196.
15. Husain, N.M.D.; Karim, M.R.; Mahmud, H.B.; Koting, S. Effects of Aggregate Gradation on the Physical Properties of Semiflexible Pavement. *Adv. Mater. Sci. Eng.* **2014**, *2014*, 1–8. [CrossRef]
16. Al-Mosawe, H.; Thom, N.; Airey, G.; Al-Bayati, A. Effect of Aggregate Gradation on the Stiffness of Asphalt Mixtures. *Int. J. Pavement Eng. Asphalt Technol.* **2015**. [CrossRef]
17. Kandhal, P.S.; Cooley, L.A. National Cooperative Highway Research Program Report 464: The Restricted Zone in the Superpave Aggregate Gradation Specification. In *National Cooperative Highway Research Program*; National Research Council: Washington, DC, WA, USA, 2001.
18. Arabani, M.; Azarhoosh, A.R. The Effect of Recycled Concrete Aggregate and Steel Slag on the Dynamic Properties of Asphalt Mixtures. *Constr. Build. Mater.* **2011**, *35*, 1–7. [CrossRef]
19. Jony, H.; Al-Rubaie, M.; Jahad, I. The Effect of Using Glass Powder Filler on Hot Asphalt Concrete Mixtures Properties. *Eng. Technol. J.* **2011**, *29*, 44–57.
20. Pereira, P.; Oliveira, J.; Picado-Santos, L. Mechanical Characterization of Hot Mix Recycled Materials. *Int. J. Pavement Eng.* **2004**, *5*, 211–220. [CrossRef]
21. Fulton, B. Use of Recycled Glass in Pavement Aggregate. In Proceedings of the 23rd Australian Road Research Board International Conference, Adelaide, Australia, 30 July–1 August 2008.
22. Finkle, I.; Ksaibati, K. Recycled Glass Utilization in Highway Construction. In Proceedings of the Transportation Research Board 86th Annual Meeting, University of Wyoming, Laramie, WY, USA, 21–25 January 2007.
23. Wong, Y.D.; Sun, D.D.; Lai, D. Value-Added Utilisation of Recycled Concrete in Hot-Mix Asphalt. *Waste Manag.* **2007**, *27*, 294–301. [CrossRef] [PubMed]
24. Paranavithana, S.; Mohajerani, A. Effects of Recycled Concrete Aggregates on Properties of Asphalt Concrete. *Resour. Conserv. Recycl.* **2006**, *48*, 1–12. [CrossRef]
25. Motter, J.S.; Miranda, L.F.R.; Bernucci, L.L.B. Performance of Hot Mix Asphalt Concrete Produced with Coarse Recycled Concrete Aggregate. *J. Civ. Eng.* **2014**, *27*. [CrossRef]
26. Pérez, I.; Pasandin, A.R.; Medina, L. Hot Mix Asphalt Using C&D Waste as Coarse Aggregates. *Mater. Des.* **2012**, *36*, 840–846.
27. Silva, C.A.R.S. Study of Recycled Aggregate from Civil Construction in Bituminous Mixtures for Urban Roads. Master's Dissertation, Federal University of Ouro Preto, Minas Gerais, Brazil, November 2009.

28. Zhu, J.; Wu, S.; Zhong, J.; Wang, D. Investigation of Asphalt Mixture Containing Demolition Waste Obtained from Earthquake-Damage Buildings. *Constr. Build. Mater.* **2012**, *29*, 466–475. [CrossRef]
29. Zulkati, A.; Wong, Y.D.; Sun, D.D. Mechanistic Performance Of Asphalt-Concrete Mixture Incorporating Coarse Recycled Concrete Aggregate. *J. Mater. Civ. Eng.* **2013**, *10*, 1299–1305. [CrossRef]
30. Arulrajah, A.J.; Piratheepan, M.W.B.; Sivakugan, N. Geotechnical Characteristics of Recycled Crushed Brick Blends for Pavement Subbase Applications. *Can. Geotech. J.* **2012**, *49*, 796–811. [CrossRef]
31. Bennert, T.; Papp, W.J.; Maher, A.; Gucunski, N. Utilization of Construction and Demolition Debris under Traffic-Type Loading in Base and Subbase Applications. *Transp. Res. Rec.* **2000**. [CrossRef]
32. Blankenagel, B.J. Characterization of Recycled Concrete for Use as Pavement Base Material. Master's Thesis, Brigham Young University, Provo, UT, USA, 20 August 2005.
33. Conceicao, F.D.; Motta, R.D.S.; Vasconcelos, K.L.; Bernucci, L. Laboratory Evaluation of Recycled Construction and Demolition Waste for Pavements. *Constr. Build. Mater.* **2011**, *25*, 2972–2979.
34. Jayakody, S.; Gallage, C.; Kumar, A. Assessment of Recycled Concrete Aggregates as a Pavement Material. *Geomech. Eng.* **2014**, *6*, 235–248.
35. Jiménez, J.R.; Ayuso, J.; Agrela, F.; López, M.; Galvín, A.P. Utilization of Unbound Recycled Aggregates from Selected CDW in Unpaved Rural Roads. *J. Resour. Conser. Recycl.* **2012**, *58*, 88–97. [CrossRef]
36. Papp, W.J.; Maher, M.H.; Bennet, T.A.; Gucunski, N. Behavior of construction and demolition debris in base and subbase applications. In *Recycled Materials in Geotechnical Applications*; Vipulanandan, C., Elton, D.J., Eds.; American Society of Civil Engineers: Reston, VA, USA, 1998; pp. 122–136. ISBN 0784403872.
37. Nataatmadja, A.; Tan, Y.L. Resilient Response of Recycled Concrete Road Aggregates. *J. Transp. Eng.* **2001**. [CrossRef]
38. Celauro, C.; Benardo, C.; Gabriele, B. Production of Innovative, Recycled and High-Performance Asphalt for Road Pavements. *Resour. Conser. Recycl. J.* **2010**, *54*, 337–347. [CrossRef]
39. Hossain, M.; Metcalf, D.G.; Scofield, L.A. Performance of Recycled Asphalt Concrete Overlays in Southwestern Arizona. *Transp. Res. Rec.* **1993**, *1427*, 30–37.
40. Rebbechi, J.; Green, M. Going Green: Innovations in Recycling Asphalt. In Proceedings of the AAPA, Queensland, QLD, Australia, 13–16 September 2015.
41. Berthelot, C.; Haichert, R.; Podborochynski, D.; Wandzura, C.; Taylor, B.; Gunther, D. Mechanistic Laboratory Evaluation and Field Construction of Recycled Concrete Materials for Use in Road Substructures. *Transp. Res. Board* **2010**. [CrossRef]
42. Wu, S.; Zhu, J.; Zhong, J.; Wang, D. Influence of Demolition Waste Used as Recycled Aggregate on Performance of Asphalt Mixture. *Road Mater. Pavement Des.* **2013**, *14*, 679–688. [CrossRef]
43. Dahir, S.A. Review of Aggregate Selection Criteria for Improved Wear Resistance and Skid Resistance of Bituminous Surfaces. *J. Test. Eval.* **1979**, *7*, 245–253.
44. Brown, E.R.; McRae, J.L.; Crawley, A.B. Effect of Aggregate on Performance of Bituminous Concrete. In *ASTM STP 1016*; American Society for Testing and Materials: West Conshohocken, PA, USA, 1989; pp. 34–63.
45. Brown, E.R.; Bassett, C.E. Effects of Maximum Aggregate Size on Rutting Potential and Other Properties of Asphalt-Aggregate Mixtures. In *Transportation Research Record*; Transportation Research Board: Washington, DC, USA, 1990; pp. 107–119.
46. Button, J.W.; Perdomo, D.; Lytton, R.L. Influence of Aggregate on Rutting in Asphalt Concrete Pavements. In *Transportation Research Record*; Transportation Research Board: Washington, DC, USA, 1990; pp. 141–152.
47. Elliot, R.P.; Ford, M.C.; Ghanim, M.; Tu, Y.F. Effect of Aggregate Gradation Variation on Asphalt Concrete Mix Properties. In *Transportation Research Record*; Transportation Research Board: Washington, DC, USA, 1991; pp. 52–60.
48. Krutz, N.C.; Sebaaly, P.E. The Effects of Aggregate Gradation on Permanent Deformation of Asphalt Concrete. In Proceedings of the Asphalt Paving Technology, Austin, TX, USA, 22–24 March 1993.
49. Oduroh, P.K.; Mahboub, K.C.; Anderson, R.M. Flat and Elongated Aggregates in Superpave Regime. *J. Mater. Civ. Eng.* **2000**, *12*, 124–130. [CrossRef]
50. Chen, J.S.; Liao, M.S. Evaluation of Internal Resistance in Hot-Mix Asphalt (HMA) Concrete. *Constr. Build. Mater. J.* **2002**, *16*, 313–319. [CrossRef]
51. Sengoz, B.; Onsori, A.; Topal, A. Effect of Aggregate Shape on the Surface Properties of Flexible Pavement, KSCE. *J. Civ. Eng.* **2014**, *18*. [CrossRef]

52. Masad, E.; Rezaei, A.; Chowdhury, A.; Harris, P. Predicting Asphalt Mixture Resistance Based on Aggregate Characteristics. In *Aggregate Resistance to Polishing and Its Relationship to Skid Resistance Project*; National Technical Information Service: Springfield, VA, USA, 2009.
53. Wu, Z.; King, W. Development of Surface Friction Guidelines for LADOTD. In *LTRC Project No. 09-2B*; Louisiana State University: Baton Rouge, LA, USA, April 2011.
54. Marinho, M.N. Analysis of Mechanical Performance of Hot Mixed Asphaltic Concrete Produced with Recycled Coarse Concrete Aggregate. Master's Dissertation, Federal University of Pernambuco, Recife, Brazil, July 2011.
55. Mohajerani, A. A Study of Relationships between Polished Aggregate Friction Value, Aggregate Crushing Value, and Point Load Strength Index. *Aust. Geomech.* **1997**, *1997*, 62–65.
56. Prowell, B.D.; Zhang, J.; Brown, E.R. Aggregate Properties and the Performance of Superpave Designed Hot Mix Asphalt. In *National Cooperative Highway Research Program Report 539*; National Research Council: Washington, DC, USA, 2005. [CrossRef]
57. Dickinson, E.J. *Bituminous Roads in Australia*; Australian Road Research Board: Melbourne, MEL, Australia, 1984.
58. Masad, E.; Al-Rousan, T.; Button, J.; Little, D. Test Methods for Characterizing Aggregate Shape, Texture, and Angularity. In *National Cooperative Highway Research Program Report 555*; National Research Council: Washington, DC, USA, 2007.
59. Guide to Pavement Technology Part 4B: Asphalt, Austroads. 2014. Available online: https://www.onlinepublications.austroads.com.au/items/AGPT04B-1 (accessed on 31 June 2014).

4

Application of Finite Layer Method in Pavement Structural Analysis

Pengfei Liu [1], Qinyan Xing [2], Yiyi Dong [2], Dawei Wang [1,3,*], Markus Oeser [1] and Si Yuan [2]

1 Institute of Highway Engineering, RWTH Aachen University, Mies-van-der-Rohe-Street 1, D52074 Aachen, Germany; liu@isac.rwth-aachen.de (P.L.); oeser@isac.rwth-aachen.de (M.O.)

2 Department of Civil Engineering, Tsinghua University, 100084 Beijing, China; xingqy@tsinghua.edu.cn (Q.X.); dyy14@mails.tsinghua.edu.cn (Y.D.); yuans@tsinghua.edu.cn (S.Y.)

3 Institute of Highway Engineering, Paul-Bonatz-Street 9–11, University of Siegen, D57076 Siegen, Germany

* Correspondence: wang@isac.rwth-aachen.de

Academic Editor: Zhanping You

Abstract: The finite element (FE) method has been widely used in predicting the structural responses of asphalt pavements. However, the three-dimensional (3D) modeling in general-purpose FE software systems such as ABAQUS requires extensive computations and is relatively time-consuming. To address this issue, a specific computational code EasyFEM was developed based on the finite layer method (FLM) for analyzing structural responses of asphalt pavements under a static load. Basically, it is a 3D FE code that requires only a one-dimensional (1D) mesh by incorporating analytical methods and using Fourier series in the other two dimensions, which can significantly reduce the computational time and required resources due to the easy implementation of parallel computing technology. Moreover, a newly-developed Element Energy Projection (EEP) method for super-convergent calculations was implemented in EasyFEM to improve the accuracy of solutions for strains and stresses over the whole pavement model. The accuracy of the program is verified by comparing it with results from BISAR and ABAQUS for a typical asphalt pavement structure. The results show that the predicted responses from ABAQUS and EasyFEM are in good agreement with each other. The EasyFEM with the EEP post-processing technique converges faster compared with the results derived from ordinary EasyFEM applications, which proves that the EEP technique can improve the accuracy of strains and stresses from EasyFEM. In summary, the EasyFEM has a potential to provide a flexible and robust platform for the numerical simulation of asphalt pavements and can easily be post-processed with the EEP technique to enhance its advantages.

Keywords: finite layer method; asphalt pavement structural analysis; EasyFEM; element energy projection; super-convergence

1. Introduction

The analysis of stress states is of considerable importance for the design, construction, maintenance, and rehabilitation of asphalt pavements in practice. In the past several decades, a lot of computer software has been developed and is increasingly used routinely in pavement design and assessment processes.

In Germany, the guidelines for the analytical design of asphalt pavement superstructures RDO Asphalt 09 [1] propose the use of layer elastic theory (LET), e.g., the program BISAR, (Shell global, The Hague, The Netherlands) as the kernel of its pavement response model to calculate stresses, strains and displacements for different temperatures at critical locations within the pavement structure. Due to its simplicity, the LET has been utilized by pavement engineers for several decades. Currently, LET has been expanded to handle some problems with the material properties of viscoelasticity and nonlinearity as well as non-uniform loads [2]. However, the features of the software application based on the LET are still limited. For example, in BISAR, each layer is assumed homogeneous, isotropic, and linear elastic; all materials are weightless (no inertial effects are considered); all layers are assumed to be infinite in lateral extent and have a finite thickness except for the sub-grade, which is assumed to be infinite; the pavement systems are loaded statically over a uniform circular area; the compatibility of strains and stresses is assumed to be satisfied at all layer interfaces [3,4]. However, the reality of asphalt pavements may be very different to the assumptions made in the BISAR, e.g., finite geometrical scale in lateral extent, non-uniform loading conditions, and inelastic material properties. These differences may result in significant deviations between the calculated and the real responses of asphalt pavements.

With the finite element method (FEM), which is a numerical analysis technique proposed for obtaining approximate solutions for a wide variety of engineering problems, these specific requirements can be met and the asphalt pavement responses can be predicted more realistically. Some representative software applications have been widely used in pavement engineering, such as CAPA-3D (Delft University of Technology, Delft, The Netherlands) and APADS 2D (Austroads, Sydney, Australia). The development history of CAPA-3D goes back to the late 1980s. Currently, CAPA-3D is a linear/nonlinear, static/dynamic finite element (FE) system for the solution of very large scale three-dimensional (3D) solid models such as those typically encountered in pavement and soil engineering. It includes several constitutive material models, e.g., linear elasticity, hyperelasticity, elastoplasticity and viscoelasticity. As such, it can simulate a very broad range of engineering materials under various loading conditions [5]. Unfortunately, due to its 3D characteristics, the high hardware demands and long execution times render it suitable, primarily for research purposes [6]. APADS 2D was developed from 2008, in the scope of the Austroads project Developments of Pavement Design Models. It applies a two-dimensional (2D) axisymmetric concept to reduce the computational time and considers the effect of multiple loads through superposition. Due to the inherent limitations of axisymmetric models, it is difficult to simulate a pavement model with a determined geometry and complex loading conditions [7]. Other general-purpose FE software, such as ABAQUS (SIMULIA, Johnston, RI, USA) and ANSYS (ANSYS, Canonsburg, PA, USA), may provide more powerful capabilities to simulate the response of asphalt pavements to a certain extent. However, the expensive costs to get valid licenses and the time-consuming training process often render it impractical to be used by a road engineer.

One proposed method to overcome the aforementioned difficulties assumes that the displacements in one geometrical direction can be represented using Fourier series. Exploiting its orthogonal properties, a problem of such a class can be simplified into a series of 2D solutions. This method is so called semi-analytical FEM and was first developed in linear analysis by Wilson [8]. The analysis was extended to an elasto-plastic body by Meissner [9] based on Wilson's work. A visco-plastic formulation was used by Winnicki and Zienkiewicz [10] to tackle material nonlinearity. To analyze the consolidation of elastic bodies subjected to non-symmetric loading, Carter and Booker [11] provided an effective solution by using the continuous Fourier series. Lai and Booker [12] successfully applied a discrete Fourier technique to analyze the stress state of solids with nonlinear behavior under 3D loading conditions. Further developments were made by Fritz [13] and Hu et al. [14], who programed simple FE codes to apply the semi-analytical FEM for the analysis of asphalt pavements. Recently, an FE code named SAFEM with more features was developed by Liu et al. [15–19]. Viscoelastic material property and dynamic analyses were integrated in this code. Here, the partial bonds between pavement layers can be considered and the infinite element was applied to reduce the influence of the boundary on the computational results.

The finite layer method (FLM) is proposed as another alternative way for analyzing the structural response of pavements, whilst reducing the computational requirements without decreasing the computational accuracy. This method combines a one-dimensional (1D) FEM discretization in the pavement depth direction with Fourier series in the two horizontal directions, which can be considered as a further development of the semi-analytical FEM. The idea is particularly suitable for multilayered structures. Furthermore, due to the use of Fourier series in two directions, the computational cost of FLM is significantly lower than that of conventional FE analysis methods. The FLM started to be applied in the analysis of elastic, horizontally layered foundations from 1979 [20], when the basic theory and application were proposed. Thereafter, the approach was further developed for the analysis of nonhomogeneous soils whose modulus increases linearly with depth [21]. An exact finite layer flexibility matrix was introduced to overcome the difficulties when the conventional finite layer stiffness approach was applied to incompressible materials [22,23]. Meanwhile, soil consolidation and surface deformation entailing the extraction of water was also investigated by using this approach [24–26]. In the next several years, researchers applied the FLM in groundwater flow models [27] and analyses of 3D Biot consolidation of layered transversely isotropic soils [28]. Recently, the FLM was applied for modeling the noise transmission through double walls [29]. analytical model, called 3D-Move uses a continuum-based FLM to compute pavement responses [30,31]. The 3D-Move model can account for important pavement response factors such as moving loads, 3D contact stress distributions (normal and shear) of any shape, and viscoelastic material characterization for the pavement layers.

Although the 3D-Move is freely available, it is not open source, and it still lacks some specific features such as the interlayer behavior for the pavement modeling. Therefore, a re-implementation of the FLM in code is necessary. As mentioned in the manual of 3D-Move, the accuracy of the calculated results is not guaranteed. Therefore, an algorithm or a technique that can provide more accurate results is of significant importance. In recent years, the research on super-convergence has increasingly drawn attention since it can provide higher order accuracy. A number of research studies have been conducted on this subject [32,33]. In recent years, a super-convergence technique named Element Energy Projection (EEP) method was proposed by Yuan et al. [34–36]. Its core assumption is to assume the energy projection theorem in the FEM mathematical theory [37] to be almost true for a single element. For various 1D problems, this assumption proves to be true and the convergent displacements and derivatives of the simplified form (i.e., with linear test/weight functions) converge at least one order higher than the conventional FEM results [38]. This method has been proven to be simple, convenient and effective. Moreover, it can be easily applied to other specific FE methods including FLM.

In the following sections, the basic theory of FLM is introduced, based on which the computer code named EasyFEM is self-developed. The EEP super-convergence strategy for EasyFEM is then described in detail to improve the accuracy of the results derived from EasyFEM. The EasyFEM is verified by comparing the results from BISAR and ABAQUS; the convergence rates between the results from ordinary EasyFEM and the EasyFEM with the post-processing with EEP are compared. The application of EasyFEM is to predict the mechanical responses of asphalt pavements. Lastly, a brief summary and outlook are provided.

2. Numerical Solution of EasyFEM

2.1. Finite Elements

A typical 3D EasyFEM model for the pavement problem was set up as shown in Figure 1.

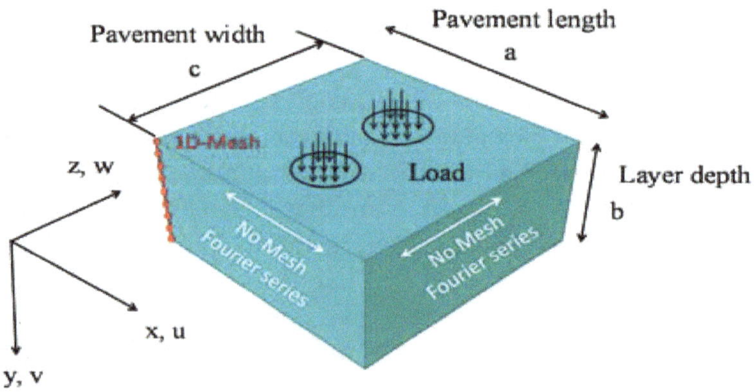

Figure 1. A typical EasyFEM model.

The system is discretized in the pavement depth direction; thus, the conventional 1D isoparametric elements of degree k can be adopted, e.g., the 3-node quadratic elements as shown in Figure 2:

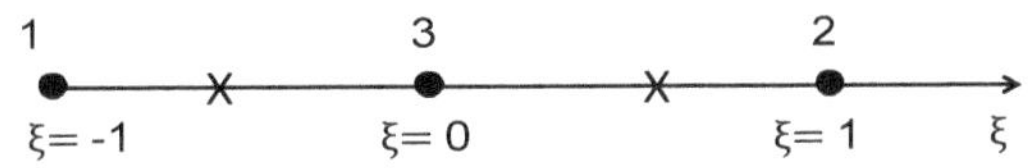

Figure 2. The natural coordinates for 1D quadratic elements.

where the shape function used was as follows:

$$\begin{aligned} N_1 &= \frac{-\xi(1-\xi)}{2} \\ N_2 &= \frac{\xi(1+\xi)}{2} \\ N_3 &= 1-\xi^2 \end{aligned} \tag{1}$$

with ξ being local coordinate.

2.2. FE Solution of Displacements

The pavement is assumed to be supported at side edges ($x = 0$, $x = a$, $z = 0$ and $z = c$) in a manner preventing all displacements in the vertical planes but permitting "unrestricted" movement in the x- and z-directions. The shape functions are re-written as the conventional 1D shape functions multiplied by the Fourier series, in which x ranges between zero and a; z ranges between zero and c. If the displacement functions with three components u, v and w are formulated in the following form, the boundary condition will meet this above-mentioned requirement:

$$\begin{aligned} d = \begin{Bmatrix} u \\ v \\ w \end{Bmatrix} &= \sum_{m=1}^{M}\sum_{n=1}^{N} \begin{bmatrix} \cos\frac{m\pi x}{a}\sin\frac{n\pi z}{c} & 0 & 0 \\ 0 & \sin\frac{m\pi x}{a}\sin\frac{n\pi z}{c} & 0 \\ 0 & 0 & \sin\frac{m\pi x}{a}\cos\frac{n\pi z}{c} \end{bmatrix} \begin{Bmatrix} u^{mn} \\ v^{mn} \\ w^{mn} \end{Bmatrix} \\ &= \sum_{m=1}^{M}\sum_{n=1}^{N} N^{mn}.d^{mn}, \\ \text{with } N^{mn} &= \begin{bmatrix} \cos\frac{m\pi x}{a}\sin\frac{n\pi z}{c} & 0 & 0 \\ 0 & \sin\frac{m\pi x}{a}\sin\frac{n\pi z}{c} & 0 \\ 0 & 0 & \sin\frac{m\pi x}{a}\cos\frac{n\pi z}{c} \end{bmatrix}, \\ d^{mn} &= \begin{Bmatrix} u^{mn} \\ v^{mn} \\ w^{mn} \end{Bmatrix} = [N_k(y)]\{d_k^{mn}\}. \end{aligned} \tag{2}$$

N^{mn} is the matrix for the Fourier series expansion in the two horizontal directions, d^{mn} is the displacement vector without the Fourier series expansion, $[N_k(y)]$ is the conventional 1D shape function applied in the FEM model, and $\{d_k^{mn}\}$ represents the corresponding nodal displacements.

Similarly, the loading function for the pavement analysis can be formulated as:

$$\begin{aligned} f &= \sum_{m=1}^{M}\sum_{n=1}^{N} p(y)\sin\frac{m\pi x}{a}\sin\frac{n\pi z}{c} = \sum_{m=1}^{M}\sum_{n=1}^{N}\{p\}^{mn}, \\ p(y) &= \sum_{s=1}^{S}\left(\frac{2P_s}{m\pi}\right)\left[\cos\frac{m\pi}{a}X_{s1} - \cos\frac{m\pi}{a}X_{s2}\right]\sum_{t=1}^{T}\left(\frac{2P_t}{n\pi}\right)\left[\cos\frac{n\pi}{c}Z_{t1} - \cos\frac{n\pi}{c}Z_{t2}\right], \end{aligned} \tag{3}$$

where m and n are the terms of Fourier series in the x and z directions, respectively. In this study, both M and N are adopted as 100, which has been previously proved to be sufficiently accurate to represent the load variation [16,17]. The product of P_s and P_t represents the tire load pressure, when $m = n$; X_{s1} and Z_{t1} are the x and z coordinates, where the tire load area starts, respectively; X_{s2} and Z_{t2} are the x and z coordinates where the tire load area ends, respectively.

At the nodes the strains are determined through displacements:

$$\varepsilon = \left\{ \begin{array}{c} \varepsilon_x \\ \varepsilon_y \\ \varepsilon_z \\ \gamma_{xy} \\ \gamma_{yz} \\ \gamma_{zx} \end{array} \right\} = \begin{bmatrix} \frac{\partial}{\partial x} & 0 & 0 \\ 0 & \frac{\partial}{\partial y} & 0 \\ 0 & 0 & \frac{\partial}{\partial z} \\ \frac{\partial}{\partial y} & \frac{\partial}{\partial x} & 0 \\ 0 & \frac{\partial}{\partial z} & \frac{\partial}{\partial y} \\ \frac{\partial}{\partial z} & 0 & \frac{\partial}{\partial x} \end{bmatrix} d = \left(L_1\tfrac{\partial}{\partial x} + L_2\tfrac{\partial}{\partial y} + L_3\tfrac{\partial}{\partial z}\right) \sum_{m=1}^{M}\sum_{n=1}^{N} N^{mn}.d^{mn}$$
$$= \sum_{m=1}^{M}\sum_{n=1}^{N} E^{mn} d^{mn} = \sum_{m=1}^{M}\sum_{n=1}^{N} B^{mn}.\{d_k^{mn}\}, \tag{4}$$

where L_1, L_2, L_3 are three different matrices of size 6×3 whose elements are either 0 or 1,

$$E^{mn} = L_1\frac{\partial N^{mn}}{\partial x} + L_2 N^{mn}\frac{\partial}{\partial y} + L_3\frac{\partial N^{mn}}{\partial z}, \tag{5}$$

and matrix

$$B^{mn} = \left[B_1^{mn}\ B_2^{mn} \cdots B_{k+1}^{mn}\right] = \left[\ E^{mn}N_1(y)\quad E^{mn}N_2(y)\quad \cdots E^{mn}N_{k+1}(y)\ \right] \tag{6}$$

is called the strain-displacement matrix.

The relation between stresses and strains is used to express the total potential energy with nodal displacements $\{d\}$:

$$\Pi(\{d\}) = \int_V \frac{1}{2}([B]\{d\})^T[D]([B]\{d\})dV - \int_V([N]\{d\})^T\{b\}dV - \int_S([N]\{d\})^T\{f\}dS\,. \tag{7}$$

The nodal displacements $\{d\}$ which correspond to the minimum of the functional Π are determined by the condition that the first variation of Π should be zero, i.e.,

$$\begin{gathered} \delta\Pi(\{d\}) = 0, \\ \delta\Pi(\{d\}) = \int_V\{\delta d\}^T[B]^T[D][B]\{d\}dV - \int_V\{\delta d\}^T[N]^T\{b\}dV - \int_S\{\delta d\}^T[N]^T\{f\}dS\,. \end{gathered} \tag{8}$$

Then, the arbitrariness of $\{\delta d\}$ leads the following equilibrium equation for FE:

$$\int_V[B]^T[D][B]dV\{d\} - \int_V[N]^T\{b\}dV - \int_S[N]^T\{f\}dS = 0, \tag{9}$$

which is commonly given in the following form:

$$\begin{gathered} [k]\{d\} = \{f\}, \\ [k] = \int_V[B]^T[D][B]dV \text{ and } \{f\} = \int_V[N]^T\{b\}dV + \int_S[N]^T\{f\}dS, begin \end{gathered} \tag{10}$$

where $[k]$ is the element stiffness matrix; and $\{f\}$ is the load vector.

From Equations (4) and (10), the stiffness matrix of one element includes:

$$\begin{aligned} I_1 &= \int_0^a\int_0^c \sin\tfrac{m\pi x}{a}\sin\tfrac{n\pi z}{c}\cdot\sin\tfrac{p\pi x}{a}\sin\tfrac{q\pi z}{c}\cdot dxdz, \\ I_2 &= \int_0^a\int_0^c \cos\tfrac{m\pi x}{a}\sin\tfrac{n\pi z}{c}\cdot\cos\tfrac{p\pi x}{a}\sin\tfrac{q\pi z}{c}\cdot dxdz, \\ I_3 &= \int_0^a\int_0^c \cos\tfrac{m\pi x}{a}\cos\tfrac{n\pi z}{c}\cdot\cos\tfrac{p\pi x}{a}\cos\tfrac{q\pi z}{c}\cdot dxdz. \end{aligned} \tag{11}$$

The integrals exhibit orthogonal properties, which ensure that:

$$I_1 = I_2 = I_3 = \left\{ \begin{array}{c} \frac{ac}{4},\ for\ m = p\ and\ n = q \\ 0,\ \ for\ m \neq p\ or\ n \neq q \end{array} \right\}. \tag{12}$$

This means that the matrix $(k^{mnpq})^e$ is diagonal. In other words, the non-zero values are only located on the diagonal, where $m = p$ and $n = q$. Thus, the stiffness matrix can be reduced to:

$$\left(k_{gk}^{mnmn}\right)^e = \frac{ac}{4}\int_{length}\left(B_g^{mn\,T} D B_k^{mn}\right)dy \qquad m = 1, 2 \ldots M;\ n = 1, 2 \ldots N, \tag{13}$$

where g and k represent the nodes of the element, respectively. The length corresponds to the length of the element.

A typical term for the force vector becomes:

$$(f^{mn})^e = \int\int\int_{vol} (N^{mn})^T \{p\}^{mn} dxdydz. \tag{14}$$

By assembling the stiffness matrix of each element to the global domain, the global linear system is achieved as follows:

$$\begin{bmatrix} K^{1111} & & & \\ & K^{1212} & & \\ & & \ddots & \\ & & & K^{MNMN} \end{bmatrix}\begin{Bmatrix} U^{11} \\ U^{12} \\ \vdots \\ U^{MN} \end{Bmatrix} + \begin{Bmatrix} F^{11} \\ F^{12} \\ \vdots \\ F^{MN} \end{Bmatrix} = 0. \tag{15}$$

Equation (15) shows that the large system of equations splits up into $M \times N$ separate problems:

$$K^{mnmn}U^{mn} + F^{mn} = 0, \tag{16}$$

where K is the global stiffness matrix; U is the global displacement vector; and F is the global loading vector.

3. EEP Super-Convergence Strategy for EasyFEM

When Equation (16) is solved for U^{mn}, both d^{mn} and d can be obtained from Equation (2). These two approximate solutions are denoted as d_h^{mn} and d_h to distinguish them from the exact solutions. It is clear, that if M and N are fixed, the accuracy of d_h mainly depends on the accuracy of d_h^{mn}. For conventional FEM models, a rather extensive mesh should be used to attain more accurate d_h^{mn}, which leads to a large number of degrees of freedom in the calculation. In this case, the newly-developed super-convergence technique EEP was introduced into EasyFEM as to improve the accuracy of the solutions for strains and stresses over the whole pavement model.

The original idea of the EEP technique came from a well-known mathematical theorem for FEM called the projection theorem [37]. Actually, the principle of minimum potential energy with respect to nodal displacements $\{d\}$ in Equation (8) can be rewritten in the same manner with respect to the continuous 1D displacement $d(y)$ as follows:

$$\delta\Pi(d(y)) = \int_V \delta d(y)^T[E]^T[D][E]d(y)dV - \int_V \delta d(y)^T[N(x,z)]^T\{b\}dV - \int_S \delta d(y)^T[N(x,z)]^T\{f\}dS = 0. \tag{17}$$

By replacing $\delta d(y)$ in Equation (17) with the test displacement $\widetilde{d}(y)$, an equivalent expression for the above principle of minimum potential energy is the virtual work equation as follows:

$$\begin{aligned} a(d(y), \widetilde{d}(y)) &= (f, \widetilde{d}(y)), \forall \widetilde{d}(y) \in S^h, \\ \text{with } a\left(d(y), \widetilde{d}(y)\right) &= \textstyle\int_V \widetilde{d}(y)^T[E]^T[D][E]d(y)dV, \\ (\mathrm{f},\ \widetilde{\mathrm{d}}(\mathrm{y})) &= \textstyle\int_V \widetilde{\mathrm{d}}(\mathrm{y})^T[\mathrm{N}(\mathrm{x},\mathrm{z})]^T\{\mathrm{b}\}\mathrm{dV} + \int_S \widetilde{\mathrm{d}}(\mathrm{y})^T[\mathrm{N}(\mathrm{x},\mathrm{z})]^T\{\mathrm{f}\}\mathrm{dS}, \end{aligned} \tag{18}$$

where $a(\cdot,\cdot)$ and $(\cdot,\cdot)$ are defined as the energy product and the linear form in mathematics, respectively, and S^h is the test space of FLM.

It can be easily understood that Equation (18) holds true both for the exact solution $d(y)$ and the approximate solution $d_h(y)$. Always taking the same test displacement for these two cases, the following projection theorem is derived for FLM:

$$a(d(y) - d_h(y), \tilde{d}(y)) = 0, \forall \tilde{d}(y) \in S^h. \tag{19}$$

This theorem holds true over the entire 3D pavement model. By assuming that it is approximately true over one element of FLM, the EEP equation was obtained:

$$a^e(d(y) - d_h(y), \tilde{d}(y)) \approx 0, \quad \forall \tilde{d}(y) \in S^h. \tag{20}$$

A typical element e of the FLM is shown in Figure 3, which is actually a small layer of the pavement with the y-coordinate ranging from y_1 to y_2.

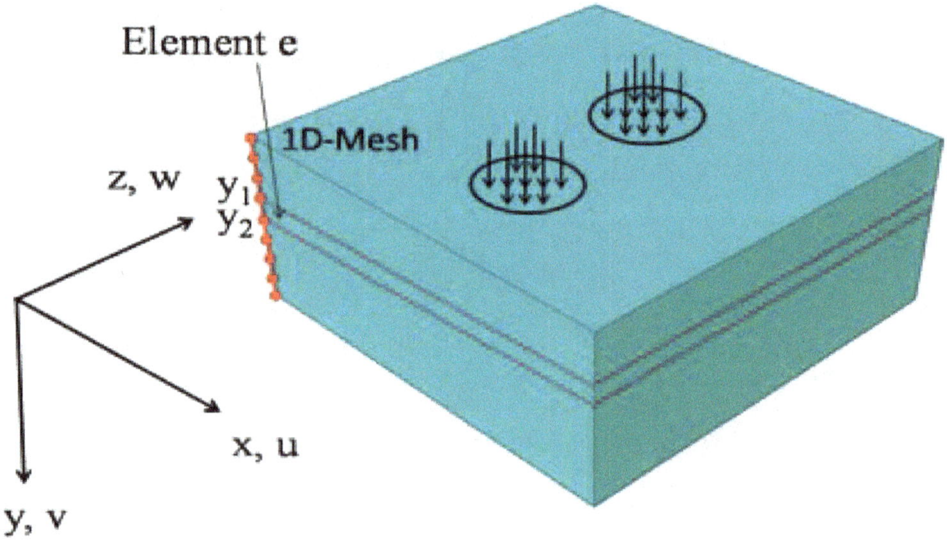

Figure 3. A typical EasyFEM element e in the pavement model.

Denote $e_h = d(y) - d_h(y)$. Then:

$$\begin{gathered} a^e\left(d(y) - d_h(y), \tilde{d}(y)\right) = \int_{V_e} \tilde{d}(y)^T [E]^T [D][E] e_h dV \\ = \sum_{m=1}^{M} \sum_{n=1}^{N} \int_{V_e} \tilde{d}^{mn^T} [E^{mn}]^T [D][E^{mn}] e_h^{mn} dV = \sum_{m=1}^{M} \sum_{n=1}^{N} a^e(e_h^{mn}, \tilde{d}^{mn}) \approx 0 \end{gathered} \tag{21}$$

Denoting $E_{13}^{mn} = L_1 \frac{\partial N^{mn}}{\partial x} + L_3 \frac{\partial N^{mn}}{\partial z}$, $E_2^{mn} = L_2 N^{mn}$, and considering the zeros from integration of the Fourier series as shown in Equation (12), Equation (21) can be further rewritten as follows:

$$\begin{gathered} \sum_{m=1}^{M} \sum_{n=1}^{N} \int_{V_e} \left(E_{13}^{mn} \tilde{d}^{mn} + E_2^{mn} \frac{\partial \tilde{d}^{mn}}{\partial y}\right)^T [D] \left(E_{13}^{mn} e_h^{mn} + E_2^{mn} \frac{\partial e_h^{mn}}{\partial y}\right) dV = \\ \sum_{m=1}^{M} \sum_{n=1}^{N} \int_{y_1}^{y_2} \left[\left(\frac{\partial \tilde{d}^{mn}}{\partial y}\right)^T A^{mn} \frac{\partial e_h^{mn}}{\partial y} + \left(\frac{\partial \tilde{d}^{mn}}{\partial y}\right)^T G^{mn} e_h^{mn} + \left(\tilde{d}^{mn}\right)^T (G^{mn})^T \frac{\partial e_h^{mn}}{\partial y} + \right. \\ \left. \left(\tilde{d}^{mn}\right)^T H^{mn} e_h^{mn}\right] dy \approx 0, \end{gathered} \tag{22}$$

where

$$\begin{aligned} A^{mn} &= \int_0^a \int_0^c (E_2^{mn})^T [D] E_2^{mn} dx dz, \\ G^{mn} &= \int_0^a \int_0^c (E_2^{mn})^T [D] E_{13}^{mn} dx dz, \\ H^{mn} &= \int_0^a \int_0^c (E_{13}^{mn})^T [D] E_{13}^{mn} dx dz, \end{aligned}$$

with integration by parts along the y-direction, the EEP can be equivalently converted into the following expression:

$$\sum_{m=1}^{M}\sum_{n=1}^{N}\int_{y_1}^{y_2}\left(-\widetilde{d}^{mn^T}A^{mn}\frac{\partial^2 e_h^{mn}}{\partial y^2}+\widetilde{d}^{mn^T}((G^{mn})^T-G^{mn})\frac{\partial e_h^{mn}}{\partial y}+\left(\widetilde{d}^{mn}\right)^T H^{mn}e_h^{mn}\right)dy + \left(\widetilde{d}^{mn^T}(A^{mn}\frac{\partial e_h^{mn}}{\partial y}+G^{mn}e_h^{mn})\right)_{y_1}^{y_2}\approx 0. \quad (23)$$

Since the test function $\widetilde{d}^{mn}$ can be arbitrarily selected for any m and n, the left side of Equation (23) is decoupled into

$$\int_{y_1}^{y_2}\widetilde{d}^{mn^T}\left(-A^{mn}\frac{\partial^2 e_h^{mn}}{\partial y^2}+\left(G^{mn^T}-G^{mn}\right)\frac{\partial e_h^{mn}}{\partial y}+H^{mn}e_h^{mn}\right)dy+\left(\widetilde{d}^{mn^T}(A^{mn}\frac{\partial e_h^{mn}}{\partial y}+G^{mn}e_h^{mn})\right)_{y_1}^{y_2}\approx 0. \quad (24)$$

Taking two linear polynomials:

$$\overline{N}_1=\frac{y_2-y}{y_2-y_1};\ \overline{N}_2=\frac{y-y_1}{y_2-y_1} \quad (25)$$

as two test functions $\widetilde{d}^{mn}$, respectively, and considering that:

$$d^{mn}(y_1)=d_h^{mn}(y_1), d^{mn}(y_2)=d_h^{mn}(y_2), \quad (26)$$

since the nodal displacements are super-convergent based on mathematical theory, one can obtain the following equation to calculate a recovered derivative solution at either of the two end-nodes of element e [39]:

$$\begin{aligned}\frac{\partial d_E^{mn}(y_1)}{\partial y} &= (A^{mn})^{-1}\left[(b,\overline{N}_1)-a^e(d_h^{mn},\overline{N}_1)-G_1 d_h^{mn}(y_1)\right],\\ \frac{\partial d_E^{mn}(y_2)}{\partial y} &= -(A^{mn})^{-1}\left[(b,\overline{N}_2)-a^e(d_h^{mn},\overline{N}_2)+G_2 d_h^{mn}(y_2)\right],\end{aligned} \quad (27)$$

where:

$$i=1,\ 2,\quad m=1,\cdots,M,\quad n=1,\cdots,N. \quad (28)$$

Substituting the derivative results calculated from Equation (28) into Equation (4), recovered strains and the corresponding stresses are obtained for the pavement model. This is the EEP super-convergence technique for FLM. When it is implemented in EasyFEM, stresses are obtained with much fewer elements and significantly less computational effort.

4. Numerical Analysis Using EasyFEM and EEP Techniques

4.1. Analytical Verification of EasyFEM

The analytical verification of the EasyFEM without EEP post-processing was carried out. The models of the asphalt pavement were created in BISAR, ABAQUS and EasyFEM, which are representative software applications developed based on LET, FEM and FLM, respectively. The pavement type is widely used in Germany according to the guidelines RStO-12 [40] and RDO-Asphalt-09 [1], as shown in Figure 4. The thicknesses of all layers except for the sub-grade were derived from RstO-12 [40]. The thickness of the sub-grade was defined to be 2000 mm in ABAQUS and EasyFEM after a previous mesh study in order to minimize the influence of the boundary on the computational results. Furthermore, the length and width of the pavement in the full-scale model were set to be 6000 mm based on the same reason. A full-scale model was created in the EasyFEM while the model in ABAQUS was one-fourth symmetrical model. The full-scale model exhibits advantages for the simulation of nonsymmetrical models with complex loading conditions in the further development of the EasyFEM. In BISAR, the thickness of the sub-grade as well as the length and width of the pavement were set to be infinite according to the LET. The pavement surface temperatures of −12.5 (winter) and 27.5 °C (summer) were assumed, and then the associated material properties were determined according to RDO-Asphalt-09 [1], as listed in Figure 4.

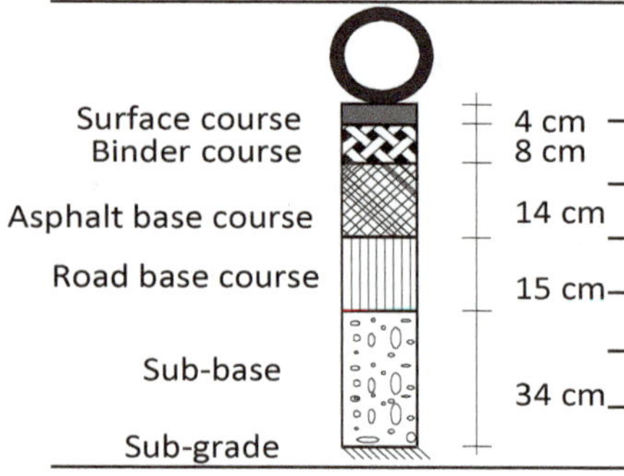

Layer	μ	Winter	Summer
		E [MPa]	E [MPa]
Surface course	0.35	22690	2902
Binder course	0.35	27283	6817
Asphalt base course	0.35	17853	4903
Road base course	0.25	10000	10000
Sub-base	0.5	100	100
Sub-grade	0.5	45	45

Figure 4. Geometrical data and material properties of the pavement.

The square load with the side length of 264 mm and uniformly distributed contact stress of 0.7 MPa was applied at the center of the full-scale pavement surface in the EasyFEM and a corresponding set-up was applied in ABAQUS. In BISAR, only circular loads can be defined and thus a uniform contact load of 0.7 MPa with a radius of 150 mm was applied. In all models, the three asphalt layers were totally bound; the two contact layers among the asphalt base course, road base course, sub-base and sub-grade were defined as being partially bound.

In order to attain a high accuracy, a very fine mesh was adopted in ABAQUS and EasyFEM after a mesh study. The mesh size in EasyFEM was uniformly set as 10 mm and thus 275 3-node quadratic elements with 551 nodes were generated. In order to reduce the computational consumption, the mesh size increases gradually from the top to the bottom and from the loading center to the boundary in the one-fourth symmetrical model in ABAQUS, as shown in Figure 5. The number of elements and nodes in ABAQUS are 162,728 and 193,601, respectively. There was no mesh generated from the BISAR.

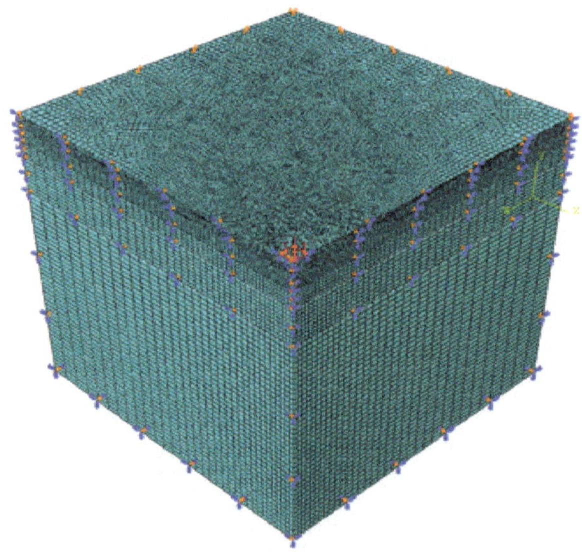

Figure 5. Mesh generated from ABAQUS.

The computational results shown in Figure 6 are derived from five series of response points offset from the loading center to the boundary.

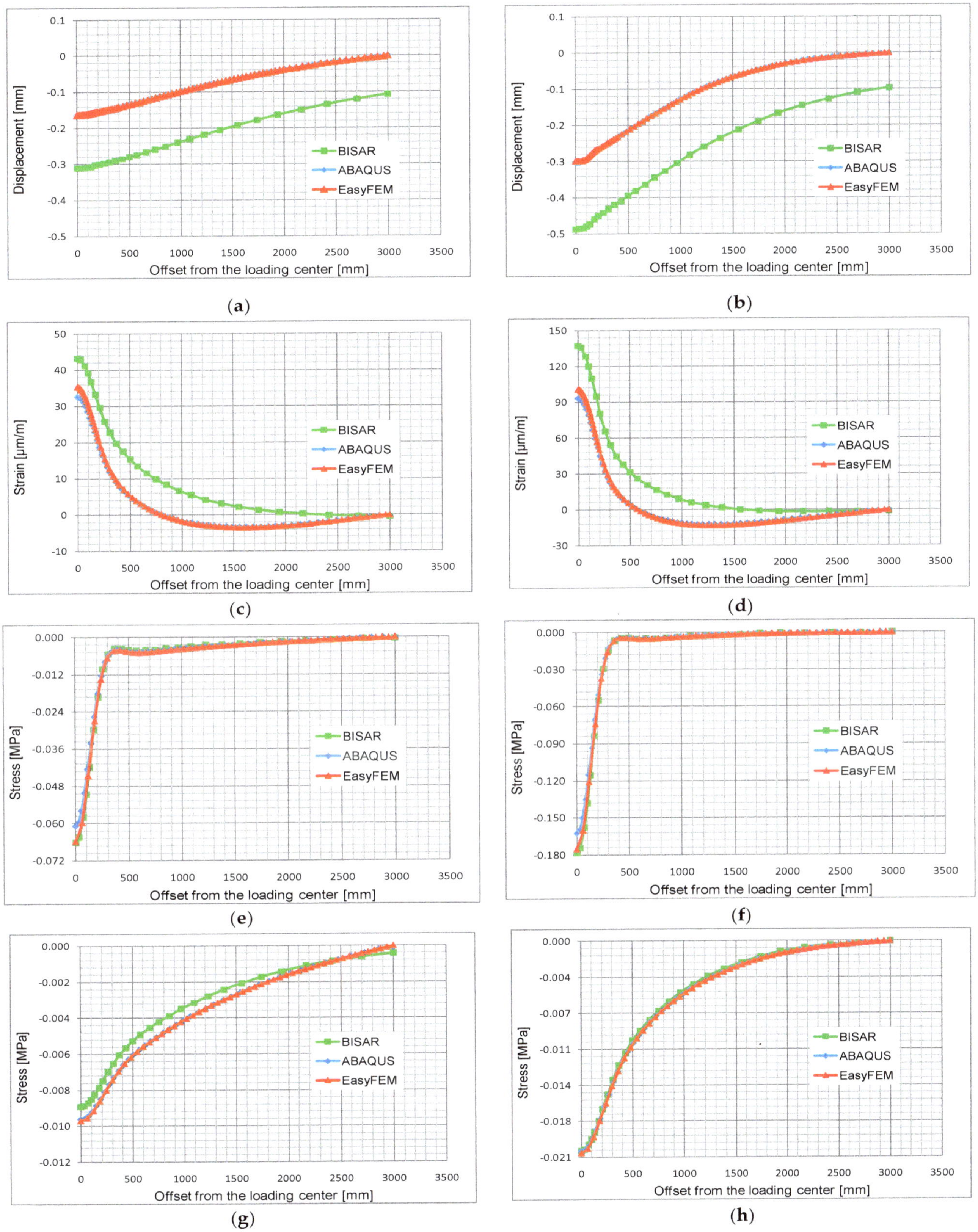

Figure 6. *Cont.*

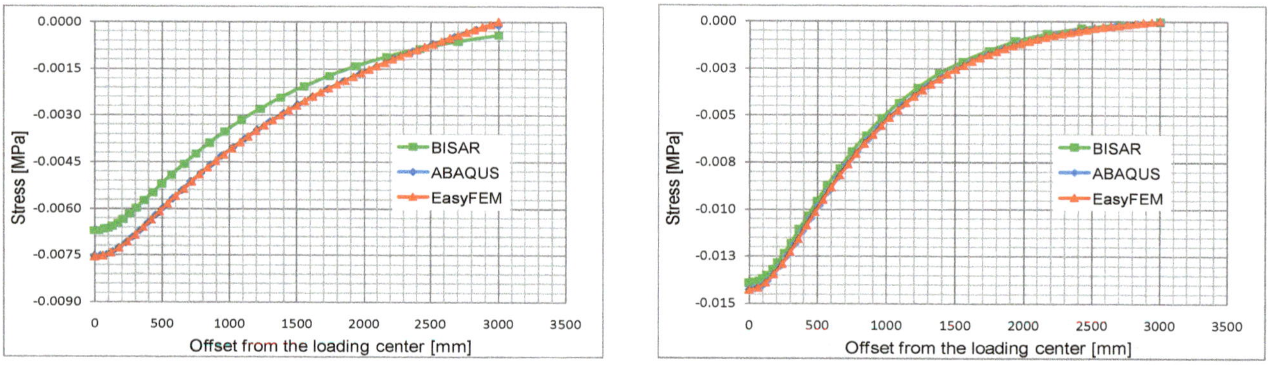

(i) (j)

Figure 6. Comparison of vertical displacement at the top of the asphalt surface in (**a**) winter, (**b**) summer; comparison of horizontal strain along transverse direction at the bottom of the asphalt base course in (**c**) winter, (**d**) summer; comparison of vertical stress at the top of the road base layer in (**e**) winter, (**f**) summer; comparison of vertical stress at the top of the sub-base layer in (**g**) winter, (**h**) summer; comparison of vertical stress at the top of the sub-grade layer in (**i**) winter, (**j**) summer.

Given these figures, it can be concluded that the results derived from ABAQUS and EasyFEM are consistent with each other, yet the results from BISAR exhibit great deviations from those from the other two software applications, especially along the first two series of response points. The underlying reasons for the differences can be attributed to the different geometrical definitions of the models in BISAR due to its intrinsic limitation (circular load, infinite length and width of the layers and infinite thickness of the sub-grade); the deviations are the largest at the response points that are closest to the load.

In Table 1, the computational results from ABAQUS and EasyFEM are considered at five critical locations directly below the loading center shown in Figure 6, where maximum compressive or tensile values may occur.

Table 1. Comparison between ABAQUS and EasyFEM based on the results at critical points below the loading center.

Points	Results	Winter		
		EasyFEM	**ABAQUS**	**Difference**
1	Vertical displacement [mm]	-1.63×10^{-1}	-1.64×10^{-1}	7.72×10^{-4} (−0.47%)
2	Horizontal strain [-]	3.54×10^{-5}	3.25×10^{-5}	2.82×10^{-6} (8.68%)
3	Vertical stress [MPa]	-6.57×10^{-2}	-6.09×10^{-2}	-4.84×10^{-3} (7.95%)
4	Vertical stress [MPa]	-9.70×10^{-3}	-9.62×10^{-3}	-8.61×10^{-5} (0.90%)
5	Vertical stress [MPa]	-7.57×10^{-3}	-7.55×10^{-3}	-1.51×10^{-5} (0.20%)
Points	**Results**	**Summer**		
		EasyFEM	**ABAQUS**	**Difference**
1	Vertical displacement [mm]	-2.99×10^{-1}	-3.02×10^{-1}	3.24×10^{-3} (−1.07%)
2	Horizontal strain [-]	1.01×10^{-4}	9.33×10^{-5}	7.39×10^{-6} (7.92%)
3	Vertical stress [MPa]	-1.75×10^{-1}	-1.62×10^{-1}	-1.27×10^{-2} (7.85%)
4	Vertical stress [MPa]	-2.06×10^{-2}	-2.04×10^{-2}	-1.75×10^{-4} (0.86%)
5	Vertical stress [MPa]	-1.43×10^{-2}	-1.43×10^{-2}	-1.01×10^{-5} (0.07%)

It can be stated that the results from both programs have a high correlation except for a slightly larger difference in the horizontal strain at the bottom of the asphalt base course and vertical stresses at the top of the sub-base. Considering the different element types and mesh algorithms in the two programs, the differences are judged to be acceptable. Moreover, the computational time of the EasyFEM is much shorter than that of the ABAQUS. Both analyses were run on a computer with an Intel Core Duo 3.4 GHz, 32 GB RAM (Santa Clara, CA, USA). On average, the computational time required by the ABAQUS is about 420 s, whereas the EasyFEM model requires 120 s. It is worth mentioning that the model in ABAQUS is one-fourth symmetrical, but the one in EasyFEM is full-scale. With code optimization, the computation time of the EasyFEM can be reduced further. In summary, the accuracy and efficiency of the EasyFEM are proved by these comparisons.

4.2. Comparison of Convergence Rate between Ordinary EasyFEM and EasyFEM with EEP Post-Processing Techniques

As discussed above, by using 275 3-node quadratic finite elements along the y-direction, EasyFEM has performed fairly well compared to ABAQUS. If the number of elements can be further reduced without decreasing the computational accuracy, the EasyFEM will be even more suitable for application in pavement design and assessment from an engineering point of view. The EEP technique for recovering derivatives makes this possible. The convergence rates between ordinary EasyFEM and EasyFEM with the EEP post-processing technique are compared in this section to show the impact of the EEP technique on the solution accuracy.

To emphasize the significant improvement brought about by the application of EEP, a series of rough meshes with 2-node linear elements was adopted in the EasyFEM. In particular, three models with different meshes were created, i.e., each layer of the asphalt pavement was divided into one, two and four elements (N_e = 1, 2 and 4), respectively; the asphalt pavement had six layers resulting in a total of $6 \times N_e = 6$, 12 and 24 linear elements in the three pavement models. EasyFEM with the EEP post-processing technique can produce strains and stresses with a higher accuracy, which is demonstrated by a number of numerical results. In Figure 7, vertical strains at the points located below the loading center and at the bottom of asphalt base course were compared between ordinary EasyFEM and EasyFEM with the EEP technique, where one, two and four linear finite elements were used in each layer along the y-direction, respectively. The results from the model with the fine mesh in ABAQUS created in Section 4.1 were taken as the reference. Since the stresses are calculated by multiplying strains with elastic constants, similar results can be obtained for vertical stresses as well. It can be seen that the EEP technique for FLM, which converges faster, can significantly improve the accuracy of vertical strains and stresses of a plain EasyFEM. With the EEP technique, the EasyFEM can offer reliable results much more quickly and efficiently and is thus suitable to be applied in pavement engineering practice.

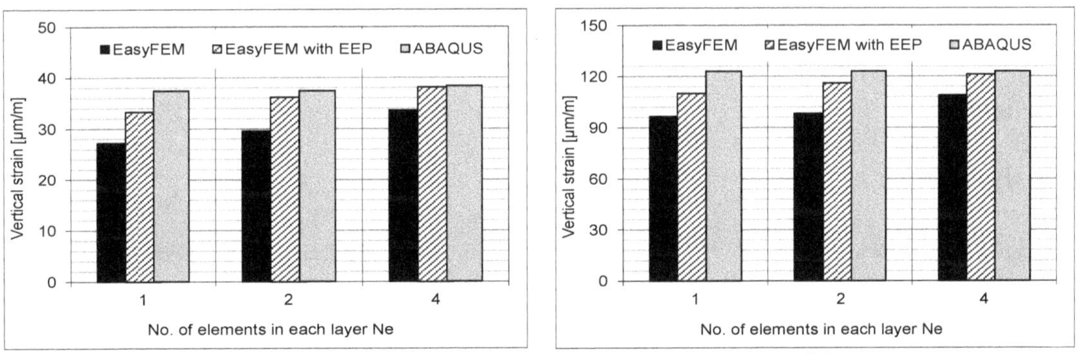

(a) **(b)**

Figure 7. Comparison of the vertical strains at the points located below the loading center and at the bottom of asphalt base course obtained with ABAQUS, ordinary EasyFEM and EasyFEM with EEP (Element Energy Projection), (**a**) winter; and (**b**) summer.

4.3. Application of EasyFEM

The EasyFEM was used to predict the mechanical responses and the predictions were compared with the data derived from field measurements on the test track of the German Federal Highway Research Institute (BASt), as shown in Figure 8a. A truck driving at a speed of approximately 30 km/h was selected for the measurement to apply the loads, as shown in Figure 8b.

(**a**)

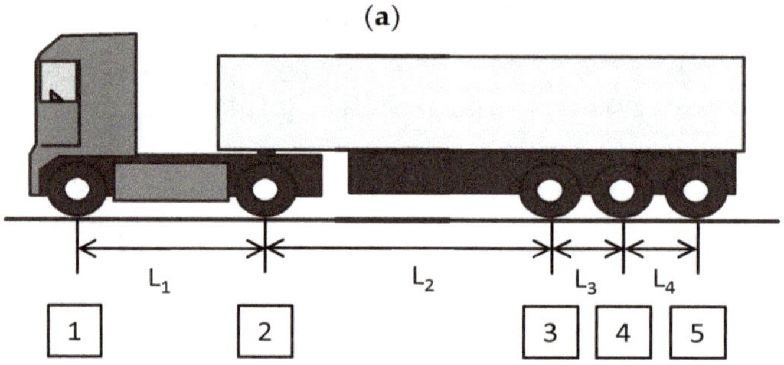

Distance	L_1	L_2	L_3	L_4	Sum
mm	3600	5860	1310	1310	12080

(**b**)

Figure 8. (**a**) The test track in BASt (the German Federal Highway Research Institute); (**b**) geometrical data and tires of the truck S23.

During the test track construction, strain gauges and pressure load cells were embedded at different depths of the test track along the center of the wheel path, which can measure strains and stresses along the corresponding directions when the truck passes. Due to the low speed of loads and the suggestion from [17], a static analysis with stationary loads is precise enough to compute the pavement responses. The material parameters of the pavement layers were derived from the laboratory tests on the specimens drilled from the test track. The thicknesses and the material properties of the test track are listed in Table 2 [17]. The length longitudinal to the direction of traffic (both directions) was defined as 20 times the loading radius to limit the time required for the computational calculation. The width of the pavement was defined to be 3750 mm.

Table 2. Layer thickness and material properties of the test track.

Layer	Thickness [mm]	μ	E [MPa]
Surface course	40	0.35	11150
Binder course	50	0.35	10435
Asphalt base course	110	0.35	6893
Gravel base layer	150	0.49	157.8
Frost protection layer	570	0.49	125.7
Sub-grade	2000	0.49	98.9

The loading parameters are listed in Table 3 [17]. All of the contact areas are assumed as squares with a side length of 264 mm. The distance between the center of the first left tire and the left edge of the test track along the traffic direction is 1100 mm.

Table 3. Loading parameters of the tire.

Axle	Wheel Number	Pressure [MPa]	Axle Load [kg]
1	2	0.522	7425
2	4	0.377	10725
3	2	0.513	7300
4	2	0.511	7275
5	2	0.519	7375

The upper three asphalt layers were totally bound. The two contact layers between the asphalt base course, gravel base layer, frost protection layer and sub-grade were defined as being partially bound. The 3-node quadratic elements with the mesh size of 10 mm were applied in the EasyFEM with the EEP technique.

The following computational values were compared with the measured data, which is used to verify the EasyFEM. The strain along the traffic direction at the bottom of the asphalt base course and the vertical tensile stress at the top of the gravel base layer from the location below the center of the left tire of each axle were used in the validation process. The values from the measurement and simulation can be seen in Figure 9. All computational strains are higher than the measured values and all computational stresses are lower than those obtained from the measurements. Due to the uncertainties and fluctuations, the error range is considered to be ±20% [17]. The computational values are all within this range. Thus, the EasyFEM is suited to simulate the response of the asphalt pavement under the traffic load with sufficient accuracy.

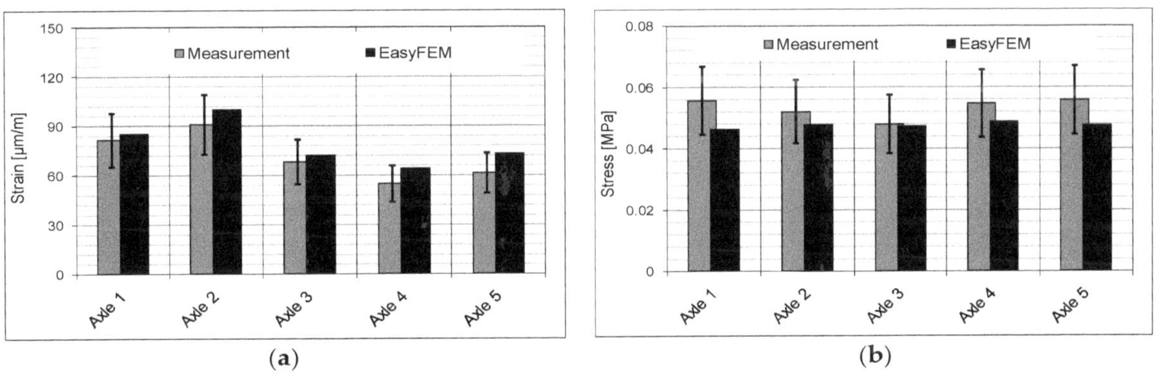

Figure 9. Comparison of the results derived from measurement and simulation. (**a**) horizontal strain along the traffic direction on the bottom of the asphalt base course; (**b**) vertical stress on the top of the gravel base layer.

5. Conclusions

Initially, the EasyFEM code is developed to predict the mechanical responses of asphalt pavements under stationary load. The accuracy of the program is verified by comparing the results to those obtained from BISAR and ABAQUS. The analytical verification showed that the pavement responses derived from EasyFEM and ABAQUS are in accordance with one another. Due to the different definitions in the models of the BISAR, some results appear to exhibit larger differences than those observed between EasyFEM and ABAQUS, which indicates that BISAR is not as flexible as the other two software applications when simulating the responses of the asphalt pavements. It should be emphasized that the computational time of the EasyFEM is much shorter than that of ABAQUS.

The EEP super-convergence strategy for FLM is introduced and applied in the EasyFEM to improve the accuracy of derivatives at two end-nodes on any element. The convergence rate of the EasyFEM with the EEP post-processing technique is much faster than without the EEP. The application of the EasyFEM proves that the current version of EasyFEM can provide a reliable prediction of the mechanical responses of asphalt pavements and thus represents a flexible and robust base for further development.

For the next step of development, various material properties, e.g., viscoelasticity and nonlinear elasticity, may be applied for asphalt layers and sub-base of the pavement, respectively. The EEP technique will be further developed for the new EasyFEM with more functions. Based on the EEP technique, adaptivity is to be included into the EasyFEM, i.e., an error tolerance for the solution is specified by the user before the calculation, and, finally, a numerical solution is obtained that satisfies the error tolerance, while the meshes used in the simulation are adaptively generated and adjusted automatically to produce a satisfactory solution with a high quality and accuracy. With these improvements, the EasyFEM should be more suited to predict the mechanical performances of the asphalt pavement.

Acknowledgments: This paper is based on a part of the research project carried out at the request of the German Research Foundation, under research project No. FOR 2089. It is also a part of the project No. 51508305 supported by the National Natural Science Foundation of China. The authors are solely responsible for the content.

Author Contributions: Pengfei Liu developed the FLM program EasyFEM; Qinyan Xing applied the EEP technique to EasyFEM; Yiyi Dong analyzed the data; Dawei Wang conceived the paper; Markus Oeser contributed programming algorithm; and Si Yuan directed the development of the EEP technique.

References

1. *Richtlinien für die Rechnerische Dimensionierung von Verkehrsflächen mit Asphaltdeckschicht: RDO Asphalt 09*; FGSV Publishing: Cologne, Germany, 2009.
2. Dong, Z.; Tan, Y. *Dynamic Response of Asphalt Pavement*; Science Press: Beijing, China, 2015.
3. *BISAR: Main Frame Computer Program, User Manual*; Shell International Oil Products: The Hague, The Netherlands, 1998.
4. Huang, H. *Pavement Analysis and Design*; Prentice Hall: New Jersey, NJ, USA, 1993.
5. Scarpas, A. CAPA-3D: A Mechanics Based Computational Platform for Pavement Engineering. Ph.D. Thesis, Delft University of Technology, Delft, The Netherlands, 2005.
6. Lytton, R.L.; Tsai, F.L.; Lee, S.I.; Luo, R.; Hu, S.; Zhou, F. Appendix Q: Finite Element Program to Calculate Stress Intensity Factor. In *NCHRP Report 669: Models for Predicting Reflection Cracking of Hot-Mix Asphalt Overlays*; The National Academies of Sciences, Engineering, and Medicine: Washington, DC, USA, 2010. [CrossRef]
7. Gonzalez, A.; Bodin, D.; Jameson, G.; Oeser, M.; Vuong, B. *Development of a Nonlinear Finite Element Pavement Response to Load Model*; Austroads Publication No. AP-T199-12; Austroads: Sydney, Australia, 2012.
8. Wilson, E.L. Structural analysis of axisymmetric solids. *AIAA J.* **1965**, *3*, 2269–2274. [CrossRef]
9. Meissner, H.E. Laterally loaded pipe pile in cohesionless soil. In Proceedings of the 2nd International Conference on Numeric Methods in Geomechanics, Virginia Polytechnical Institute and State University, Blacksburg, VA, USA, 20–25 June 1976.
10. Winnicki, L.A.; Zienkiewicz, O.C. Plastic (or visco-plastic) behaviour of axisymmetric bodies subjected to non-symmetric loading-semi-analytical finite element solution. *Int. J. Num. Meth. Eng.* **1979**, *14*, 1399–1412. [CrossRef]
11. Carter, J.P.; Booker, J.R. Consolidation of axi-symmetric bodies subjected to non axi-symmetric loading. *Int. J. Numer. Anal. Meth. Geomech.* **1983**, *7*, 273–281. [CrossRef]
12. Lai, J.Y.; Booker, J.R. Application of discrete Fourier series to the finite element stress analysis of axi-symmetric solids. *Int. J. Num. Meth. Eng.* **1991**, *31*, 619–647. [CrossRef]
13. Fritz, J.J. Flexible Pavement Response Evaluation Using the Semi-Analytical Finite Element Method. *Int. J. Mater. Pavement Des.* **2002**, *3*, 211–225.
14. Hu, S.; Hu, X.; Zhou, F. Using Semi-Analytical Finite Element Method to Evaluate Stress Intensity Factors in Pavement Structure. In Proceedings of the RILEM International Conference on Cracking in Pavements, Chicago, IL, USA, 16–18 June 2008; pp. 637–646.

15. Liu, P.; Wang, D.; Oeser, M.; Chen, X. Einsatz der Semi-Analytischen Finite-Elemente-Methode zur Beanspruchungszustände von Asphaltbefestigungen. *Bauingenieur* **2014**, *89*, 333–339.
16. Liu, P.; Wang, D.; Otto, F.; Hu, J.; Oeser, M. Application of Semi-Analytical Finite Element Method to Evaluate Asphalt Pavement Bearing Capacity. *Int. J. Pavement Eng.* **2016**. [CrossRef]
17. Liu, P.; Wang, D.; Hu, J.; Oeser, M. SAFEM – Software with Graphical User Interface for Fast and Accurate Finite Element Analysis of Asphalt Pavements. *JTE* **2017**, *45*, 1–15. [CrossRef]
18. Liu, P.; Wang, D.; Oeser, M. The Application of Semi-Analytical Finite Element Method to Analyze Asphalt Pavement Response under Heavy Traffic Loads. *JTTE* **2017**. [CrossRef]
19. Liu, P.; Wang, D.; Oeser, M. Application of semi-analytical finite element method coupled with infinite element for analysis of asphalt pavement structural response. *JTTE* **2015**, *2*, 48–58. [CrossRef]
20. Cheung, Y.K.; Fan, S.C. Analysis of pavements and layered foundations by finite layer method. In *S.A.E. Preprints, Proceedings of the Third International Conference on Numerical Methods in Geomechanics, Aachen, Germany, 2–6 April 1979*; Wittke, W., Balkema, A.A., Eds.; Society of Automotive Engineers: Rotterdam, The Netherlands, 1979; pp. 1129–1135.
21. Rowe, R.K.; Booker, J.R. Finite layer analysis of nonhomogeneous soils. *JEMD* **1982**, *108*, 115–132.
22. Small, J.C.; Booker, J.R. Finite layer analysis of layered elastic materials using a flexibility approach, 1, Strip loadings. *Int. J. Numer. Meth. Eng.* **1984**, *20*, 1025–1037. [CrossRef]
23. Small, J.C.; Booker, J.R. Surface deformation of layered soil deposits due to extraction of water. In Proceedings of the Ninth Australasian Conference on the Mechanics of Structures and Materials, Sydney, Australia, 29–31 August 1984; pp. 33–38.
24. Booker, J.R.; Small, J.C. Finite layer analysis of consolidation I. *Int. J. Numer. Anal. Meth. Geomech.* **1982**, *6*, 151–171. [CrossRef]
25. Booker, J.R.; Small, J.C. Finite layer analysis of consolidation II. *Int. J. Numer. Anal. Meth. Geomech.* **1982**, *6*, 173–194. [CrossRef]
26. Booker, J.R.; Small, J.C. Finite layer analysis of viscoelastic layered materials. *Int. J. Numer. Anal. Meth. Geomech.* **1986**, *10*, 415–430. [CrossRef]
27. Smith, S.S.; Allen, M.B.; Puckett, J.; Edgar, T. *The Finite Layer Method for Groundwater Flow Models*; Wyoming Water Research Center: Laramie, WY, USA, 1992.
28. Zai, J.M.; Mei, G.X. Finite layer analysis of three dimentional Biot consolidation. *CJGE* **2002**, *24*, 31–33.
29. Díaz-Cereceda, C.; Poblet-Puig, J.; Rodríguez-Ferran, A. The finite layer method for modelling the sound transmission through double walls. *J. Sound Vib.* **2012**, *331*, 4884–4900. [CrossRef]
30. Siddharthan, R.V.; Yao, J.; Sebaaly, P.E. Pavement Strain from Moving Dynamic 3-D Load Distribution. *J. Transp. Eng.* **1998**, *124*, 557–566. [CrossRef]
31. Siddharthan, R.V.; Krishnamenon, N.; Sebaaly, P.E. Pavement Response Evaluation using Finite-Layer Approach. *Transp. Res. Rec.* **2000**, *1709*, 43–49. [CrossRef]
32. Zienkiewicz, O.C.; Zhu, J. The super-convergence patch recovery (SPR) and a posteriori error estimates, Part 1: The recovery technique, Part 2: Error estimates and adaptivity. *Int. J. Num. Method. Eng.* **1992**, *33*, 1331–1382. [CrossRef]
33. Babuska, I.; Strouboulis, T.; Upadhyay, C.S.; Gangaraj, S.K. A posteriori estimation and adaptive control of the pollution error in the h-version of the finite element method. *Int. J. Num. Method. Eng.* **1995**, *38*, 4207–4235. [CrossRef]
34. Yuan, S.; Du, Y.; Xing, Q.; Ye, K. Self-adaptive one-dimensional nonlinear finite element method based on element energy projection method. *Appl. Math. Mech.* **2014**, *35*, 1223–1232. [CrossRef]
35. Yuan, S.; He, X. A self-adaptive strategy for one-dimensional FEM based on EEP method. *Appl. Math. Mech.* **2006**, *27*, 1461–1474. [CrossRef]
36. Yuan, S.; Xing, Q.; Wang, X.; Ye, K. Self-adaptive strategy for one-dimensional FEM based on EEP method with super-convergence order. *Appl. Math. Mech.* **2008**, *29*, 591–602. [CrossRef]
37. Strang, G.; Fix, G. *An Analysis of the Finite Element Method*; Prentice-Hall: London, UK, 1973.
38. Zhao, Q.H. The Mathematical Analysis on Element Energy Projection Method. Ph.D. Dissertation, Hunan University, Changsha, China, 2007. (In Chinese)
39. Yuan, S.; Xiao, J.; Ye, K. EEP Super-Convergent Computation in FEM Analysis of FEMOL Second Order ODEs. *Eng. Mech.* **2009**, *26*, 1–9.
40. *Guidelines for the Standardization of the Upper Structure of Traffic Areas: RStO-12*; FGSV: Cologne, Germany, 2012.

5

Ultrasonic Techniques for Air Void Size Distribution and Property Evaluation in Both Early-Age and Hardened Concrete Samples

Shuaicheng Guo [1], Qingli Dai [1,*], Xiao Sun [1], Ye Sun [2] and Zhen Liu [1]

[1] Department of Civil and Environmental Engineering, Michigan Technological University, Houghton, MI 49931-1295, USA; sguo3@mtu.edu (S.G.); xiaos@mtu.edu (X.S.); zhenl@mtu.edu (Z.L.)

[2] Department of Mechanical Engineering and Engineering Mechanics, Michigan Technological University, 1400 Townsend Drive, Houghton, MI 49931-1295, USA; yes@mtu.edu

* Correspondence: qingdai@mtu.edu

Academic Editor: Faris Ali

Abstract: Entrained air voids can improve the freeze-thaw durability of concrete, and also affect its mechanical and transport properties. Therefore, it is important to measure the air void structure and understand its influence on concrete performance for quality control. This paper aims to measure air void structure evolution at both early-age and hardened stages with the ultrasonic technique, and evaluates its influence on concrete properties. Three samples with different air entrainment agent content were specially prepared. The air void structure was determined with optimized inverse analysis by achieving the minimum error between experimental and theoretical attenuation. The early-age sample measurement showed that the air void content with the whole size range slightly decreases with curing time. The air void size distribution of hardened samples (at Day 28) was compared with American Society for Testing and Materials (ASTM) C457 test results. The air void size distribution with different amount of air entrainment agent was also favorably compared. In addition, the transport property, compressive strength, and dynamic modulus of concrete samples were also evaluated. The concrete transport decreased with the curing age, which is in accordance with the air void shrinkage. The correlation between the early-age strength development and hardened dynamic modulus with the ultrasonic parameters was also evaluated. The existence of clustered air voids in the Interfacial Transition Zone (ITZ) area was found to cause severe compressive strength loss. The results indicated that this developed ultrasonic technique has potential in air void size distribution measurement, and demonstrated the influence of air void structure evolution on concrete properties during both early-age and hardened stages.

Keywords: air void structure; ultrasonic scattering measurement; inverse analysis; transport ability; compressive strength; dynamic modulus

---

1. Introduction

As an important phase existing in the concrete, the air void content can significantly affect the physical properties of concrete, including the compressive strength [1], elastic modulus [2,3], permeability [4], and long-term freeze-thaw durability [5]. During the concrete construction, the air void content needs to be properly designed and controlled for appropriate property. Also, the influence of air void content on the concrete properties development during both early-age and hardened stages needs to be investigated.

The ultrasonic scattering method has been used to analyze hardened concrete air void characteristics by different researchers. For instance, Zhang [6] studied the microstructure evolution of cementitious material with the ultrasonic methods. Based on the ultrasonic monitoring results, the influences of mineral admixture type, steel fiber content, and curing temperature on the microstructure development were discussed. Zhu [7] analyzed the relationships between the ultrasonic wave speed and the existing air voids in cement paste. Particularly, the air voids can obviously decrease the compressional wave speed, while the influence of air void content on the shear wave speed was insignificant. Liu [8] analyzed the early-age concrete microstructure evolution with the measured ultrasonic pulse speed. The influences of the water/cement ratio, mixtures, and aggregates gradation on the microstructure were analyzed. Lai [9] applied the ultrasonic surface wave to detect the honeycomb defects existed in concrete during early stage. The results demonstrated that the honeycomb can cause obvious spectra dispersion of the surface wave. Punurai [10] analyzed the volume content of capillary pores and air voids in hardened cement pastes with ultrasonic measurement. The results were comparable to the traditional petrographic methods based on American Society for Testing and Materials (ASTM) C457 [11]. Sun [12] employed the log-normal distribution for the description of air void size distribution, and obtained the inversed air void size distribution based on the scattering attenuation calculation. Yim [13] quantified the air void size distribution by ultrasonic attenuation calculation with the Roney equation [14]. The three-dimensional characteristics of air voids were determined through the inverse analysis, with the result verified by the ASTM C457 test [11]. Guo [15] compared different distribution functions for the description of hardened concrete air void size distribution. Based on the verification with the ASTM C457 test, the combination of normal distribution for small air voids and log-normal distribution for large air voids performs better. Currently, the application of ultrasonic measurements on the air void content evolution during early stages has not been conducted to the best of the authors' knowledge.

Besides the application on concrete air void structure analysis, the ultrasonic scattering technique has also been employed to investigate concrete properties evolution during both early-age and hardened stages. Boumiz [16] applied ultrasonic measurements to study the properties of early-age cement pastes and mortars. Particularly, the elastic modulus and Poisson's ratio were determined acoustically. Jerome [17] monitored the setting process of mortar samples with the ultrasonic technique. The shear wave velocity was found more appropriate to determine the final setting time than the compression wave velocity. The method was further applied on the concrete samples [18], and the effect of aggregate on the setting process was also analyzed. Subramaniam [19] studied the steel reinforced concrete with the ultrasonic measurement. The wave reflection factor at the steel–concrete interface was calculated and its relationship to strength gain was also analyzed. Qixian [20] predicted the dynamic modulus of concrete with the measured surface wave velocity. The calculated dynamic modulus was in good accordance with the measured results through the resonance method. Trtnik [21] proposed the TG (Trtnik Gregor) parameter for the strength gain prediction, which is the peak value ratio between the high frequency and low frequency range of the transmission signal in the frequency domain. The TG parameter was found to be strongly linearly correlated to the strength gain during the first eight hours. The TG parameter also became stable after the first several curing hours; thus, it cannot be used for the strength gain prediction during the early stages (1–28 days). The ultrasonic pulse velocity (UPV) has also been used to monitor the hardening behavior. The UPV-based models were found to be especially suitable for the early age property prediction [22].

The influence of the air void structure on the concrete properties has already been widely studied [23], but the study on the influence of the air void structure evolution of concrete performance is relatively limited. Wong [24] studied the influence of entrained air voids on the concrete transport ability. It is reported that the entrained air void can decrease or increase the concrete transport ability, depending on saturation level. Gutmann [25] studied the strength loss due to the entrained air bubbles in concrete. The compressive strength was found to decrease, with [26] analyzing the effect of air void clustering on concrete strength loss. It is found that the strength loss was mainly caused by the total air content, rather than the air void clustering; this is contradictory to the findings of Hover [27]. Currently, research on how the transport ability is evaluated during curing age is limited, and the effect of air void clustering on compressive strength is still not clear.

This paper aims to monitor the air void structure evolution and the change of concrete properties during the early-age and hardened stages. Three concrete samples with the same mix design but different air entrainment agent dosages were prepared. The ultrasonic scattering measurements were conducted during early-age and hardened stages to detect the air void structure evolution. The experimental attenuation was obtained based on the wave spectral ratio of the first and the second transmission waves in the frequency domain. The method is following the former study [15] on hardened concrete samples by employing the combination of normal and log-normal distribution. The theoretical attenuation was then calculated with the obstacle scattering theory, by integrating the pre-determined cement paste, aggregate attenuation, and air void distribution effects. The inverse analyses were conducted to determine air void distribution parameters by achieving the minimum error between theoretical and experimental attenuations. The ultrasonic measured air void distribution was then verified with the ASTM C 457 [11] test results at Day 28. The influence of air void structure evolution on concrete properties were also evaluated in this study, including transport ability, compressive strength, and dynamic modulus. The transport ability was evaluated with the electrical resistivity measurement. The experimental measurements on compressive strength and ultrasonic parameters were fitted with linear relations at early stages. The dynamic modulus predicted with ultrasonic parameters was also compared with ASTM C 215 measurement of hardened concrete.

2. Attenuation-Based Ultrasonic Scattering Measurement Method

2.1. Theoretical Attenuation

The obstacle scattering theory built by Ying [28] was applied for the theoretical ultrasonic scattering calculation. This theory is based on a motion equation, which describes the wave function in homogeneous solid materials. Particularly, the wave function can be obtained by solving the motion equation, and represented with the Bessel function. The total energy loss can be obtained by the integration of the scattered wave and represented as the item scattering cross section γ^{Sca}. In this research, both the aggregates and the air voids are considered as spherical obstacles. Then, assuming the obstacles inside concrete do not interact with each other [12], the total theoretical attenuation can be depicted as Equation (1).

$$\alpha_f = (1-\phi)\alpha_{a,f} + \frac{1}{2}\sum_{i=1}^{m1} n_{i1}\gamma_{i1,f}^{Sca} + \frac{1}{2}\sum_{i=1}^{m2} n_{i2}\gamma_{i2,f}^{Sca} \quad (1)$$

where α_f is the total concrete sample attenuation coefficient (Nepper/m); ϕ is the combined volume fraction of the air voids and aggregate (%); $\alpha_{a,f}$ is the attenuation coefficient of the viscoelastic cement paste (Nepper/m); n_{i1} and n_{i2} are the numbers of a certain size aggregate and air void per volume respectively; $\gamma_{i1,f}^{Sca}$ and $\gamma_{i2,f}^{Sca}$ represent the corresponding scattering cross sections of the aggregate and air voids respectively.

2.1.1. Attenuation Generated by Cement Paste and Aggregates

The attenuation generated by the cement paste matrix $\alpha_{a,f}$ can be directly measured with the ultrasonic scattering measurement on the separately prepared cement paste with the same water/cement ratio. The aggregate size distribution $G(a_{i1})$ can be determined with the designed aggregate aggregation. Then, the corresponding amount of a certain size aggregate per unit volume n_i can be deduced as shown in Equation (2). Introducing Equation (2) into Equation (1), the theoretical attenuation generated by the aggregate can be determined as depicted in Equation (3).

$$n_{i1} = \left[\left[G(a_{i1}) - G(a_{i1-1})\right]\phi_1\right] \Big/ \frac{4}{3}\pi a_{i1}^3 \quad (2)$$

$$\alpha_{f,aggregate} = \frac{1}{2}\sum_{i=1}^{m1}\left[\left[G(a_{i1}) - G(a_{i1-1})\right]\phi_{i1}\gamma_{i1,f}^{Sca}\right] \Big/ \frac{4}{3}\pi a_{i1}^3 \quad (3)$$

where ϕ_1 is the volume fraction of the total aggregate (%); a_{i1}, ϕ_{i1} and n_{i1} represent the size, volume fraction and count per volume of a certain type aggregate; $\alpha_{f,aggregate}$ is the theoretical attenuation of the aggregate (Nepper/m).

2.1.2. Attenuation Generated by Air Voids

The normal and log-normal distribution are desirable to characterize the size distribution of small air voids and large air voids respectively [15]. The combination of the normal and log-normal distribution function is depicted in Equation (4). The detailed normal and log-normal distributions are shown as Equations (5) and (6) respectively

$$f = \phi_{21} \times f_{norm}(r_{norm}; \mu_{norm}, \sigma_{norm}) + \phi_{22} \times f_{log-norm}\left(r_{log-norm}; \mu_{log-norm}, \sigma_{log-norm}\right) \quad (4)$$

where f represents the distribution function for the whole air void structure; r_{norm}, μ_{norm}, σ_{norm} are the radius of the air void, mean value, and standard deviation of the norm size distribution, respectively; $r_{log-norm}$, $\mu_{log-norm}$, $\sigma_{log-norm}$ represent the corresponding parameters for the log-normal distribution. The volume fractions of small air voids and large air voids are ϕ_{21} and ϕ_{22}, respectively ($\phi_2 = \phi_{12} + \phi_{22}$).

$$f_{norm}(r_{norm}; \mu_{norm}, \sigma_{norm}) = \left(1/\sqrt{2\pi\sigma_{norm}^2}\right)\exp\left[-(r_{norm} - \mu_{norm})^2/2\sigma_{norm}^2\right] \quad (5)$$

$$f_{log-norm}\left(r_{log-norm}; \mu_{log-norm}, \sigma_{log-norm}\right) = \left(1/\sqrt{2\pi\sigma_{log-norm}^2}\right)\exp[-\left(\ln r_{log-norm} - \mu_{log-norm}\right)^2 \Big/ 2\sigma_{log-norm}^2 \quad (6)$$

where $f_{norm}(r_{norm}; \mu_{norm}, \sigma_{norm})$ and $f_{log-norm}\left(r_{log-norm}; \mu_{log-norm}, \sigma_{log-norm}\right)$ represent the normal distribution and log-normal distribution separately, and r_{norm}, μ_{norm}, σ_{norm} are the radius of the air void, mean value, and standard deviation of the norm size distribution, respectively, while $r_{log-norm}$, $\mu_{log-norm}$, $\sigma_{log-norm}$ represent those parameters for the log-normal distribution.

Then, the corresponding cumulative function for a certain size air void can be represented as $F_2(a_{i2})$ through the integration of distribution function $f_2(a_{i2})$. Similar to the analysis of aggregate attenuation above, the amount of a certain size air void per unit volume n_{i2} can be deduced as shown in Equations (7) and (8). Particularly, the air void size distribution function is divided by $F_2(a_{N2}) - F_2(a_{12})$ for normalization, where a_{N2} and a_{12} represent the largest and smallest air void size respectively.

$$\phi_{i2} = \left[F_2(a_{i2}) - F_2(a_{i2-1})\right]\phi_2 \Big/ \left[F_2(a_{N2}) - F_2(a_{12})\right] \quad (7)$$

$$n_{i2} = \left[F_2(a_{i2}) - F_2(a_{i2-1})\right]\phi_2 \Big/ \left[F_2(a_{N2}) - F_2(a_{12})\right]\frac{4}{3}\pi a_{i2}^2 \quad (8)$$

where ϕ_2 is the volume fractions of the air void ($\phi_1 + \phi_2 = 1 - \phi$).

Introducing Equation (9) into Equation (1), the total theory attenuation can be linked to the air voids' size distribution and aggregate gradation as illustrated in Equation (9).

$$\alpha_{f,airvoid} = \frac{1}{2}\sum_{i=1}^{m2}\left[F_2(a_{i2}) - F_2(a_{i2-1})\right]\phi_{i2}\gamma_{i2,f}^{Sca} \Big/ \left[F_2(a_{N2}) - F_2(a_{12})\right]\frac{4}{3}\pi a_{i2}^3 \quad (9)$$

2.2. Experimental Setup and Experimental Attenuation Measurement

The basic setup for ultrasonic scattering measurement is demonstrated in Figure 1. The two Olympus 5077 transducers (Olympus Co., Center Valley, PA, USA) are located on the opposite sides of the concrete specimen (Prepared by the authors in Benedict Lab, Houghton, MI, USA) as the pulser and receiver separately. The couplant B2 from Olympus (Center Valley, PA, USA) was also used to mitigate the coupling effect between the specimen surface and ultrasonic transducers. The ultrasonic wave centered at 500 kHz frequency is applied in this study, which is appropriate for the detection of the air voids and has been used in the concrete study by different researchers [29,30]. Ultrasonic waves generated by the excitation voltage can transmit through the specimen and keep being reflected at the transducer–specimen interface. The wave signal captured by the receiver transducer is recorded by an oscilloscope (Olympus Co., Center Valley, PA, USA) and displayed on a computer (Dell Co., Round Rock, TX, USA). Particularly, the first two transmission waves $S_1(t)$ and $S_2(t)$, as shown in Figure 1, are used for the experimental attenuation calculation.

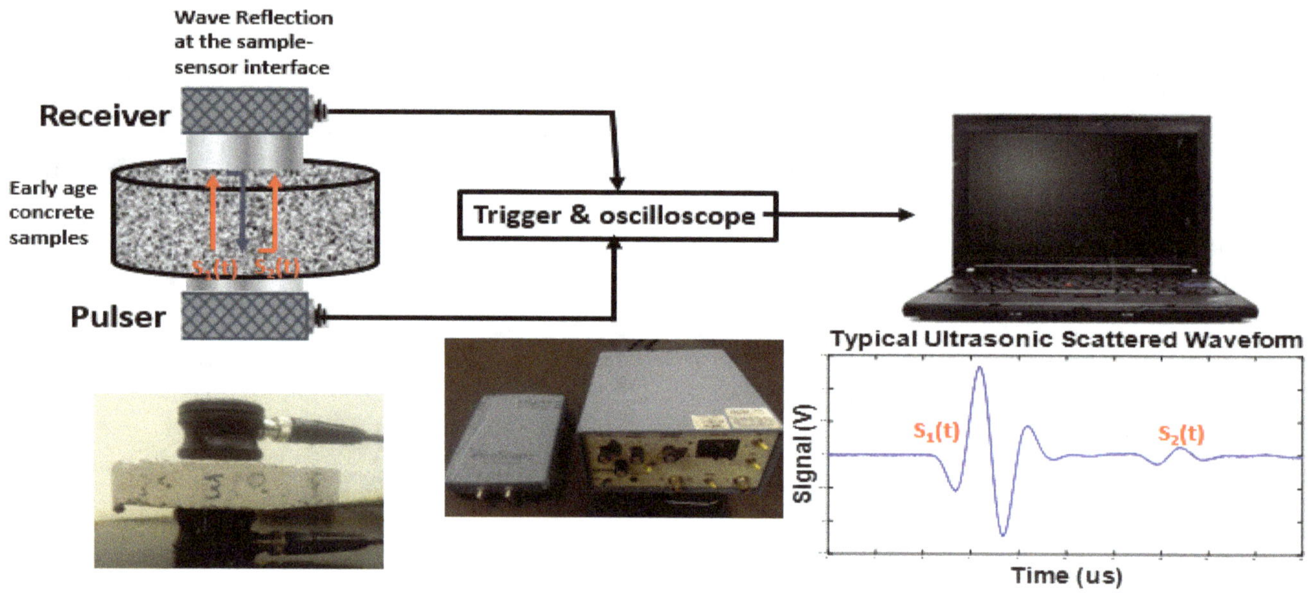

Figure 1. Demonstration of the ultrasonic scattering measurement of air void size distribution in concrete samples.

The first and second transmission waves referred to as S_1 and S_2 were first converted from the time domain to the frequency domain by Fast Fourier Transform on MATLAB (R2015a, MathWorks Co., Natick, MA, USA), as shown in Figure 2. It is clear that the amplitude of S_2 is significantly reduced along the whole frequency range compared to the first transmission wave S_1. The experimental attenuation is estimated from Equation (10) proposed by Sears [31]. The irrelevant attenuation caused by the wave diffraction was eliminated by using the diffraction correlations as shown in Equation (10).

$$\alpha(f) = \frac{1}{2d}\left[\ln\left(\frac{s_1(f)}{s_2(f)}\right) - \ln\left(\frac{D(s;d)}{D(s;3d)}\right)\right] \tag{10}$$

where $\alpha(f)$ is the experimental attenuation of the specimen, which is expressed with wave frequency f; d is the specimen thickness; and $s_1(f)$ and $s_2(f)$ are the first and second transmission waves respectively. $D(s;d)$ and $D(s;3d)$ are proposed by Rogers and Buren [32], which are the simplified closed-form expressions for the diffraction correction.

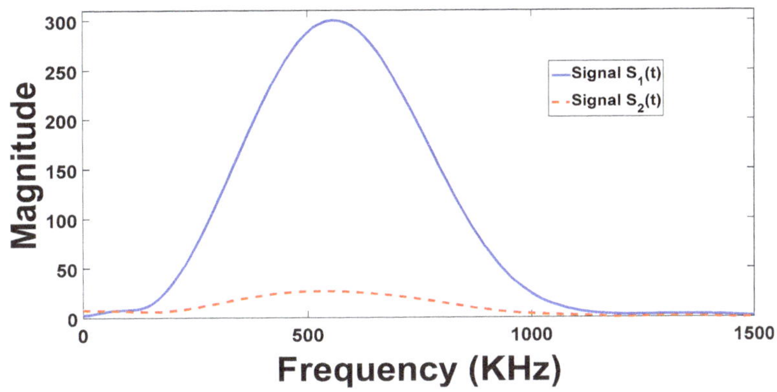

Figure 2. Magnitude spectrum of the cement paste sample ultrasonic results.

Before the investigation of ultrasonic attenuation on the concrete specimen, the attenuation measurement from the ultrasonic test setup was validated with a thin polymethylmethacrylate sample. The sample was placed between the "pulser" and "receiver" transducers, and the couplant was also used to reduce the coupling effect. The first and second transmission waves were recorded and converted to the frequency spectrum with Fourier transformation for attenuation calculation. The attenuation was calculated with Equation (10) and the measurement results are shown in Figure 3. The relationship between ultrasonic attenuation and wave frequency is demonstrated by the fitting curve of $y = 1.725 \times 10^{-5} \times x + 10.958$, where x and y represent the wave frequency (Hz) and wave attenuation (Np/m), respectively. The transmission speed can also be measured based on the arrival time of the first transmission signal and the sample thickness. The measured transmission speed is 2717 m/s; the corresponding attenuation per wavelength can then be calculated as 0.0469 NP. The results are close to the measured values (2782 m/s and 0.045 NP) in reference [10], and the published values (2750 m/s and 0.041 NP) in reference [33], which demonstrate the accuracy of attenuation measurement from this system.

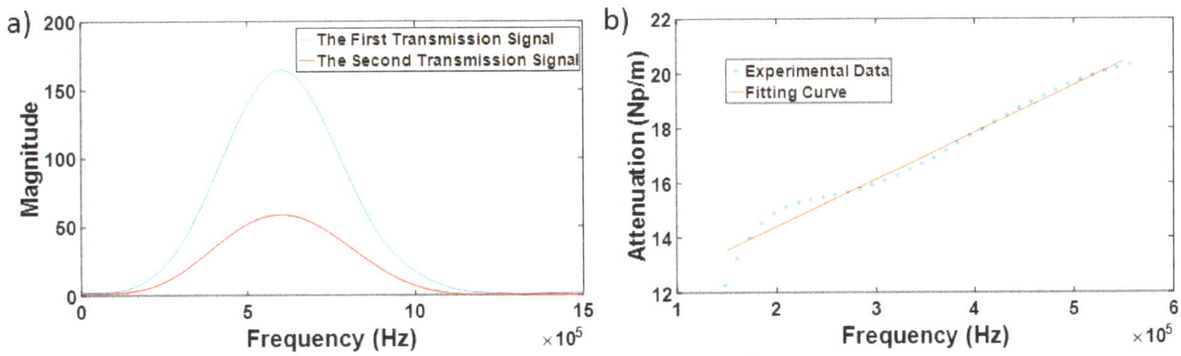

Figure 3. The attenuation measurement results with a polymethylmethacrylate material sample. (**a**) The magnitude spectrum of the first and second transmission wave after Fourier transformation; (**b**) The curve fitting based on the experimentally measured ultrasonic attenuation.

2.3. Inverse Analysis to Determine Air Void Distribution Functions

According to Equation (1), the total theoretical attenuation includes the viscoelastic property of cement paste and the obstacle effect of the aggregate and air voids. The cement paste prepared with the same water/cement ratio was measured with the same method as shown in Figure 1. The corresponding cement paste attenuation was used as the viscoelastic matrix attenuation $\alpha_{a,f}$ as shown in Equation (1). Based on Equation (6), the attenuation generated by the aggregate can be calculated with the designed aggregate gradation. Then, the total theoretical attenuation can be determined and compared with the experimental attenuation estimated from Equation (10). The air voids' size distribution parameters can be obtained by achieving the minimum error between experimental and theoretical attenuations.

3. Ultrasonic Measurement of Air Void Distribution in Early-Age Concrete

3.1. Sample Preparation and Early-Age Property Measurement

Three different types of concrete samples were prepared based on the ASTM C192/C192M-16a [34] during this study. To focus on the influence of the air entrainment agent on the air void structure, the samples were prepared with the same mix design but different air entrainment agent dosages. MasterAir AE 200 type air entrainment agent (BASF Co., Florham Park, NJ, USA) was chosen in this study. The detailed mix design for three types of concrete are shown in Table 1. The volume percentage of size three-quarters (19 mm), one-half (12.5 mm), three-eighths (9.5 mm), No. 4 (4.75 mm), No. 8 (2.36 mm) and No. 16 (1.18 mm) coarse aggregate are 5%, 25%, 25%, 20%, 16% and 9% respectively. Particularly, the fine aggregates belong to the No. 8 (2.36 mm) sieve size. After mixing, the specific gravity of each sample was measured and used for the air void volume fraction calculation based on the ASTM C138 [35]. Then, the fresh concrete was placed into cylinder molds (Humboldt Mfg. Co., Elgin, IL, USA) with 10.16 cm (4 in) diameter and 20.32 cm (8 in) height. After demolding at 24 h, the concrete samples were submerged into water and cured at room temperature for 28 days. Before the tests at each curing age, the samples were first air dried for 12 h to remove the extra water. Besides the ultrasonic scattering measurement, the compressive strength and electrical resistivity tests were also conducted at Day 3, 7 and 28. The compressive strength test was conducted based on the ASTM C39/C39M-16 [36] standard. The bulk electrical resistivity of the concrete sample was measured with the Rcon2 [37] equipment from Giatec Scientific Co. (Ottawa, ON, Canada). Considering the transmission ability of the ultrasonic wave, the concrete cylinder was cut with a diamond saw to obtain slices of around 2.54 cm (1 in) in height for measurement. The selection of thickness partly depended on the penetration ability of the ultrasonic source; the first two signals will not overlap with each other at the thickness of 2.54 cm.

Table 1. Lab concrete mixture proportional design for early-age concrete studies.

Sample	Cement Content (kg/m^3)	Coarse Aggregates (kg/m^3)	Fine Aggregates (kg/m^3)	Water (kg/m^3)	Glass Particle (kg/m^3)	Design w/c	Air entrainment Agent (mL/m^3)
Type 1	334.6 (*sg* * = 3.15)	1110.7 (*sg* = 2.75)	666.4 (*sg* = 2.65)	148.4	74.0 (*sg* = 2.5)	0.45	0
Type 2	334.6 (*sg* = 3.15)	1110.7 (*sg* = 2.75)	666.4 (*sg* = 2.65)	148.4	74.0 (*sg* = 2.5)	0.45	167.5
Type 3	334.6 (*sg* = 3.15)	1110.7 (*sg* = 2.75)	666.4 (*sg* = 2.65)	148.4	74.0 (*sg* = 2.5)	0.45	335.0

* *sg* is short for specific gravity.

3.2. Attenuation Measurement of Cement Paste for Different Ages

As mentioned before, the ultrasonic attenuation of the cement paste first needs to be determined as attenuation for the corresponding viscoelastic matrix in concrete, especially during the early ages. The cement paste samples with the same water/cement ratio (0.45) as concrete samples were separately prepared and no air entrainment agent was added during preparation. Then, the air voids in the separately prepared cement paste are mainly capillary pores due to cement hydration. The attenuation of cement paste is mainly generated from its viscoelastic property, and the contribution from the capillary air void structure is limited. The air void structure of the separately prepared cement paste and that of the cement paste in concrete is different, as the air void structure in concrete can be further affected by the added air entrainment agent, air voids entrapped by the aggregate, and the existence of the Interfacial Transition Zone (ITZ) area. Also, due to the relatively small volume fraction (around

25%) of cement paste inside concrete, the influence of the intrinsic capillary air void content of the cement paste on the concrete air void structure is further limited. The influence of the entrained air voids, entrapped air voids, and the air voids in the ITZ area on the ultrasonic attenuation, are all considered during the measurement in this study.

The cement paste properties are highly impacted by the hydration process. In this study, the ultrasonic scattering measurement was conducted at Day 1, 3, 7, 14, and 28 for both the cement paste and concrete specimen. The calculated cement paste attenuation on the frequency domain based on Equation (10) is shown in Figure 4. A parabolic equation was then used for the curve fitting between the attenuation and corresponding frequency. The fitted equations are depicted as Equations (11)–(14). These fitted equations served as the background attenuation for further inverse analysis of the characteristics of concrete air voids. As the center attenuation used in this study is 500 kHz, the signal is more stable during the range of [150, 550] kHz and this range was applied for the further analysis. It can be observed from Figure 3 that the attenuation of the cement paste matrix decreased with the curing age on the whole frequency domain.

$$\alpha_{a,f} = -222.8110e - 6 \times f^2 + 259.3297e - 3 \times f + 23.8052 \quad (11)$$

$$\alpha_{a,f} = -371.3589e - 6 \times f^2 + 362.8550e - 3 \times f - 10.4897 \quad (12)$$

$$\alpha_{a,f} = -275.0380e - 6 \times f^2 + 280.7362e - 3 \times f - 5.5200 \quad (13)$$

$$\alpha_{a,f} = -129.7629e - 6 \times f^2 + 146.8296e - 3 \times f + 20.7148 \quad (14)$$

where f is the frequency range (KHz) and $\alpha_{a,f}$ is the attenuation (Np/m).

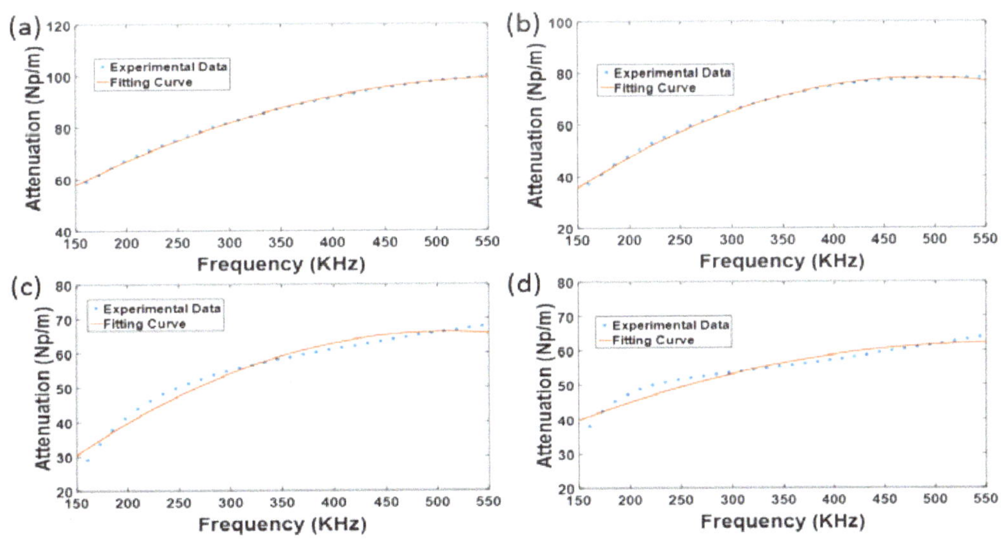

Figure 4. The curve fitting of the early-age cement paste attenuation with frequency. (**a**) Curve fitting at Day 1; (**b**) Curve fitting at Day 3; (**c**) Curve fitting at Day 7; (**d**) Curve fitting at Day 14.

3.3. Determination of Air Void Distribution at Early Stages

The inverse analysis was conducted on MATLAB code by comparing the theoretical attenuation and experimental attenuation estimated based on Equations (1) and (10) respectively. The fitted attenuation equations for cement paste shown in Section 3.2 were applied as the background attenuation $\alpha_{a,f}$. The attenuation generated by the aggregate were calculated with Equation (3), depending on the designed aggregate gradation. Based on the description equation for the air void size distribution shown as Equation (4), the attenuation generated by air voids can also be determined with

Equation (9). Combining all these three parts, the total theoretical attenuation can be determined and compared with the experimental attenuation determined with the methods mentioned in Section 2.2. Three parameters need to be determined for both the small and large air void size distribution: the volume fraction, average size, and the corresponding standard deviation. The air void size distribution parameters can be obtained by achieving the minimum error between experimental and theoretical attenuation.

The comparison between the experimental and theoretical attenuation in the frequency domain for Sample Type 1 was demonstrated in Figure 5, where (a), (b), (c) and (d) depict the comparison at Day 1, 3, 7, and 14 respectively. The corresponding inversed air void size distribution and volume fraction were shown in Figure 6, where (a) and (b) represent the change of the air void volume fraction and evolution of air void size distribution with curing ages. Particularly, the air void volume fraction at Day 0 (fresh stage) was measured by the ASTM C138 test, which is in good agreement with the ultrasonic scattering measurement results.

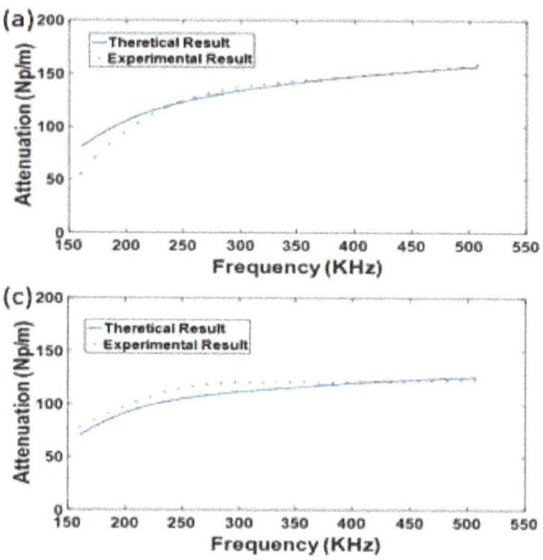

Figure 5. The comparison between theoretical and experimental attenuation during early stages (**a**) The comparison at Day 1; (**b**) The comparison at Day 3; (**c**) The comparison at Day 7; (**d**) The comparison at Day 14.

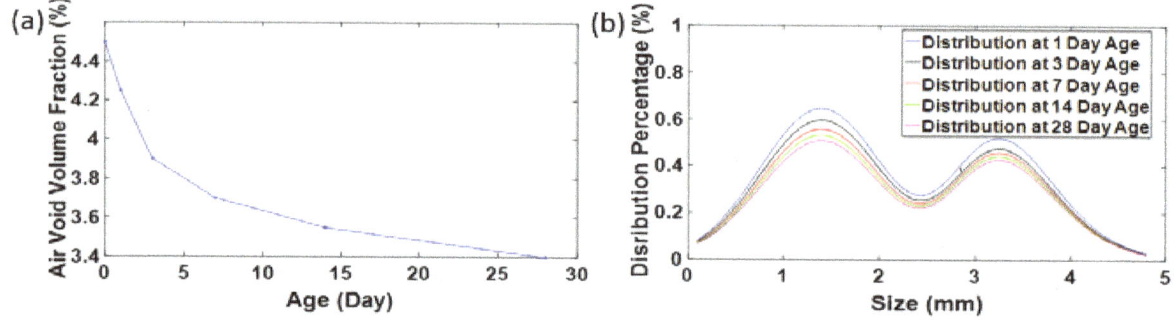

Figure 6. Air voids evolution for Sample Type 1. (**a**) The change of air void content with curing age measured with ultrasonic scattering (at early stages Days 1–28) and the ASTM C138 (Day 0); (**b**) Air void size distribution evolution with curing time.

Similar to the inverse analysis methods used for Sample Type 1, the inverse analysis results were conducted for Sample Types 2 and 3 separately. The changes of air void volume fractions with curing ages are represented in Figures 7a and 8a. Similar to the results shown in Figure 6a, the decreasing trend of the air void volume fraction was observed for all these three samples. Particularly, the air void volume fraction change after Day 14 is much slower compared to the volume change during the first 14 days. The corresponding air void size distribution evolution for Sample Types 2 and 3 are shown in Figures 7b and 8b respectively. The results demonstrate that the volume fraction of both small and large air voids decrease with the curing ages. As mentioned above, the air void volume fraction was observed to decrease with curing age. This phenomenon is in accordance with the results that the air void volume fraction can decrease during hydration age [38], especially on the ITZ area. It is also clear that the decreasing rate of the air void volume fraction declines with the curing age, which is in accordance with the cement hydration speed [39].

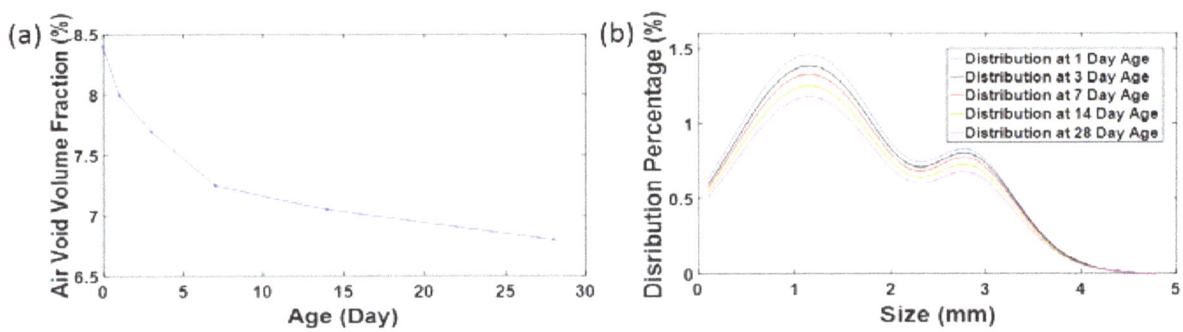

Figure 7. Air voids evolution for Sample Type 2. (**a**) The change of air void content with curing age measured with ultrasonic scattering (at early stages Day 1–28) and the ASTM C138 (Day 0); (**b**) Air void size distribution evolution with curing time.

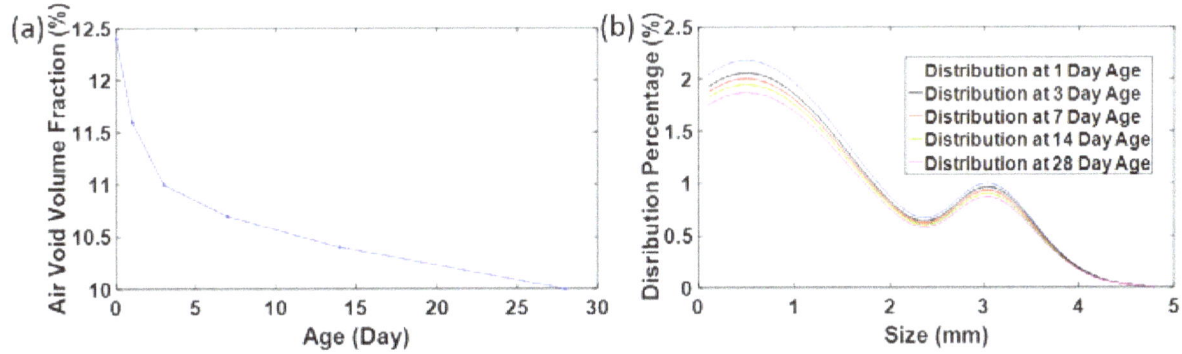

Figure 8. Air voids evolution for Sample Type 3. (**a**) Comparison of the air void content measurement with ultrasonic scattering (at early stages Day 1–28) and the ASTM C138 (Day 0); (**b**) Air void size distribution evolution with curing time.

3.4. The Correlation between Compressive Strength and Utlrasonic Parameter

Besides the analysis of air void volume shrinkage, the compressive strength gain during early age can also be possibly investigated with ultrasonic parameters. The cement hydration process can lead to the increase in concrete stiffness and the decline of viscosity [39]. The rising concrete stiffness means higher transmission speed. The decreasing viscosity can also generate a drop in cement attenuation, as demonstrated in Figure 3. In addition, the shrinkage of the air void volume fraction, as depicted in Figures 6–8, can also lead to higher transmission speed and lower attenuation. Hence, the ultrasonic transmission speed and attenuation are in positive and negative correlations with the compressive strength during the concrete early-age. The ultrasonic transmission speed is determined with the arrival time of Signal $S_1(t)$ and the sample thickness. The calculated ultrasonic transmission speeds are shown in Table 2. Similar to the study [21], this study also applied the peak value ratios for the strength gain prediction. As shown in Figure 2, the peak value of both the first and second transmission waves are located around 500 kHz. Hence, the ultrasonic attenuation value at 500 kHz was chosen to represent the change of attenuation with time, as depicted in Table 2.

Table 2. Properties of concrete samples during the early stages.

Sample	Compressive Strength (MPa)			Ultrasonic Transmission Speed (m/s)			Ultrasonic Transmission Attenuation at 500 KHz (Np/m)			Electrical Resistivity (Ω·m)		
	Day 3	Day 7	Day 28	Day 3	Day 7	Day 28	Day 3	Day 7	Day 28	Day 3	Day 7	Day 28
Type 1	35.8	42.4	48.7	4468.2	4507.8	4612.9	128.1	116.7	105.3	47.3	51.2	57.3
Type 2	29.0	37.7	45.1	4243.9	4311.7	4488.4	134.7	120.8	112.1	49.1	54.6	75.1
Type 3	20.1	25.1	30.0	3788.9	3824.1	4024.2	155.0	141.2	129.4	47.5	87.6	116.5

The compressive strength analysis, applied to the measured results of all three sample types, was applied for the correlation analysis to better demonstrate the relationship between ultrasonic parameters and the strength gain during early stages. The relationship between compressive strength with ultrasonic speed and attenuation is demonstrated in Figure 9a,b respectively. Both of these correlations were evaluated with the linear fitting, and the fitted results are demonstrated as Equations (15) and (16), respectively. The corresponding correlation coefficient between the fitted and measured results are 93.23% and 98.20%, respectively, which demonstrate that the change of ultrasonic attenuation is in stronger linear correlation with the strength gain during early stages.

$$f_c = 0.0291 \times v_p - 88.96 \quad (15)$$

where f_c is the compressive strength (MPa) and v_p is the ultrasonic transmission speed (m/s).

$$f_c = -0.61 \times \alpha_{a,f} + 112.59 \quad (16)$$

where $\alpha_{a,f}$ is the ultrasonic attenuation (Np/m).

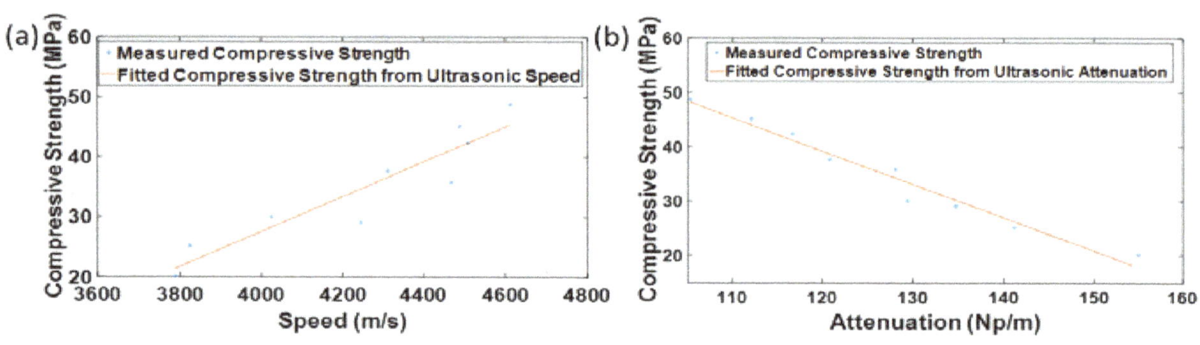

Figure 9. The comparison between the fitted and measured compressive strength. (**a**) The fit between the compressive strength and ultrasonic speed; (**b**) The fit between compressive strength and ultrasonic attenuation.

3.5. The Correlation between Transport Property and Air Void Structure

The back-calculation results demonstrate that the air void content can decrease with the curing age. The shrinkage of air voids can affect the transport ability of concrete. In this study, the transport property was evaluated with the electrical resistivity measurement; the measured results are shown in Table 2. It is clear that the electrical resistivity increases with the curing age for all three sample types. The raising of electrical resistivity can be generated from the decreasing of ion concentration in pore solution and the decline of concrete connectivity. Based on reference [40], the ion concentrations in concrete pore solution become relatively stable after Day 7. Hence the obvious increase in electrical resistivity from Day 7–28 can be mainly generated from the declining connectivity. This result is in accordance with the air void shrinkage during the curing age, as shown in Figures 6–8. In addition, the electrical resistivity can also increase with the air entrainment agent dosage as shown in Table 2. This result demonstrates that the small air voids generated by the added air entrainment agent are mainly isolated in the cement matrix. The small air voids can serve as an insulator [24], and reduce the conductivity of concrete.

4. The Air Void Size Distribution and Dynamic Modulus Analysis of the Hardened Stage

In this study, the estimated air void size distribution based on ultrasonic scattering measurement was compared with the ASTM C457 results for verification at the hardened stage. Further studies on the prediction of the dynamic modulus of hardened concrete were also conducted.

4.1. Compare Measured Air Void Size Distribution with ASTM C457 at Hardened Stages

The obtained air void size distribution and volume fraction from the ultrasonic scattering measurement were further compared with the ASTM C457 test on Day 28. The ASTM C457 tests were conducted with the flatbed scanning methods [41]. The cylinder samples were cut into slabs (7.62 cm (3 in) × 10.16 cm (4 in)) and the surface was carefully polished with grinder. The surfaces were then painted with black marker pen and powdered with calcium silicate. The prepared samples are shown in Figure 10. After processing, the white spots represent air voids existing in concrete specimens. Comparing these three samples, it is clear that the whitened areas increase with the added air entrainment agent dosage. To obtain the size distribution and volume fraction of air voids, the prepared samples were scanned and analyzed with the BubbleCounter [41] embedded in ImageJ software [42].

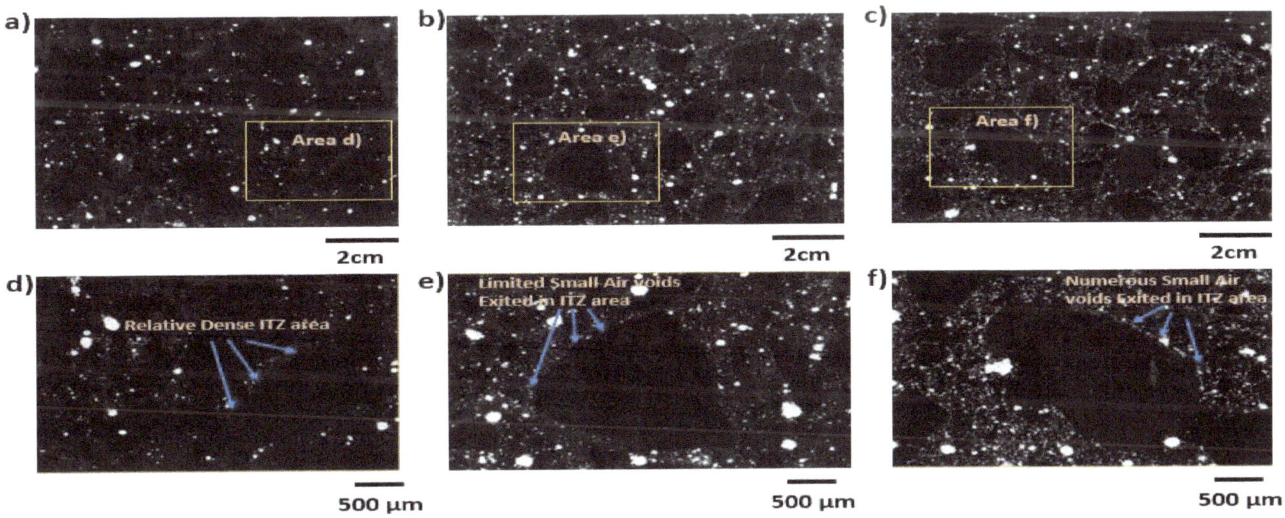

Figure 10. Scanned pattern of concrete slabs from the ASTM (American Society for Testing and Materials) C457 Test. (**a**) The scanned results of Sample Type 1; (**b**) The scanned results of Sample Type 2; (**c**) The scanned results of Sample Type 3d; (**d**) The demonstration of the Interfacial Transition Zone (ITZ) area (Area **d**)) in Sample Type 1; (**e**) The demonstration of the ITZ area (Area **e**)) in Sample Type 2; (**f**) The demonstration of the ITZ area (Area **f**)) in Sample Type 3.

The comparison between the ultrasonic scattering measurement results and the ASTM C457 results is shown in Figure 11a–c, respectively. The corresponding correlation coefficients on the whole size range are 88.86%, 96.88%, and 90.75% for Samples 1–3 separately. The distribution parameters are shown in Table 3. It is clear that the average size and standard deviation for the small air voids decrease and increase respectively with the air entrainment agent content. The air void fraction obtained with ultrasonic scattering techniques is around 3.4%, 6.8%, and 10.0% for Samples 2–4, which are very close to the ASTM C 457 measured values of 3.4%, 6.9%, and 10.1% for tested samples. The size distribution of the small air voids (0.1–1.2 mm) are shown in Figure 12, where a–c represent the non-entrained concrete (Sample Type 1), entrained concrete with regular entrainment dosage (Sample Type 2), and entrained concrete with double entrainment dosage (Sample Type 3), respectively. On the range of small air voids, the correlation coefficient between the ultrasonic scattering measurement results and ASTM C457 measured results are 93.98%, 93.75%, and 71.33%, for Sample Type 1, 2, and 3, respectively. The closeness of both the size distributions curve and volume fraction between these two results demonstrates the feasibility of the ultrasonic scattering measurement method.

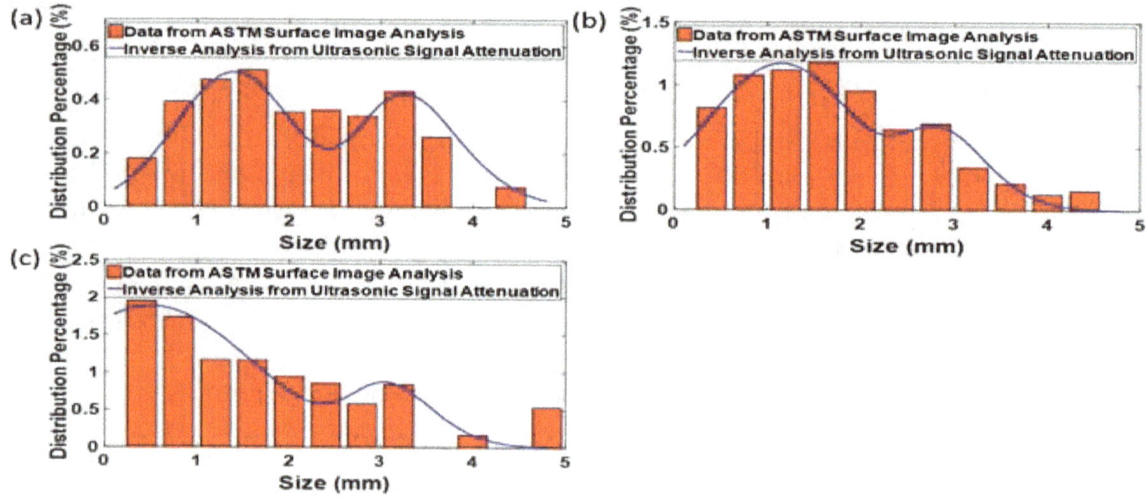

Figure 11. Comparison between the air void size distribution measurement with ultrasonic scattering and ASTM C 457 methods. (**a**) The comparison for Sample Type 1; (**b**) The comparison for Sample Type 2; (**c**) The comparison for Sample Type 3.

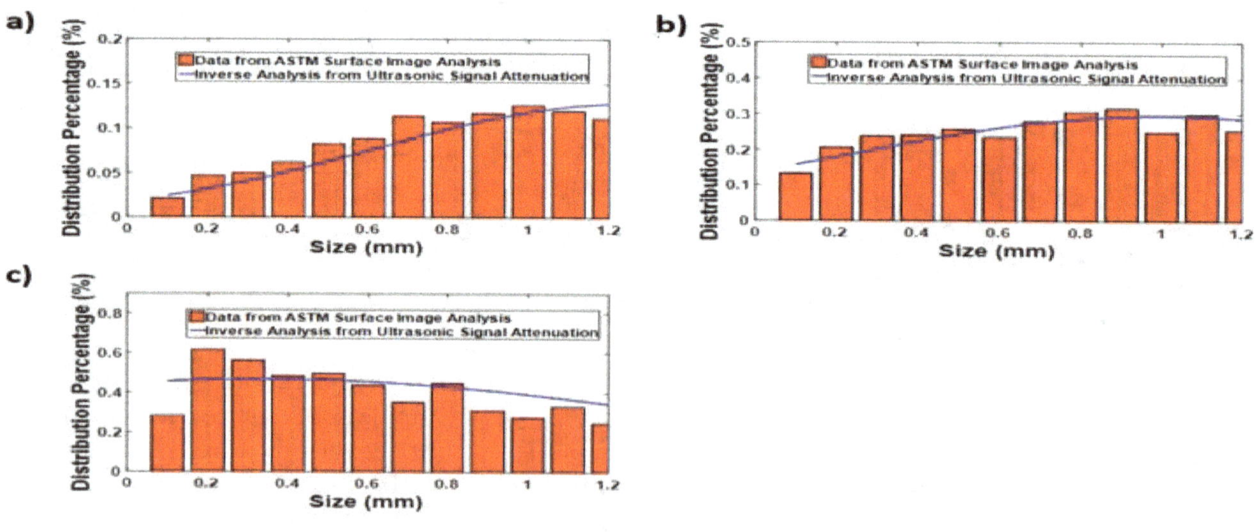

Figure 12. Comparison of the air void size distribution in the small size range. (**a**) The comparison of the experimental result and the theoretical result on the small-size air voids range for Sample Type 1; (**b**) The comparison of the experimental result and the theoretical result on the small-size air voids range for Sample Type 2; (**c**) The comparison of the experimental result and the theoretical result on the small-size air voids range for Sample Type 3.

Table 3. Inverse analysis of parameters for combined distributions.

Sample	Small Air Voids			Large Air Voids			Total air Void Content (%) ϕ
	Volume Fraction (%) ϕ_1	Average Size(mm) μ_{norm}	Standard Deviation σ_{norm}	Volume Fraction (%) ϕ_2	Average Size(mm) $\mu_{log-norm}$	Standard Deviation $\sigma_{log-norm}$	
Sample Type 1	2	1.2	6.3×10^{-4}	1.4	3.2	3×10^{-7}	3.4
Sample Type 2	5.4	0.8	8×10^{-4}	1.4	2.8	2×10^{-7}	6.8
Sample Type 3	8.6	0.3	1.1×10^{-3}	1.4	3.0	2×10^{-7}	10.0

As shown in these figures, the volume fractions of both small air voids and large air voids decrease with the curing age. Particularly, a peak shift from about 1.2 mm to about 0.3 mm can be observed for small air void distribution from Sample Type 1 to Type 3. In contrast, the peaks of the large air void distribution are relatively stable among these three samples. Also, the volume fraction of small air voids increases significantly with the added air entrainment agent, while the volume fraction of the large air voids is almost the same among these three samples. These differences indicate that almost all of the air voids introduced by AE 200 air entrainment agent are smaller than 1.2 mm.

4.2. *Analysis of Compressive Strength Loss and Air Void Content*

As shown in Table 2, the added air entrainment agent can lead to obvious compressive strength loss. Compared to the Type 1 sample, the compressive strength of Types 2 and 3 samples decrease by 7.39% and 38.40%, respectively. In addition, the air void content of Types 2 and 3 Samples increase by 3.4% and 6.6% separately, as compared to the Type 1 sample. Hence, for every 1% increase in air void content, the compressive strength decreases by 2.17% and 5.82% for Types 2 and 3 samples, respectively. It is reported that the compressive strength loss rate was in the range of [2%, 6%] for a 1% increase in air void content [26]. Though the strength loss rate of Types 2 and 3 samples are both in this range, the loss rate of the Type 3 sample is still much higher than the value of the Type 2 sample. Hence, the correlation between compressive strength and air void content cannot be simply treated as a linear relationship. It is considered that the air void distribution can affect the compressive strength, and not only the air void content.

In particular, the distributions of small air voids along the ITZ area are demonstrated in Figure 10d–f for Type 1, 2, and 3 samples, respectively. It is clear that the ITZ area of the Type 1 sample is relatively dense and small air voids can seldom be detected. Similarly, only limited small air voids can be found in the ITZ area of the Type 2 sample, as shown in Figure 10e. For the Type 3 sample, numerous small air voids can be found along the ITZ area as depicted in Figure 10f. The existence of small air voids in the ITZ area can lead to a weaker bonding strength between the cement paste and aggregate, which can be the main reason for the severe strength loss in sample Type 3. This finding demonstrates that clustered air voids can play an important role for the compressive strength loss.

4.3. *Dynamic Modulus Analysis with Ultrasonic Parameters on Hardened Samples*

The dynamic Young's modulus of elasticity was usually applied to evaluate the freeze-thaw durability of concrete [43,44], which can be determined with the resonance frequency based on ASTM C215-14 [45]. During the resonant frequency tests, the concrete sample supported by the soft plastic materials was struck with an impactor at the center of one end of the surface, while the resultant signal was captured by an accelerator on the opposite site. The resonant frequency was then determined with the captured signal, after which the dynamic modulus can be calculated. In this study, the dynamic modulus was determined with the longitudinal resonance frequency as shown in Equation (17).

$$E_{Dynamic} = D \times M \times n^2,\ D = 5.093\left(\frac{L}{d^2}\right) \text{ for a cylinder and } 4\left(\frac{L}{bt}\right) \text{ for a prism.} \quad (17)$$

where $E_{Dynamic}$ is the Young's dynamic modulus of elasticity; M is the concrete mass; n is the measured resonant frequency; L is the sample length; d is the cylinder diameter; b and t are the cross-section dimensions of the concrete beam.

The determination of the dynamic modulus with resonance frequency is based on empirical equations as shown in Equation (17). The dynamic modulus can also be calculated with the sound speed through a theoretical equation [20,46] as shown in Equation (18).

$$E_p = \frac{(1+\vartheta)(1-2\vartheta)}{(1-\vartheta)} v_p^2 \rho_{concrete} \quad (18)$$

where E_p is the concrete dynamic Young's modulus; ϑ is the dynamic Poisson's ratio of concrete; v_p is the longitudinal transverse speed of concrete; $\rho_{concrete}$ is the mass density of concrete samples.

The dynamic Poisson's ratio of concrete is in the range of [0.21, 0.28] based on reference [20]. Similarly, the average dynamic Poisson's ratio was determined as 0.24 in reference [47]. Hence, the Poisson's ratio chosen as 0.24 is used in this study. The longitudinal ultrasonic wave speed is determined based on the arrival time of signal $S_1(t)$ and the sample thickness. The mass density is obtained as the sample mass/volume ratio. The longitudinal dynamic modulus can then be determined with Equation (18). The analyzed dynamic modulus based on Equation (18) was then compared with the measured results from ASTM C215 in Table 4. The relative error between the predicted and measured results are all within 5%, which demonstrate the feasibility of this method.

Table 4. Properties of concrete samples during the early stages .

Sample	Mass Density at Hardened Stage (kg/m^3)	Ultrasonic Transmission Speed (m/s)	Predicted Dynamic Modulus (GPa)	Measured Dynamic Modulus (GPa)	Relative Errors (%)
Type 1	2.30	4612.9	41.5	43.3	4.2%
Type 2	2.23	4488.4	39.2	37.5	1.6%
Type 3	2.16	4024.2	31.6	31.7	0.3%

5. Conclusions

This study analyzed the change of the air void structure and concrete property during the early stage. The ultrasonic scattering method was applied to measure the air void size distribution and to analyze the properties of both early-age and hardened concrete. The results were favorably compared with the ASTM C457 test results for validation. Based on the measured results, the change of the air void structure during the early age, and its relationship to concrete transport ability and the compressive strength were also analyzed.

During the early stages, the decreasing trend of the cement paste attenuation can be observed with the curing ages. The air void volume fraction obtained from the ultrasonic scattering measurement was found to decrease in both the small and large air void range with the curing ages. These phenomena are caused by the change in cement paste volume during the continuous hydration process. The results obtained from the ultrasonic scattering measurement were further verified with the ASTM C 457 test results on Day 28. The correlation coefficient between the ultrasonic scattering back-calculated values and the ASTM C457 measured results on the whole size range are 88.86%, 96.88%, and 90.75%, for Sample Types 1–3, respectively. The relatively high correlation coefficients further indicate the feasibility of the ultrasonic scattering measurement methods.

The electrical resistivity of concrete samples was found to increase with the curing age, which is in accordance with the air void shrinkage during the hydration process. In addition, the correlation between ultrasonic parameters and compressive strength was evaluated. The dynamic modulus was analyzed with the ultrasonic transmission speed, and the results are in good agreement (within 5% error) with the resonance-measured results. The relationship between the air void content and strength loss was also analyzed. The clustered air voids along the ITZ area can cause more severe compressive strength loss.

This paper provides an ultrasonic scattering method to measure the air void size distribution and volume fraction during both early-age and hardened stages. The correlation between concrete properties and ultrasonic parameters was also analyzed. Thus, the ultrasonic techniques can be further developed to measure concrete air void evolution and property development during early stages for concrete construction inspection.

Acknowledgments: The authors would like to acknowledge the Michigan Tech Transportation Institute Initiative Fund program for partial financial support. The first author acknowledges the financial support from China Scholarship Council under No. 201406370141. The authors would like to thank Gerald Anzalone at Michigan Technological University for assistance in the ASTM C457 air void measurement.

Author Contributions: Qingli Dai and Shuaicheng Guo conceived and designed the experiments; Shuaicheng Guo and Xiao Sun performed the experiments and analyzed the data; Ye Sun, Zhen Liu and Shuaicheng Guo contributed the analysis tools; Shuaicheng Guo and Qingli Dai wrote the paper.

References

1. Sturrup, V.; Vecchio, F.; Caratin, H. Pulse Velocity as a Measure of Concrete Compressive Strength. *Spec. Publ.* **1984**, *82*, 201–228.
2. Du, L.; Folliard, K.J. Mechanisms of air entrainment in concrete. *Cem. Concr. Res.* **2005**, *35*, 1463–1471. [CrossRef]
3. Hansen, T.C. Influence of aggregate and voids on modulus of elasticity of concrete, cement mortar, and cement paste. *Am. Concr. Inst. J. Proc.* **1965**, *62*, 193–216.
4. Kearsley, E.; Wainwright, P. Porosity and permeability of foamed concrete. *Cem. Concr. Res.* **2001**, *31*, 805–812. [CrossRef]
5. Cordon, W.A. *Freezing and Thawing of Concrete-Mechanisms and Control*; American Concrete Institute: Farmington Hills, MI, USA, 1966.
6. Zhang, Y.; Zhang, W.; She, W.; Ma, L.; Zhu, W. Ultrasound monitoring of setting and hardening process of ultra-high performance cementitious materials. *NDT E Int.* **2012**, *47*, 177–184.
7. Zhu, J.; Kee, S.H.; Han, D.; Tsai, Y.T. Effects of air voids on ultrasonic wave propagation in early age cement pastes. *Cem. Concr. Res.* **2011**, *41*, 872–881. [CrossRef]
8. Liu, Z.; Zhang, Y.; Jiang, Q.; Sun, G.; Zhang, W. In situ continuously monitoring the early age microstructure evolution of cementitious materials using ultrasonic measurement. *Constr. Build. Mater.* **2011**, *25*, 3998–4005. [CrossRef]
9. Lai, W.; Wang, Y.H.; Kou, S.C.; Poon, C.S. Dispersion of ultrasonic guided surface wave by honeycomb in early-aged concrete. *NDT E Int.* **2013**, *57*, 7–16.
10. Punurai, W.; Jarzynski, J.; Qu, J.; Kurtis, K.E.; Jacobs, L.J. Characterization of entrained air voids in cement paste with scattered ultrasound. *NDT E Int.* **2006**, *39*, 514–524.
11. ASTM C457/C457M-12. *Standard Test Method for Microscopical Determination of Parameters of the Air-Void System in Hardened Concrete*; ASTM International: West Conshohocken, PA, USA, 2012.
12. Sun, Y.; Yu, X.B.; Liu, Z.; Liu, Y.; Tao, J. Advanced ultrasonic technology for freezing damage prevention of concrete pavement. *Int. J. Pavement Res. Technol.* **2013**, *6*, 86–92.
13. Yim, H.J.; Kim, J.H.; Bang, Y.L.; Kwak, H.G. Air voids size distribution determined by ultrasonic attenuation. *Constr. Build. Mater.* **2013**, *47*, 502–510. [CrossRef]
14. Roney, R.K. *The Influence of Metal Grain Structure on the Attenuation of an Ultrasonic Acoustic Wave*; California Institute of Technology: Pasadena, CA, USA, 1950.
15. Guo, S.; Dai, Q.; Sun, X.; Sun, Y. Ultrasonic scattering measurement of air void size distribution in hardened concrete samples. *Constr. Build. Mater* **2016**, *113*, 415–422. [CrossRef]
16. Boumiz, A.; Vernet, C.; Tenoudji, F.C. Mechanical properties of cement pastes and mortars at early ages: Evolution with time and degree of hydration. *Adv. Cem. Based Mater.* **1996**, *3*, 94–106. [CrossRef]
17. Carette, J.; Staquet, S. Monitoring the setting process of mortars by ultrasonic P and S-wave transmission velocity measurement. *Constr. Build. Mater* **2015**, *94*, 196–208. [CrossRef]
18. Carette, J.; Staquet, S. Monitoring the setting process of eco-binders by ultrasonic P-wave and S-wave transmission velocity measurement: Mortar vs. concrete. *Constr. Build. Mater* **2016**, *110*, 32–41. [CrossRef]

19. Subramaniam, K.V.; Shaw, C.K.; Subramaniam, K.V. Ultrasonic technique for monitoring concrete strength gain at early age. *Mater. J.* **2002**, *99*, 458–462.
20. Qixian, L.; Bungey, J. Using compression wave ultrasonic transducers to measure the velocity of surface waves and hence determine dynamic modulus of elasticity for concrete. *Constr. Build. Mater* **1996**, *10*, 237–242. [CrossRef]
21. Trtnik, G.; Gams, M. Ultrasonic assessment of initial compressive strength gain of cement based materials. *Cem. Concr. Res.* **2015**, *67*, 148–155. [CrossRef]
22. Carette, J.; Staquet, S. Monitoring and modelling the early age and hardening behaviour of eco-concrete through continuous non-destructive measurements: Part II. Mechanical behaviour. *Cem. Concr. Compos.* **2016**, *73*, 1–9. [CrossRef]
23. Neville, A.M. *Properties of Concrete*; Longman Group UK Limited: Harlow, UK, 1995.
24. Wong, H.; Pappas, A.M.; Zimmerman, R.W.; Buenfeld, N.R. Effect of entrained air voids on the microstructure and mass transport properties of concrete. *Cem. Concr. Res.* **2011**, *41*, 1067–1077. [CrossRef]
25. Gutmann, P.F. Bubble characteristics as they pertain to compressive strength and freeze-thaw durability. *MRS Proc.* **1987**, *114*, 128–130.
26. Vosahlik, J.; Riding, K.; Esmaeily, A.; Billinger, A.; Mcleod, H. Effects of Air Void Clustering on Concrete Compressive Strength. *ACI Mater. J.* **2016**, *113*, 759–767. [CrossRef]
27. Hover, K.C. Some recent problems with air-entrained concrete. *Cement. Concr. Aggreg.* **1989**, *11*, 67–72.
28. Ying, C.; Truell, R. Scattering of a plane longitudinal wave by a spherical obstacle in an isotropically elastic solid. *J. Appl. Phys.* **1956**, *27*, 1086–1097. [CrossRef]
29. Demčenko, A.; Visser, H.; Akkerman, R. Ultrasonic measurements of undamaged concrete layer thickness in a deteriorated concrete structure. *NDT E Int.* **2016**, *77*, 63–72.
30. Larose, E.; Hall, S. Monitoring stress related velocity variation in concrete with a 2×10^{-5} relative resolution using diffuse ultrasound. *J. Acoust. Soc. Am.* **2009**, *125*, 1853–1856. [CrossRef] [PubMed]
31. Sears, F.M.; Bonner, B.P. Ultrasonic attenuation measurement by spectral ratios utilizing signal processing techniques. *IEEE Trans. Geosci. Remote Sens.* **1981**, *2*, 95–99. [CrossRef]
32. Rogers, P.H.; van Buren, A.L. An exact expression for the Lommel-diffraction correction integral. *J. Acoust. Soc. Am.* **1974**, *55*, 724–728. [CrossRef]
33. Cheeke, J.D.N. *Fundamentals and Applications of Ultrasonic Waves*; CRC Press: New York, NY, USA, 2012.
34. ASTM C192/C192M-16a. *Standard Practice for Making and Curing Concrete Test Specimens in the Laboratory*; ASTM International: West Conshohocken, PA, USA, 2016.
35. ASTM C138/C138M-16a. *Standard Test Method for Density (Unit Weight), Yield, and Air Content (Gravimetric) of Concrete*; ASTM International: West Conshohocken, PA, USA, 2016.
36. ASTM C39/C39M-16. *Standard Test Method for Compressive Strength of Cylindrical Concrete Specimens*; ASTM International: West Conshohocken, PA, USA, 2016.
37. Giatec Scientific Inc. *RCON2 User Manual*; Giatec Scientific Inc.: Ottawa, ON, Canada, 2016.
38. Scrivener, K.L.; Crumbie, A.K.; Laugesen, P. The interfacial transition zone (ITZ) between cement paste and aggregate in concrete. *Interface Sci.* **2004**, *12*, 411–421. [CrossRef]
39. Mehta, P.K. *Concrete. Structure, Properties and Materials*; Prentice-Hall: Englewood Cliffs, NJ, USA, 1986.
40. Vollpracht, A.; Lothenbach, B.; Snellings, R.; Haufe, J. The pore solution of blended cements: A review. *Mater. Struct.* **2016**, *49*, 3341–3367. [CrossRef]
41. Peterson, K.K.W.; Anzalone, G.C.; Nezami, S.; Oh, C.Y.S.; Lu, H. Robust Test of the Flatbed Scanner for Air-Void Characterization in Hardened Concrete. *Evaluation* **2016**, *44*, 1–16. [CrossRef]
42. Schneider, C.A.; Rasband, W.S.; Eliceiri, K.W. NIH Image to ImageJ: 25 years of image analysis. *Nat. Methods* **2012**, *9*, 671–675. [CrossRef] [PubMed]
43. Martins, L.; Vasconcelos, G.; Lourenço, P.B.; Palha, C. Influence of the Freeze-Thaw Cycles on the Physical and Mechanical Properties of Granites. *J. Mater. Civ. Eng.* **2015**, *28*, 04015201. [CrossRef]
44. ASTM C666/C666M-15. *Standard Test Method for Resistance of Concrete to Rapid Freezing and Thawing*; ASTM International: West Conshohocken, PA, USA, 2015.
45. ASTM C215-14. *Standard Test Method for Fundamental Transverse, Longitudinal, and Torsional Resonant Frequencies of Concrete Specimens*; ASTM International: West Conshohocken, PA, USA, 2014.

46. Kinsler, L.E.; Frey, A.R.; Coppens, A.B.; Sanders, J.V. Fundamentals of acoustics. In *Fundamentals of Acoustics*, 4th ed.; Kinsler, L.E., Frey, A.R., Coppens, A.B., James, V., Sanders, J.V., Eds.; Wiley-VCH: Weinheim, Germany, 1999; p. 560.
47. Teller, L. Elastic properties. *Signif. Tests Prop. Concr. Concr. Aggreg.* **1956**, *169*, 94.

6

Technologies and Principles of Hot Recycling and Investigation of Preheated Reclaimed Asphalt Pavement Batching Process in an Asphalt Mixing Plant

Henrikas Sivilevičius, Justas Bražiūnas *and Olegas Prentkovskis

Department of Mobile Machinery and Railway Transport, Faculty of Transport Engineering, Vilnius Gediminas Technical University, Plytinės g. 27, LT-10105 Vilnius, Lithuania; henrikas.sivilevicius@vgtu.lt (H.S.); olegas.prentkovskis@vgtu.lt (O.P.)

* Correspondence: justas.braziunas@vgtu.lt

Abstract: More and more recycled asphalt mixtures with high reclaimed asphalt pavement (RAP) content are used in road pavement. Having determined and evaluated RAP composition (aged bitumen content and aggregate gradation) and properties, a suitable recycling agent and virgin materials are selected in the design process. The gradation of hot mix asphalt (HMA) mixture that is recycled in an asphalt mixing plant (AMP) shall correspond to its optimal gradation set out in its job-mix formula (JMF). When RAP is recycled in an AMP, inevitable systematic and random errors of performed technological operations and inhomogeneity of virgin materials and RAP have a significant influence. These factors influence the variation of components quantities of recycled hot mix asphalt (RHMA) mixture and deviations from JMF. In this study, the principles of asphalt pavement hot recycling are systematized, which allows analysis of the factors of components' interaction influencing the results of the recycling process. The paper also presents and analyses asphalt recycling technologies in AMP and their comparative analysis. During the season of asphalt mixture production in 2014, statistical parameters were calculated according to the data obtained from one of the companies, which collected and systematized RAP batch masses, when before batching it was pre-dried and pre-heated in an additional dryer. These parameters of batch mass and RAP content in RHMA position and variation were used when evaluating the accuracy and precision of the recycling process in AMP. The obtained data showed that when RHMA mixtures are produced in a modern batch-type AMP, RAP is batched accurately, but not precisely enough.

Keywords: asphalt pavement; recycling principles; asphalt mixing plant (AMP); reclaimed asphalt pavement (RAP); batching errors

---

1. Introduction

Production of asphalt can incorporate reclaimed materials from the deconstruction of road surfaces. This aids in reducing production costs and saves natural resources, bitumen, and aggregates. However, recycling can only be justified if the performance and longevity of the produced pavement is equal or better than that of traditional mixtures. Mixtures containing RAP have not always demonstrated such performance. Premature cracking due to the aged bitumen is one of the main reasons that agencies are reluctant to increase allowed RAP content in the mixtures [1,2]. Another reported problem is the variability of RAP material, which does not allow to securely assume that the produced mixture properties will reflect laboratory design [3]. Finally, problems with production technology have been reported, including excessive emissions and issues with consistency of mixture that has resulted in poor pavement performance [4,5].

These reported problems, however, can be accounted for and the desired performance for mixes containing RAP can be ensured by taking comprehensive measures in three broad areas:

- Materials. It must be ensured that the reclaimed materials correspond to the requirements, they are homogeneous, with a low moisture content, and without contamination.
- Mix design should be performed according to best practices, depending on recycled asphalt content and its properties to ensure the reduction of stiffness of the aged binder, as well as the desired performance properties for the entire life-cycle of the pavement [6].
- The production technology must allow for incorporating the desired RAP content by heating it to the necessary temperature and ensuring homogeneous blending with virgin materials without creating emissions [7].

Most laboratory studies have demonstrated evidence that high RAP pavements can achieve the desired pavement performance and longevity that are equal to conventional asphalt [8–10]. Adequate mix design procedures can even allow for creating mixtures containing 100% recycled asphalt [11,12]. It has also been confirmed that adequate full-scale RAP management, production, and mix design procedures can lead to excellent field results. The research by [13,14] provides evidence that adequate management and processing of RAP can ensure homogeneity of RAP equal or better than that of virgin aggregates. A comprehensive study by National Center for Asphalt Technology (NCAT) [15] evaluated the long-term performance of 18 sites across North America with pavements containing 30% RAP and concluded that the performance is similar to pavement constructed from virgin materials with no RAP. Thus, it becomes evident that both adequate design procedures and full-scale production technologies are currently available and allow for high RAP production [16]. So, why is it not being done routinely?

One of the reasons is that, although a lot of research is being done in laboratory, independent scientific research on production technologies and other full-scale operations and the means to improve them is scarce. This is likely because of the high research costs and many available (and constantly changing) technologies. At the same time, as reported by [17], a large portion of the failures of mixtures containing RAP have been caused by the "use of unprocessed RAP and hot-mix plants that were not designed to handle high RAP contents." Thus, a clear, full-scale research is necessary. The pavement performance is equally dependent on design procedures and consistent production and only adequate accounting for both of these factors can ensure satisfactory pavement performance [18].

This article aims to link different features of a full-scale production to allow for a theoretical evaluation of the entire production process chain. Only a comprehensive approach like this can permit analysing the "bottle necks", which preclude increasing the reclaimed asphalt content in the mixture or to find the defective link in the chain in the case of unsatisfactory performance of the pavement. Such an approach is also necessary when aiming at a cleaner production. The entire chain of operations must be analysed considering the long term consequences. This encompasses undertaking daily, weekly, and yearly analyses. This article will present the theoretical basis of the asphalt production chain and then focus on the asphalt production with regards to ensuring the desired mixture homogeneity.

The aim of the study is to present the analysis of mandatory actions of asphalt pavement hot recycling based on scientific principles and to investigate the parameters of technological operations, performed in batch AMP with additional RAP drying and heating.

2. Theoretical Formulation of the Problem

2.1. Principles of Asphalt Pavement Hot Recycling

When developing asphalt pavement recycling methods and technologies, the most important principles of this relatively new and promising method of recycling road pavement by reusing its asphalt materials shall be known and applied. We present these principles based on our personal theoretical knowledge and practical experience, as well the analysis of the findings of the investigations conducted by other researchers. The scheme of twelve principles is presented in Figure 1.

THE PRINCIPLES OF ASPHALT PAVEMENT HOT RECYCLING

1. Necessity to break up an asphalt pavement course into lumps or slabs
2. Necessity to separate components of the composite physically
3. Decomposing of mechanically or thermo-mechanically impacted RAP through matrix, i.e. a bitumen film and air voids
4. Not damaging the mineral components separable from each other and a bitumen binder
5. Water content removal from the recycled material of asphalt pavement
6. Necessity to blend mechanically: free or/and forced movement of separate components in space
7. Formation of the largest possible surface of aged bitumen films to be coated with a rejuvenator
8. Capability of a rejuvenator to coat all particles uniformly and to diffuse into aged bitumen films
9. Retardation of bitumen oxidized ageing inthe technological process of recycling
10. Failure to remove aged bitumen and/or some of the aggregate fractions from the recycled asphalt mixture
11. Compliance of the designed recycled hot mix asphalt (RHMA) quality indicators with the design specification requirements
12. Compliance of *in-plant* or *in-place* (on the road) recycled asphalt mixture composition and properties with the job-mix formula (JMF)

Figure 1. The system of scientific principles of asphalt pavement hot recycling.

The first principle (necessity to break up an asphalt pavement course into lumps, or slabs). The asphalt of the reclaimed pavement containing the matrix and inserts of the composite shall be mechanically, thermally, or hydrothermally crushed (broken-up) into separate components. This enables to improve its whole structure and properties by incorporating some materials, not only those in the surface course of the road pavement [19,20]. It is rather complicated and ineffective to incorporate rejuvenating additives into a compacted and dense pavement through air voids from the top. A ravelled aged asphalt course of the road pavement, which is not crushed into small particles, would not be uniformly recycledthroughout the whole thickness of the course.

The second principle (necessity to separate physically the components of the composite). The asphalt of the recycled pavement shall be downsized to its granular particles (RAP). Aggregate particles of different sizes (solid phase) are separated by breaking-up the films of the bitumen bonding them (liquid phase). As adhesion is stronger than cohesion, thermo-mechanically impacted RAP frequently decompose not only at the border of bitumen and particles, but in the course of volumetric bitumen and air voids. Particles are separated from each other in order that when they are changing their position in space, they allow rejuvenating agents (binder or/and mineral aggregates) to permeate through open gaps, which would distribute uniformly among the obtained particles of the recycled asphalt mixture [21].

The third principle (decomposing of mechanically or thermo-mechanically impacted RAP through matrix, i.e., a bitumen film and air voids). The RAP of aged pavement shall be broken up in order that the material that binds asphalt (the variation system of a bitumen binder and mineral fillers or volumetric bitumen films, which cement the composite material into a conglomerate) decomposed. During its recycling process, surfaces may merge with microcracks occurring in the aged road pavement and air voids. It would be desirable that the surfaces were formed in a matrix, not in inserts (coarse and fine aggregate), when asphalt decomposes (is milled, broken-up or crushed). In the contact zone of a milling machine cutter or a stone breaker (crusher) with reclaimed asphalt, new uncoated surfaces form when stone and fine aggregate particles split, which increases the area of all particles of the mixture and the necessary summative amount of the recycled bitumen. When asphalt pavement is cold-milled or its broken lumps or slabs are crushed in a stonebreaker, some coarse and fine aggregate particles inevitably split up (break up). When heated road pavement is milled with a mobile recycler, the probability of particle splitting decreases. If the degree of aged mineral aggregate fractioning is unknown and it is not taken into consideration when batching additional virgin mineral aggregates or an asphalt mixture, RHMA with a higher content of fine particles is obtained.

The fourth principle (not damaging the mineral components separable from each other and a bitumen binder). The methods and technologies of separating solidified and liquefied phases of asphalt pavement from each other shall ensure that aged bitumen and mineral particles are suitable for the use in a recycled asphalt mixture. Thermal weakening of the matrix (bitumen or asphalt binding material) by reducing the viscosity of aged bitumen before decomposing to the strength that is lower than that of inserts (particles of coarse and fine aggregate) decomposes reclaimed asphalt through matrix, not inserts, due to decreasing adhesion and cohesion. Splitting and crumbling of aggregate particles shall be avoided. When reclaimed asphalt is heated, the properties of aged bitumen should not worsen due to overheating. Usually, RAP that has been heated in the RAP channel goes to a hot RAP storage bin when it has reached its desired temperature (100–120 °C) at the end of circulation line. When it is stored in the bin, it will maintain a warm temperature. Chemical solvents that are used for bitumen recovery, but not for the improvement of its properties, are not suitable for separating the components of RAP from each other.

The fifth principle (water content removal from the reclaimed material of asphalt pavement). The RAP that was milled or crushed and sieved in a stonebreaker almost always contains a certain amount of water. It is likely to become a wet hot-in-place recycled asphalt pavement. When stockpiled, precipitation water increases the water content of this material. Therefore, it is recommended to store RAP that is prepared for use in a stockhouse or in a shelter. This water shall be removed by applying the laws of heat transfer in some equipment [22,23]. Wet and cold RAP, usually sized up to 32 mm, can be dried and heated up in an AMP in the following two ways: first, cold RAP is mixed with superheated virgin mineral materials, takes over part of their heat, dries and heats up; second, RAP is heated up in a parallel additional dryer (in-dryer) using the heat that is radiated by the burner. When the road pavement is recycled with a mobile recycling equipment (hot-in-place recycling), it is heated up by infrared burners from the top [24]. When water films are removed from the surface of RAP particles and microcracks, the adhesion of the recycled summative (aged and virgin bitumen) bituminous binder and mineral aggregates increases, which influences the strength of the recycled asphalt.

The sixth principle (Necessity to blend mechanically: free or/and forced movement of separated components in space). All or most of the particles of different sizes of the reclaimed asphalt shall be separated from each other so that when mixed, they could change their position in relation to each other and their location in a mix or a flow. When during initial decomposition road pavement structure courses are cold-milled or flat asphalt lumps or their slabs (pieces) are crushed (pulverized), and particles are obtained. When granular asphalt is heated up during the second decomposition, viscosity, adhesion, and cohesion of the old bitumen are reduced. Therefore, when exposed to mixing forces weaker than those of milling, particles partially break up and mineral particles redistribute in space, thereby opening new unbound surfaces that could be exposed to rejuvenators (additives). When the heated-up asphalt pavement is milled or broken, and when the films of the reclaimed old liquefied bitumen binder melt sufficiently, adhesion (bonding of bitumen molecules with mineral particles) and cohesion (interaction of bitumen molecules) are reduced. This makes the separation of particles from each other easier and takes less mechanical power, but requires more heat.

The seventh principle (formation of the largest possible surface of aged bitumen films to be coated with a rejuvenator). Particles of granular mixture obtained from the heated-up, and the mixed RAP is coated with extremely viscous oriented inner bitumen films of a certain thickness with a higher content of asphaltenes [25]. Due to adsorption into open voids of particles and microcracks (splits), and due to stronger adhesion, they barely melt and decompose. These particles are also coated with outer bitumen films, which weakly interact with the surface of particles as they do not directly contact with them. Volumetric bitumen films make up an outer layer, in which the content of liquid tars and oils reduces during the road use. At the temperature lower than that of the oriented bitumen they melt and become a viscous liquid that can flow in separate drops and at the molecular level mix with the rejuvenators recycling the old bitumen or virgin more liquefied bitumen [26,27]. Volumetric bitumen films oil the surfaces of mobile particles and facilitate the mixing process of the recycled asphalt mixture. When mixing the recycled asphalt mixture, more and more new surfaces of aged particles, which can absorb the batched rejuvenators or a virgin liquefied bitumen, open [28,29]. It takes more energy for finer particles to separate from each other. If they are not separated, rejuvenators, which improve aged bitumen, cannot pass and part of the mixture remains unrecycled.

The eighth principle (capability of a rejuvenator to coat all particles uniformly and to diffuse into aged bitumen films). The rejuvenators improving the properties of the aged bitumen in the recycled asphalt mixture shall uniformly distribute on the surface of each mineral particle, coated with oriented and volumetric bitumen films [7,30]. The thickness of a newly formed film of the rejuvenator shall ensure that upon completion of its diffusion into the double coating of aged bitumen, the properties of its summative bitumen are restored to the appropriate structure and properties of the recycled asphalt mixture. The deeper that a rejuvenator or a virgin bitumen penetrates into the double-coating system of reclaimed bitumen films due to diffusion after the technological process of recycling, the more complete this process is and the less asphalt properties change in the beginning of its exploitation [31–33]. The long duration of the diffusion of the rejuvenator into aged bitumen does not allow the consideration of the process of asphalt recycling as complete [34]. The duration of diffusive mixing of reclaimed old and virgin bitumen in films should be as short as possible, as only upon completion of this process the new structure of the recycled asphalt structure forms and its required (rated) resilience is obtained. Classically approach, Fick's law and Arrhenius equation describes this theory [33].

The ninth principle (retardation of bitumen oxidized ageing in the technological process of recycling). When the aged structure decomposes at high temperature of asphalt recycling process, unbound surfaces of aged volumetric bitumen contact with ambient air. The components of aged bitumen intensively oxidize and vapour even more due to ageing, thereby worsening the properties of summative bitumen [35,36]. Therefore, the temperature of the recycled asphalt mixture should not be too high and the duration of works should not be too long. When stored in a silo of an AMP and transported in a truck for a considerable period of time, laid, and compacted slowly, the structure and properties of the hot-recycled asphalt mixture deteriorate.

The tenth principle (failure to remove aged bitumen and/or some of the aggregate's fractions from the recycled asphalt mixture). The gradation of aggregates in an asphalt mixture complies with a certain optimal content of bituminous binder, which ensures its best properties [37]. When reclaimed asphalt is being recycled, the properties of aged bitumen are improved by admixing a rejuvenator or virgin bitumen, which always increases the content of summative bitumen and makes it not optimal. In order to optimize the content of summative bitumen, aggregates of appropriate fractions shall be added [38]. It is practically impossible to remove aged bitumen and (or) some aggregate fraction from recycled asphalt. It is possible only to add rejuvenators. Virgin mineral aggregates, which improve the gradation absorb part of the aged volumetric bitumen, transferring from the surface of the granular particles of reclaimed asphalt (during hot-in-plant recycling). When asphalt pavement is hot-in-place recycled with a mobile Remixer, the particles of additional HMA mixture are coated with virgin bitumen and transfer part of it to aged particles. It is likely that during the mass exchange process, aged bitumen is transferred to new particles of an additional HMA mixture. This way, a reversible double coating of new particles is obtained: exogenous coating of aged bitumen and indigenous coating of new bitumen films. These two different hot recycling technologies impact on the different formation of the structure of double-coating of a bitumen film, which coats aged and virgin particles, the thickness of the coatings, diffusion processes, and the properties of the recycled asphalt.

The eleventh principle (compliance of the designed RHMA quality indicators with the design specification requirements). RHMA is designed using the RAP, virgin mineral aggregates, virgin bitumen, or any other rejuvenator, when their factual properties and composition are identified and evaluated. The proportion of virgin bituminous binding materials and aggregates is selected taking into account the composition of RAP, its homogeneity and the properties of aged bitumen [39–41]. The aim is to obtain such physical and mechanical properties of RHMA, which are not worse than those of the mixtures made of purely virgin materials (for mixtures of the mix group) and the type of bituminous mixture and which comply with the specifications set out in European Standart (norm) EN 13108-1, ..., EN 13108-7 and EN 13108-20. The lower the homogeneity of RAP is, the less amount of it can be batched in the manufactured recycled HMA mixture [42]. RAP homogeneity is evaluated according to EN 13108-20 and TRA ASFALTAS 08.

The twelfth principle (compliance of in-plant or in-place recycled asphalt mixture composition and properties with the JMF). The recycled hot bituminous mixture manufactured from RAP and virgin materials in an AMP or in a mobile Remixer shall meet Factory Production Control requirements, as specified in EN 13108-21. Therefore, the errors of technological parameters are minimized during the recycling process, which enables to deviate the factual composition of the recycled HMA mixture, temperature, physical, and mechanical indicators from the JMF within permitted (tolerance) limits [43].

2.2. Model of Technologies of RAP Recycling

Such asphalt pavement recycling principles are entirely or partially applied in hot-mix recycling process of RAP in an AMP. The recycling process may be divided into sequential technological operations (Figure 2).

In conventional AMP constructed to recycle aged RAP, additional technological equipment to feed, store, supply (transport), batch, dry, heat up, and mix RAP with virgin materials be used (Figure 3). Recently, the increasing content of RAP has been observed. The structure of the equipment designed to recycle RAP depends on the maximum permitted RAP percentage content, the methods of drying and heating up, the place of batching in AMP, as well as mixing with virgin materials. When a small percentage (up to 10% or 20% of RAP content from the total mass of recycled asphalt mixture) is used, a batch-type AMP without additional heating of RAP is the most suitable one (Figure 3D). AMP of this type enables the delivery RAP into its four types of equipment: 1—hot elevator; 2—hot aggregate bin (compartment of bypass); 3—hot aggregate scale bin; and, 4—mixer. To batch the increased RAP percentage content (from 20% to 50%), it is dried and heated up in an additional cylinder dryer, incorporated into a batch-type AMP (Figure 3C).

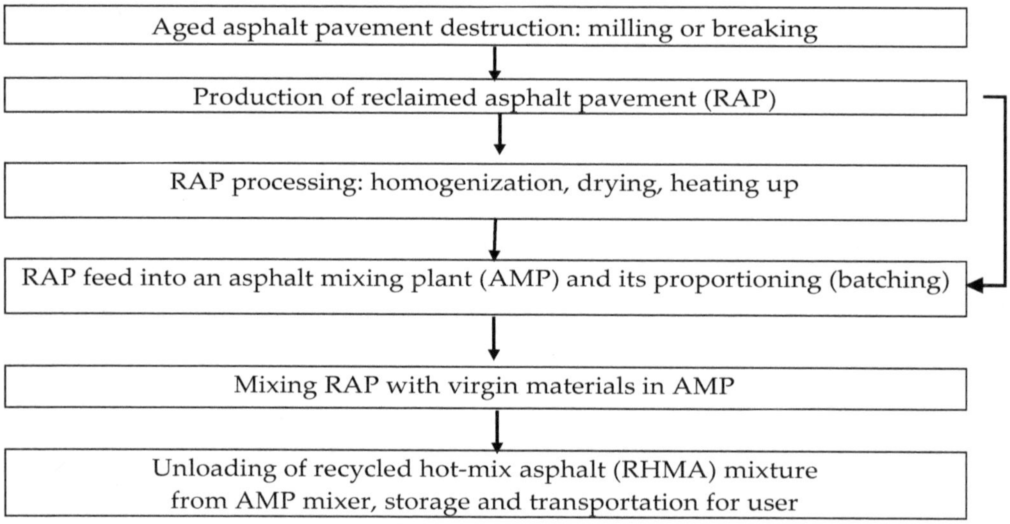

Figure 2. Production operations of the in-plant recycling process of aged asphalt pavement.

RAP can also be recycled in a continuous AMP with integrated drum-mix (Figure 3A) or double barrel dryers-mixers (Figure 3B). When using this equipment, 50% or more of wet and cold particles of RAP could be batched. All of the asphalt recycling technologies presented in Figure 3 have both advantages and disadvantages.

Recent theoretical and practical achievements enable the recycling of up to 100% of RAP, of course, we need to evaluate that the studies were conducted in ideal and lab-controlled conditions [44,45]. RAP inhomogeneity has a huge impact on its maximum permitted percentage in a recycled RHMA. It is only in rare cases that RAP is sieved into two or more different size fractions. A lack of an efficient practical method, which enables to homogenize RAP prepared for use, does not allow the use of high RAP content in a recycled hot-mix asphalt (RHMA) mixture. As a rule, according to the standards it is allowed use less RAP content in a recycled asphalt mixture of the pavement surface course than in a binder or a base course [42].

When RHMA mixture is produced in an AMP, cold and wet RAP is dried and heated up to break them down into separate component particles, which could absorb thin films of a rejuvenator on their surface. RAP may dry out and heat up in two technological ways of heat transfer: first, (Figure 3A,B,D technologies) without preheating the reclaimed material [46], when RAP take over some of the heat of aggregates (superheated virgin materials, superheating temperature) in contact with superheated aggregates; second, (Figure 3C technology) the flow of RAP is dried and heated up (up to 100–110 °C) in a separate parallel dryer, where a mixture of heated up gas and air is emanated from a burner [47].

The most important technological requirements for the production of RHMA mixtures are as follows:

- the conformity of its gradation (content of mineral filler, fine aggregate and coarse aggregate and bituminous binder content with JMF; and,
- the conformity of the RHMA mixture temperature to the type of used bituminous binder grade.

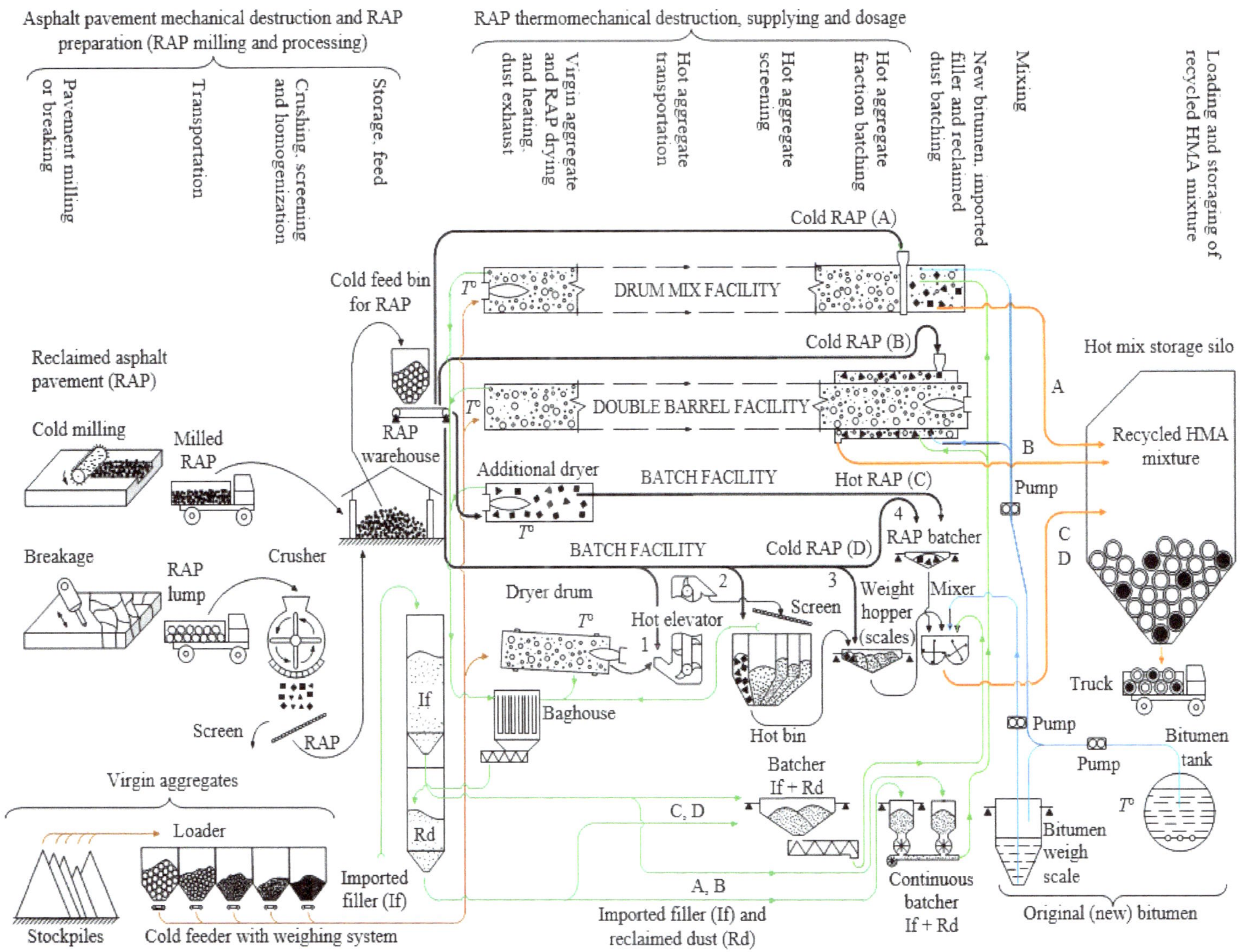

Figure 3. The model of asphalt pavement recycling operations and equipment used in batch plant (**C**,**D**), drum mix (**A**) and a double barrel-drum mix (**B**) when reclaimed asphalt pavement (RAP) is supplied and dosed to different places.

Technical requirements TRA ASFALTAS 08 drafted in compliance with the above-mentioned standards requires that the max allowed content of RAP K_i should be calculated according to Equations (1) and (2) taking into account the homogeneity of RAP. Such homogeneity is determined by the range indicators for individual properties (a_i). K_i shall be computed by the general permissible deviations $N_{adm,i}$ depending on the property of RAP and the type of recycled HMA mixture:

$$K_i = \frac{0.50N_{adm,i}}{a_i} \times 100 \quad (1)$$

or

$$K_i = \frac{0.33N_{adm,i}}{a_i} \times 100, \quad (2)$$

where a_i is the range between the max $x_{i,max}$ and the min $x_{i,min}$ values of RAP i-property indicator estimated for the sample (min sample size $n = 5$).

With respect to all of the properties of base and surface-base course mixtures, Equation (1) shall apply. It shall be also applied for the softening point of binder course and surface course mixtures. With respect to all other properties, Equation (2) shall apply [42].

RAP is not only homogeneous, but it is also dosed with systemic and random errors of a certain magnitude in the AMP that increase the deviations of the RHMA component content from JMF and variation [38].

Derivation of the procedure for calculation of the maximum allowable RAP content is as follows. ASTM D 4460 gives equations for calculating standard deviation values for quantities determined from calculations involving two other values. From these equations, the following formula for calculating the standard deviation of a blend of two materials can be derived:

$$\sigma_m = \sqrt{\alpha^2\sigma_a^2 + (1-\alpha)^2\sigma_b^2 + \left(\overline{X}_a^2 + \overline{X}_b^2\right)\sigma_\alpha^2}, \quad (3)$$

where σ_m—standard deviation of the mixture; σ_a—standard deviation of component "a"; σ_b—standard deviation of component "b"; α—proportion of component "a" in the mixture; $\overline{X}_a^2$— mean value for component "a"; $\overline{X}_b^2$—mean value for component "b"; and, σ_α—standard deviation of the proportions.

We can rewrite this for percent passing for a selected sieve for HMA mixtures consisting of a blend of new HMA materials with RAP [48]:

$$\sigma_{PM} = \sqrt{w_R^2\sigma_{PR}^2 + w_N^2\sigma_{PN}^2 + \left(\overline{P}_R^2 + \overline{P}_N^2\right)\sigma_W^2}, \quad (4)$$

where σ_{PM}—standard deviation of percent passing for a selected sieve for the mixture with RAP (i.e., RHMA); w_R—weight fraction of RAP in the mixture; σ_{PR}—standard deviation of percent passing for the selected sieve for the RAP; w_N—weight fraction of new materials (new HMA) in the mixture $= (1 - w_R)$; σ_{PN}—standard deviation of percent passing for the selected sieve for the new HMA; $\overline{P}_R$—mean value for RAP% passing for the selected sieve; $\overline{P}_N$—mean value for new HMA% passing for the selected sieve; and, σ_w—standard deviation of the weight fractions, also called "batching variability".

Equation (3) can be solved for the maximum amount of RAP that can be added to new material without increasing the standard deviation for percent passing on the selected sieve above a selected maximum value by application of the quadratic equation:

$$Max.RAP\% = \frac{-b + \sqrt{b^2 - 4ac}}{2a}100\%, \quad (5)$$

where $Max.RAP\%$—maximum amount of RAP that can be added to the mix, weight%; $a = \sigma_{PR}^2 + \sigma_{PN}^2$; $b = -2\sigma_{PN}^2$; $c = \sigma_{PN}^2 + \left(\overline{P_{PN}^2} + P_N^2\right)\sigma_w^2 - \sigma_{PM/Max}^2$; and, $\sigma_{PM/Max}^2$ is the maximum allowable standard deviation for percent passing for the selected sieve.

Because asphalt is temperature sensitive, good preheating of RAP is helpful to eliminate the agglomerates of RAP to promote the dispersion of RAP during the hot recycling process [49]. When applying the first heat transfer method, superheating temperature of virgin mineral materials (aggregate) is calculated according to the following empirical formula applied in practice [50]:

$$T_{v.a.} = \frac{T_{RHMA} - S \cdot T_{RAP}}{1 - S} + \frac{4 \cdot S \cdot W_{RAP}}{1 - S}(637 - T_{RAP}), \quad (6)$$

where $T_{v.a.}$—superheating temperature of virgin mineral materials (aggregates), °C; T_{RHMA}—required temperature of recycled hot mix asphalt, which depends on the grade of the bitumen binder, °C; S—ratio of RAP batching mass in the recycled hot mix asphalt mixture, in units; T_{RAP}—temperature of RAP, °C; and, W_{RAP}—moisture (water content) of RAP, in units.

This formula (1) contains four variables, two of which (T_{RHMA}, S) show the temperature of RAP and component composition, whereas other two (T_{RAP}, W_{RAP})—show RAP temperature and moisture. Experimental investigations showed that RAP may absorb from 0.7 to 6.5% of moisture. The average made up 3.40%. Therefore, it is recommended to store it in warehouses. It is likely that not all humidity is removed from RAP in a parallel drum. The quantity and temperature of a batched.

RAP, which depends on its transfer type impacted by the production technology of RHMA in AMP, allows to calculate the temperature at which virgin aggregates shall be heated. According to formula (6), it was identified how the temperature required for heating virgin aggregates varies when the moisture of RAP varies from 0 to 8%, and its quantity in RHMA varies from 5 to 50% (Figure 4). When making calculations, it was assumed that the initial temperature of RAP is equal to ambient temperature of 15 °C (Figure 4a). RAP was heated in a parallel drum to the working temperature of 100°C (Figure 4b), and it also contained some residual temperature.

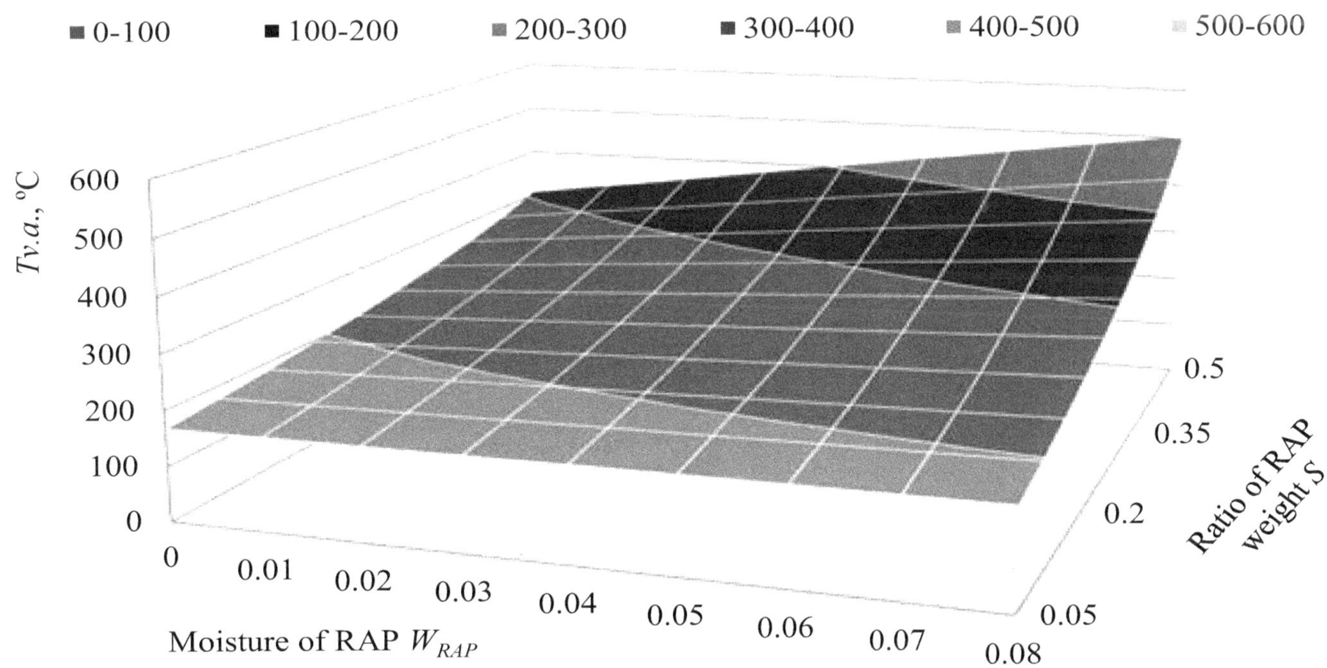

(a)

Figure 4. *Cont.*

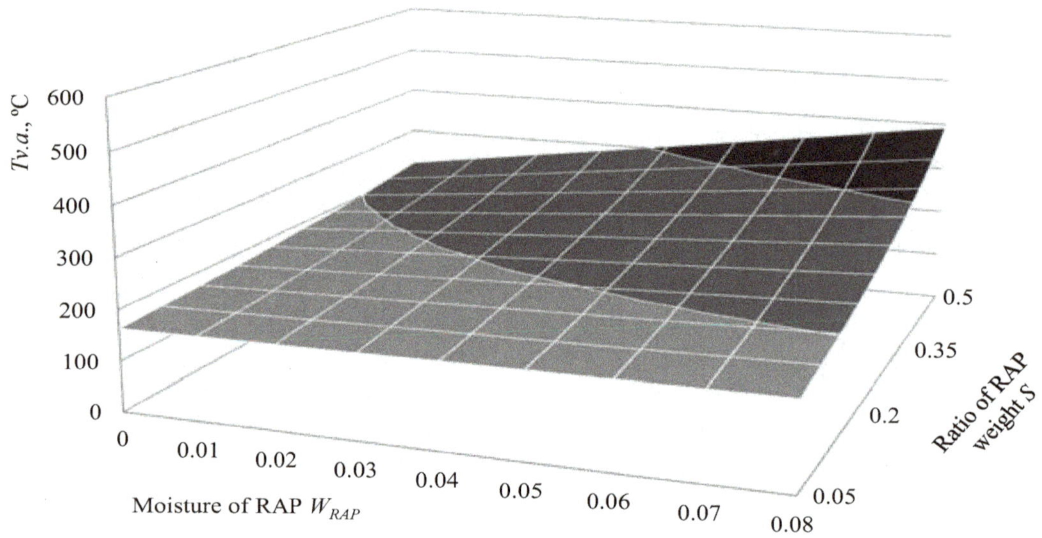

(**b**)

Figure 4. Dependence of superheating temperature of virgin mineral materials on the ratio of RAP weight and moisture of RAP when T_{RAP} is as follows: (**a**)—15 °C; (**b**)—100 °C.

It was estimated that if RAP content in RHMA mixture is 50% and RAP moisture is 2%, then the temperature required to heat virgin aggregates when producing 160 °C RHMA with RAP heated in a parallel drum is 262 °C, and with non-heated RAP-even 354 °C. The difference made up 92 °C.

3. RAP Batching Accuracy and Precision

RHMA mixture of different RAP content was produced in one asphalt producing company in a batch-type asphalt mixing plant (AMP), Ammann Euro A 240, the highest capability of which was Q_{max} = 240 t/h, and the highest mass of one mix batch was 3000 kg. RAP feeding, transporting, drying, and heating, as well as batching devices were mounted in addition (Figure 5).

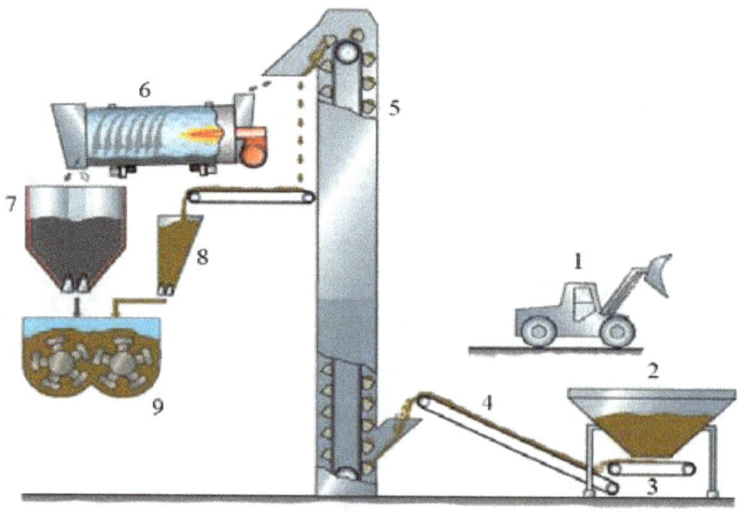

Figure 5. The technological scheme of RAP hot and cold recycling in batch AMP: 1—wheel loader; 2—RAP feeder; 3—belt scale; 4—collecting conveyor; 5—RAP elevator; 6—parallel drum; 7—recycling buffer silo with weighing appliance and recycling scale; 8—cold recycling addition via buffer silo and belt scale; 9—mixer.

3.1. Methods

The following mixture for a road pavement surface and base courses of two types was produced in a batchAMP with preheated RAP: with 25% and 50% of RAP, AC 16 PD and AC 22 PS, respectively. The grade of AC 16 DP according to the national standard TRA ASFALTAS 09 is asphalt concrete (AC); maximum size of particles—16mm; purpose—for pavement single ply surface course (PD). The grade of AC 22 PS is also asphalt concrete (AC); maximum size of particles—22 mm; purpose—for pavement base course (P); also, purpose—for the heavy load (S). To investigate the accuracy and precision of RAP batching, technological data about RHMA mixtures produced in 2014, which were stored in an APM handling computer, were used. AMP handling software presents rated (target) and factual data on the name and number of job-mix formula (JMF), weight, mixing time, and temperature of each RHMA mix batch. Masses of all mineral materials, bitumen, additives, and RAP weighed batches are presented as well.

For statistical purposes, only the data when most of RHMA mixture was produced were used: 10-day AC 16 PD mixture production data and 12-day AC 22 PS grade hot-mix asphalt (HMA) mixture production data. Quite large sample sizes, presented in Table 1, allowed the study to obtain rather dependable investigation data. RHMA mix batches of less than 2800 kg (outliers) were eliminated from each sample size. Statistical investigation was conducted on the data of 3041 mix batches, for the production of which approximately 3400 t of RAP was used and approximately 9000 t of RHMA mixture were produced.

Table 1. Sample sizes of the data used in the investigation

Type of RHMA	Days	Total Sample Size	Sample Size Without Outliers
AC 16 PD	10	1424	1324
AC 22 PS	12	1617	1569

Statistical data were calculated through the use of Statgraphics Centurion XVII software package (Statpoint Technologies, Inc., Warrenton, VA, USA) for statistical processing, which allowed to identify the accuracy and precision of RAP batching in a batch-type AMP. AMP handling computer data of each working day's lot were grouped and classified by days and grades of RHMA mixture. Total weight Q_{batchi} of RHMA mix batch and weighed RAP batch mass q_{RAPi} were calculated. RAP percentage content in each asphalt mix batch was calculated according to this data:

$$x_{RAPi} = \frac{q_{RAPi}}{Q_{batchi}} \times 100 \tag{7}$$

where x_{RAPi}—RAP content in i RHMA mix batch, %; q_{RAPi}—RAP batch mass, kg, Q_{batchi}—RHMA mix batch mass, kg; i—i RHMA mix batch (i = 1, 2, . . . N, N—RHMA mix batches' number).

Statistical indices of the produced RHMA mixture of both grades RAP batch mass position and variation were calculated, and diagrams and histograms were drawn (Figures 6 and 7).

The accuracy of RAP batching was identified by comparing average RAP content deviation from JMF in the produced RHMA mixture:

$$\Delta q_{RAP} = \frac{\bar{q}_{RAP} - q_{RAP,JMF}}{q_{RAP,JMF}} \times 100, \tag{8}$$

where $\bar{q}_{RAP}$—average of RAP batches mass, kg; $q_{RAP,JMF}$—target RAP batch mass according to JMF, kg.

Standard deviation was used to evaluate the precision of RAP batching (Figure 7).

Normal distribution of both sample sizes of RAP batch mass was tested, which enabled the calculation of the average precision of RAP batching process. Normal distribution of experimental data was investigated by applying skewness and kurtosis. The values of the coefficient of skewness *sk* and coefficient of kurtosis *ku* calculated from data of sample size were compared with their respective standard deviations s_{sk} and s_{ku}:

$$s_{sk} = \sqrt{\frac{6n(n-1)}{(n-2)(n+1)(n+3)}}, \tag{9}$$

$$s_{ku} = \sqrt{\frac{24n(n-1)^2}{(n-3)(n-2)(n+3)(n+5)}}, \tag{10}$$

where n—the sample size (number of measurements). When $|sk| < 3s_{sk}$ and $|ku| < 5s_{ku}$, it can be considered that the normality hypothesis of empirical data is accepted. Otherwise, the raised hypothesis is rejected or accepted as doubtful.

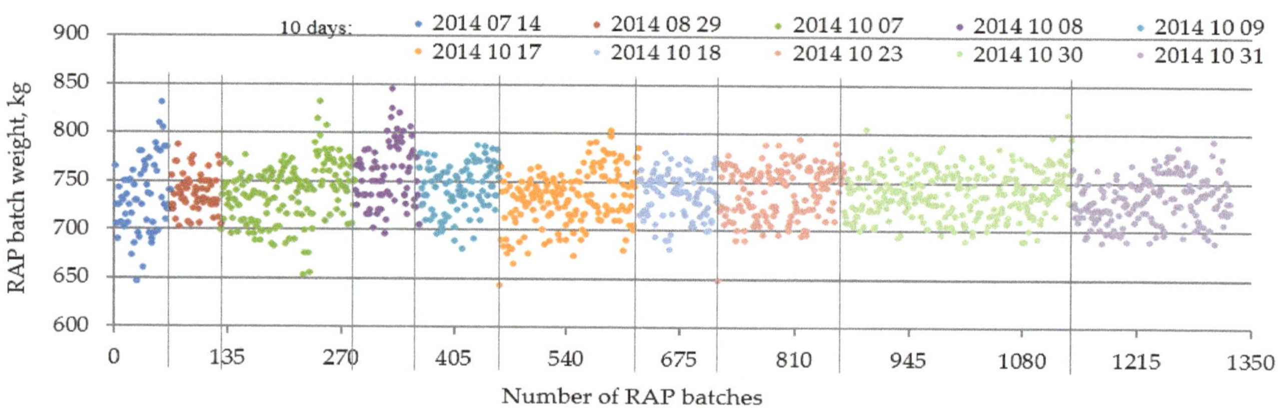

(a)

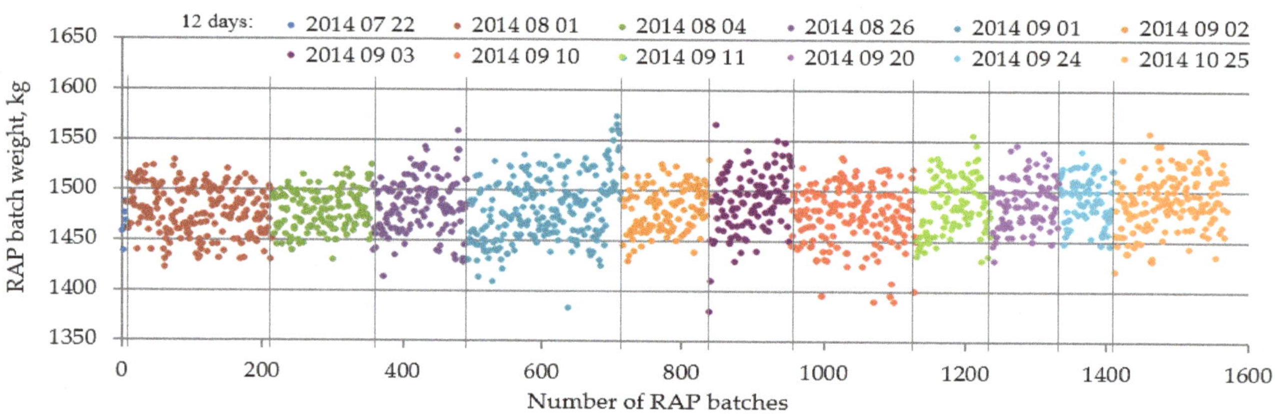

(b)

Figure 6. Distribution of RAP separate batch masses. When its job mix formula (JMF) is as follows: (**a**)—750 kg; (**b**)—1500 kg.

3.2. Materials

RHMA mixtures were produced according to various JMF's. To produce AC 16 PD grade mixture, imported fillers (IF), crushed gravel sifting, reclaimed dust (RD), natural sand, various fraction crushed gravel, and 100/150 grade road bitumen were used. AC 22 PS grade RHMA mixture was produced from 0/2 mm and 0/5 mm fraction natural sand, 5/8 mm, 8/11 mm and 11/22 mm fraction gravel, and 50/70 grade road bitumen. Additives were not used for RHMA mixture production. The batch masses of all materials' finite batching in one recycled HMA mix batch, the mass of which according to JMF is 3000 kg, are presented in Table 2.

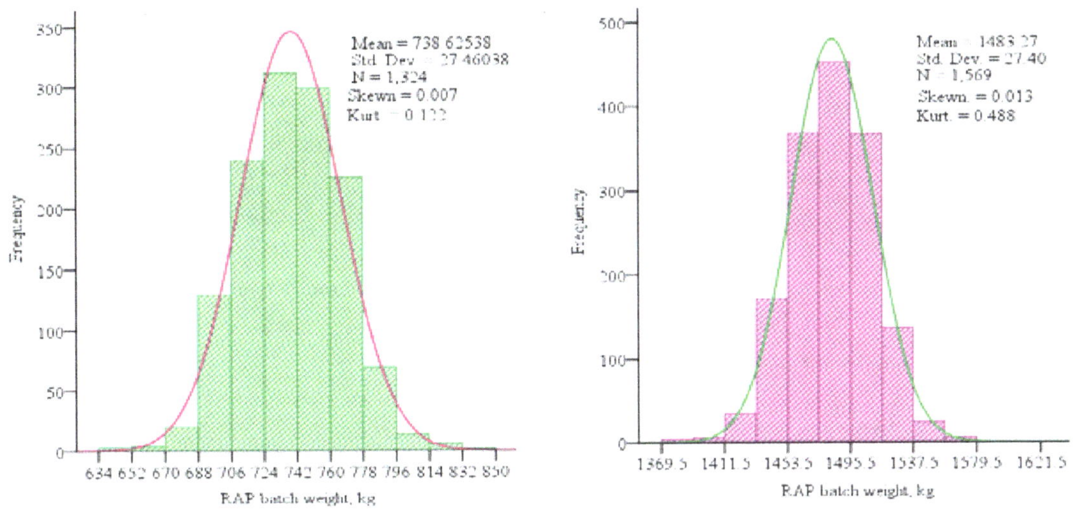

(a) (b)

Figure 7. Distribution histograms, theoretical curves of normal distribution and statistical parameters of RAP batch masses in HMA batches of mixtures: (**a**)—AC 16 PD (750 kg); (**b**)—AC 22 PS (1500 kg).

Table 2. RAP composition, %.

Particle Size, mm	Subsample Number					Average
	I	II	III	IV	V	
>11.2	6.3	7.5	8.4	8.7	7.5	7.7
>2	69.4	73.4	71.2	68.5	68.7	70.6
0.063–2	30.4	26.5	28.6	31.3	31.2	29.2
<0.125	0.5	0.3	0.2	0.3	0.3	0.3
<0.063	0.2	0.1	0.2	0.2	0.1	0.2
Bitumen	4.78	4.73	4.86	4.81	4.43	4.72

To produce RHMA mixture, crushed and sieved RAP, the size of which was not larger than 32 mm, were used. RAP was prepared according to the third principle and technological scheme (Figure 3). When designing the composition of RHMA mixture and determining JMF, RAP grading and hot fraction were evaluated. Several sample units were taken from the adjacent places of the sheltered stock of RAP and prepared for use. The material of several sample units was put into a subsample. In total, five subsamples were taken this way, each of which was tested separately in a laboratory. RAP investigation data are presented in Table 3.

Table 3. RHMA mixture batch composition according to JMF (finite batching materials' batch mass in kg in one RHMA mix batch of 3000 kg).

RHMA Mixture Group	RAP	IF	RD	Hot Aggregate Fractions, mm						Virgin Bitumen
				0/2	2/5	5/8	8/11	11/16	11/32	
AC 16 PD	750	99	99	612	219	411	384	306	-	120
AC 22 PS	1500	-	-	141	87	297	144	-	780	51

The mean of five subsamples was taken as RAP average composition. RAP homogeneity was estimated by standard deviation.

4. Results and Discussion

The aim of the study was to identify how accurately and precisely RAP is batched when producing an asphalt mixture. The experiment data show that when producing RHMA mixture AC 16 PD (with 25% RAP) for pavement single ply surface course, the batched quantity of RAP into an AMP mixer was on average $\overline{q}_{RAP} = 738.63$ kg. According to Equation (8), batching accuracy $\Delta q_{RAP} = \frac{738.63-750}{750} \times 100 = -1.52\%$. RAP batch mass varied from 643 kg to 845 kg. Range $R = q_{max} - q_{min} = 202$ kg (Figure 6a). Standard deviation showing batch mass variation was $s_{q_{RAP}} = 27.46$ kg and coefficient of variation was $V_q = 3.72\%$. The specifications do not regulate AMP batching tolerances of bitumen, hot and cold mineral materials. They are specified in ASTM Standard Specification for mixing plants for hot-mixed, hot-laid bituminous paving mixture. According to the Specifications above, the automatic proportioning system shall be capable of consistently delivering materials within the full range of a batch size of the following tolerances (Table 4).

Table 4. Tolerances of batch masses of AMP batchers' materials according to ASTM D 995-95b.

Materials	Total Batch Mass of a Paving Mix, %
Batch aggregate component	±1.5
Mineral filler	±0.5
Bituminous material	±0.1
Zero return (aggregate)	±0.5
Zero return (bituminous material)	±0.1

The electric circuits for the above delivery tolerances of each cut-off interlock shall be capable of providing the total span for the full allowable tolerance for maximum batch size. Tolerance controls shall be automatically or manually adjustable to provide spans that are suitable for less than full-size batches. The automatic controls and interlock cut-off circuits shall be capable of being consistently coordinated with the batching scale or meter within an accuracy of 0.2% of the nominal capacity of the said scale or meter throughout the full range of the batch sizes according to ASTM D995-95b.

Having investigated all 1569 asphalt mix batches of grade AC 22 PS RHMA mixture (with 50% RAP content) produced in 12 days for the base course, it was identified that RAP is batched rather accurately. When the mean of RAP of all the batches is $\overline{q}_{RAP} = 1483.27$ kg, and $q_{RAP,JMF} = 1500$ kg, then according to the mean RAP is batched rather accurately: $\Delta q_{RAP} = \frac{1483.27-1500}{1500} \times 100 = -1.11\%$ and error is less than 1.5%, i.e., lower than tolerances. The value of batching preciseness is standard deviation $s_{q_{RAP}} = 27.40$ kg, and variation coefficient $V_q = 1.85\%$.The maximum mass of a RAP batch is 1631 kg and the minimum mass is 1380 kg. Range $R = q_{max} - q_{min} = 251$ kg (Figure 6b).

Taking into consideration the specification that not only RAP batch mass mean is $\overline{q}_{RAP}$, but its weighed batch mass q_{RAP} differed from rated value $q_{RAP,JMF}$ does not exceed the permitted tolerances (±1.5%), when $q_{RAP,JMF} = 750$ kg, the amount of weighed RAP shall be not less than 738.75 kg and not more than 761.25 kg. When $q_{RAP,JMF} = 1500$ kg, the minimum permitted batch mass of RAP may be 1477.5 kg, and the maximum batch mass 1522.5 kg. In fact, the diagrams (Figure 7) above show that the mass of quite a lot of RAP separate batches does not meet tolerance specifications. Only 29% of them (392 out of 1324) are within the range of permitted tolerances, when the estimated $q_{RAP,JMF} = 750$ kg and 51% (803 out of 1569) when it was $q_{RAP,JMF} = 1500$ kg.

The normality of data distribution was verified according to skewness and kurtosis. Empirical coefficient of skewness of RAP batch masses sk were 0.007 and 0.013, and the coefficient of kurtosis ku were 0.122 and 0.488, respectively (Figure 7). They were compared to the critical values which depend only on sample size n. When RAP batch sample sizes n were 1324 and 1569, standard deviations of skewness and kurtosis were calculated as follows: $s_{sk} = 0.067$, $s_{sk} = 0.062$ and $s_{ku} = 0.134$, $s_{ku} = 0.123$ (AC 16PD and AC 22PS mixtures, respectively). The estimated s_{sk} values were multiplied by 3, which resulted in $3s_{sk} = 0.202$ and $3s_{sk} = 0.185$, respectively. Standard deviation of kurtosis s_{ku} was multiplied by 5, which resulted in $5s_{ku} = 0.672$ and $5s_{ku} = 0.617$, respectively.

When $|sk| < 3s_{sk}$ and $|ku| < 5s_{ku}$, it could be stated that the hypothesis of normality of the empirical data is confirmed. The normality of sample size data distribution can be verified using more complex methods, such as Kolmogorov and Pearson. Null hypothesis can be verified according to the Bartlett criterion, where the available data is close to normal distribution.

Thus, Bartlett's criterion was used to compare the variances of various sample sizes according to normal distribution. This criterion is convenient to compare two or more variances of the normal distribution sample sizes when these sample sizes are unequal. If statistics B is less than $\chi^2_{\alpha,\nu}$, RAP batch mass variation is statistically equal. Therefore, the average calculated value of sample variance $\bar{s}_q^2 = 752.52$ kg and general standard deviation $\bar{s}_q = 27.43$ kg. Statistics $B = 0.0094$ is less than critical $\chi^2_{\alpha,\nu} = 3.84$ with the selected $\alpha = 0.05$ significance level and degrees of freedom $\nu = l-1$ (where l—the number of comparable variances (samples). Therefore, it could be concluded that both (with 25% and 50% RAP content) batch mass variation (standard deviation $s_q = 27.46$ kg and $s_q = 27.40$ kg) are statistically equal.

The physical and mechanical properties of a recycled HMA mixture are influenced by RAP percentage content in each mix batch of RHMA mixture. This parameter allows to determine the proportion of RAP in the whole mix batch. When hot fractions, IF and RD batches are weighed in AMP, their mass like that of RAP has a certain variation, which influences the variation of RHMA mixture batch masses (Figure 8). Therefore, the mass of RHMA mixture batches always differs from JMF, i.e., from 3000 kg and is of stochastic type.

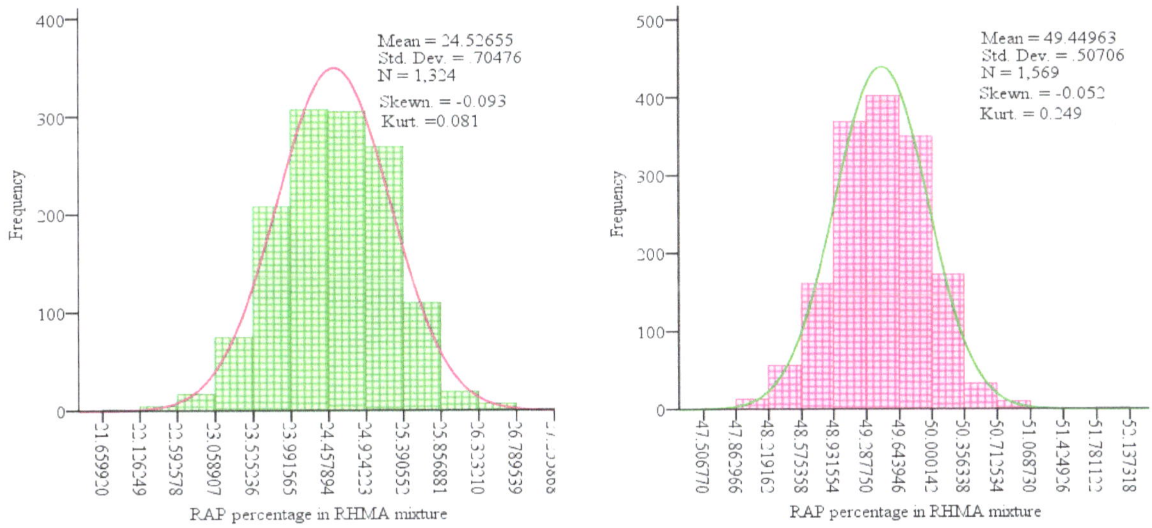

(**a**) (**b**)

Figure 8. Normal distribution curves and statistical indices of histogram of RAP percentage content in recycled hot mix asphalt (RHMA) mixture, when its content according to JMF had to be: (**a**)—25%; (**b**)—50%.

Average values $\bar{x}_{RAP}$ of RAP percentage Equation (7) of both sample sizes out of 1324 (with 25% RAP) and 1569 (with 50% RAP) were approximate to JMF (Figure 8). $\bar{x}_{RAP_{25}} = 24.53\%$, $\bar{x}_{RAP_{50}} = 49.45\%$. Standard deviation s_x of RAP percentage content was rather low: in mix batches with 25% of RAP content—0.705%; in mix batches with 50% of RAP content—0.507%, and their variation coefficients were $vc_{x25} = 2.87\%$ and $vc_{x50} = 1.03\%$, respectively.

Average values $\bar{x}_{RAP}$ of RAP percentage content Equation (7) of both sample sizes out of 1324 (with RAP percentage content in RHMA mix batches were approximate to those designed in JMF; however, histograms show that most of the data is grouped below the value of JMF (Figure 8). There was a tendency of lack of RAP content in RHMA mixture.

When asphalt pavement is hot-mix in-plant recycled, the sequence of technological operations depends on the structure of an AMP and additional equipment for feeding, transporting, and proportioning or batching of RAP. The model constructed by the authors contains two types of technological operations. First, hot-mix recycling in a batch plant, when RAP is dried and heated up due to the contact with superheated virgin mineral materials, i.e., conductive heat transfer occurs. Second, when RAP is dried and heated up in an additional parallel dryer, i.e., convective heat transfer occurs. These two recycling technologies are popular in the European Union countries. The model also contains hot-mix recycling in drum-mix and double-barrel plant technologies, when RAP dry out and heat up due to the heat exchange between them and superheated virgin mineral materials: conductive heat transfer occurs. These technologies are more popular in the USA.

When up to 10% or 20% of RAP is used, such aged asphalt pavement material may be supplied to four batch AMP places: the bottom part of a bucket hot elevator with aggregate, hot fractions' unsieved mineral materials bypass hot bin, hot mineral material fractions batcher's weight hopper, forced mixing two-shaft mixer. Such four RAP feeding places into AMP, which were selected in advance, have both advantages and disadvantages.

The aim is to consume less heat energy (gas or liquid fuel), required to remove water content in RAP and to reduce the pollution of ambient air. Ready-to-use RAP is stored in a closed warehouse, which protects the material from precipitation water. When the conductive heat transfer method is applied to vaporize the water content in RAP and to reduce the viscosity of aged bitumen, the superheated temperature of virgin mineral materials mostly depends on RAP moisture (water content) and its percentage in RHMA. When the value of these parameters increases, the required temperature of heating and drying of virgin aggregates in a drum increases, which is calculated according to the presented empirical formula.

The composition of HMA mixture with less than 100% of RAP developed in a laboratory as accurately and precisely as possible shall be guaranteed when producing it in an AMP. Inevitable systematic and random errors of batching RAP increase the deviations of RHMA mixture component content from JMF and their variation. The size of errors depends on AMP structure, technical condition, RAP percentage in RHMA and AMP operator's actions when setting the technological parameters of the recycling process and handling them.

5. Conclusions

The application of these twelve road asphalt pavement hot recycling principles allows us to expect the best possible properties and pavement of the course with reclaimed asphalt pavement. These principles may be used for asphalt pavement hot-in-place surface recycling and hot-mix in-plant recycling.

According to JMF, RAP content was 25% (750 kg batch mass, 10 days, $n = 1324$ batches) and during the investigation, it was 50% (1500 kg batch mass, 12 days, $n = 1569$ batches). The mean of 738.6 kg of all the AMP batches with its required content of 25% mass was less than 750 kg, i.e., by 1.52%, and with the required RAP content of 50%, the mean of 1483.3 kg was less than 1500 kg, i.e., by 1.1%. The accuracy of RAP batching are almost met or completely met the tolerances of the mean specified in ASTM D995-95b on the batch aggregate component ($\pm 1.5\%$) content. Due to the high variation of the batch mass of separate RAP, the tolerance of $\pm 1.5\%$ was met only by 29.6% and 51.5%, respectively, of all RAP batches. The accuracy and compliance of batch masses with tolerances increases when a higher percentage of RAP is batched.

The standard deviation mean calculated according to two standard deviations ($s_q = 27.46$ kg and $s_q = 27.40$ kg) was 27.4 kg. The variation coefficient of RAP batch mass was 3.72% (when RAP batch mass according to JMF was 750 kg) and approximately two times less (1.85%), when according to JMF RAP batch mass was 1500 kg.

RAP batch mass deviations from JMF and their variation influence RAP percentage content in a recycled HMA mixture. When the mass of an asphalt mix batch inevitably varies, the variation of

RAP percentage in RHMA mixture increases even more. When RAP content according to JMF was 25% in RHMA, the mean of its percentage content was 24.53%, and when according to JMF it was 50%, in fact, it was 49.45%. In both cases, RAP percentage content in RHMA mixture varies according to normal distribution. Standard deviations of RAP percentage content in RHMA mixture were 0.705% and 0.507%, and statistically did not diverge.

Acknowledgments: The authors would like to acknowledge Fegda JSC for providing part of materials and laboratory equipment required for the research work.

Author Contributions: Henrikas Sivilevičius systematized hot regeneration principles, submitted the model of asphalt pavement recycling operations (Figure 3) and wrote the introduction. Justas Bražiūnas conceived and designed the experiments, wrote and edited the paper. Olegas Prenkovskis processed and analyzed the data. All authors read and approved the final manuscript.

References

1. Tran, N.H.; Taylor, A.; Willis, R. *Effect of Rejuvenator on Performance Properties of HMA Mixtures with High RAP and RAS Contents*; Report Number 12-05; National Center for Asphalt Technology: Auburn, AL, USA, 2012.
2. Zaumanis, M.; Mallick, R.B. Review of very high-content reclaimed asphalt use in plant-produced pavement: State of the art. *Int. J. Pavement Eng.* **2015**, *16*, 39–55. [CrossRef]
3. Solaimanian, M.; Savory, E. Variability Analysis of Hot-Mix Asphalt Concrete Containing High Percentage of Reclaimed Asphalt Pavement. *Transp. Res. Rec.* **1996**, *1543*, 89–96. [CrossRef]
4. Howard, I.L.; Cooley, L.A., Jr.; Doyle, J.D. *Laboratory Testing and Economic Analysis of High RAP Warm Mixed Asphalt*; Mississippi Department of Transportation: Jackson, MS, USA, 2009.
5. Bloomquist, D.; Diamond, G.; Oden, M.; Ruth, B.; Tia, M. *Engineering and Environmental Aspects of Recycled Materials for Highway Construction*; Western Research Institute: Laramie, WY, USA, 1993.
6. Apeagyei, A.K.; Clark, T.M.; Rorrer, T.M. Stiffnes of high-RAP asphalt mixtures: Virginia's experience. *J. Mater. Civ. Eng.* **2013**, *25*, 747–754. [CrossRef]
7. Rad, F.Y.; Sefidmazgi, N.R.; Bahia, H. Application of diffusion mechanism. Degree of Blending between fresh and recycled asphalt pavement binder in dynamic shear rheometer. *Transp. Res. Rec. J. Transp. Res. Board.* **2014**, *2444*, 71–77. [CrossRef]
8. Miró, R.; Valdés, G.; Martínez, A.; Segura, P.; Rodriguez, C. Evaluation of high modulus mixture behaviour with high reclaimed asphalt pavement (RAP) percentages for sustainable road construction. *Const. Build. Mater.* **2011**, *25*, 3854–3862. [CrossRef]
9. Mogawer, W.; Bennert, T.; Daniel, J.S.; Bonaquist, R.; Austerman, A.; Booshehrian, A. Performance characteristics of plant produced high RAP mixtures. *Road Mater. Pavement Des.* **2012**, *13*, 183–208. [CrossRef]
10. Bražiūnas, J.; Sivilevičius, H. The bitumen batching system's modernization and its effective at the asphalt mixing plant. *Transport* **2010**, *25*, 325–335. [CrossRef]
11. Zaumanis, M.; Mallick, R.B.; Frank, R. 100% recycled hot mix asphalt: A review and analysis. *Resour. Conserv. Recycl.* **2014**, *92*, 230–245. [CrossRef]
12. Santos, L.G.D.P.; Baptista, A.M.D.C.; Capitão, S.D. Assessment of the Use of Hot-Mix Recycled Asphalt Concrete in Plant. *J. Transp. Eng.* **2010**, *136*, 1159–1164. [CrossRef]
13. Estakhri, C.; Spiegelman, C.; Gajewski, B.; Yang, G.; Little, D. *Recycled Hot-Mix Asphalt Concrete in Florida: A Variability Study*; International Center for Aggregates Research: Austin, TX, USA, 1999.
14. West, R.; Willis, J.R.; Marasteanu, M. *Improved Mix Design, Evaluation, and Materials Management Practices for Hot Mix Asphalt with High Reclaimed Asphalt Pavement Content*; NCHRP Report 752; NCHRP: Washington, DC, USA, 2013.
15. National Center for Asphalt Technology (NCAT). LTPP Data Shows RAP Mixes Perform as Well as Virgin Mixes. In *Asphalt Technology News*; NCAT: Auburn, AL, USA, 2009; Volume 21.
16. Diefenderfer, S.; Nair, H. Evaluation of Production, Construction, and Properties of High Reclaimed Asphalt Pavement Mixture. *Transp. Res. Rec. J. Transp. Res. Board.* **2014**, *2445*, 75–82. [CrossRef]
17. Bonaquist, R. Can I run more RAP? *HMAT Hot Mix Asph. Technol.* **2007**, *12*, 11–13.
18. Mohammad, L.; Wu, Z.; Zhang, C.; Khattak, M.; Abadie, C. Variability of Air Voids and Mechanistic Properties of Plant-Produced Asphalt Mixtures. *Transp. Res. Rec. J. Transp. Res. Board.* **2004**, *1891*, 85–97. [CrossRef]

19. Ali, H.; Grzybowski, K. Life cycle of hot in-place pavement recycling. Case study. *Transp. Res. Rec. J. Transp. Res. Board.* **2012**, *2292*, 29–35. [CrossRef]
20. Tahmoorian, F.; Samali, B.; Tam, V.W.Y.; Yeaman, J. Evaluation of mechanical properties of recycled material for utilization in the asphalt mixtures. *Appl. Sci.* **2017**, *7*, 763. [CrossRef]
21. Kriz, P.; Grant, D.L.; Veloza, B.A.; Gale, M.J.; Blahey, A.G.; Brownie, J.H.; Shirts, R.D. Blending and diffusion of reclaimed asphalt pavement and virgin asphalt binders. *Road Mater. Pavement Des.* **2014**, *15*, 78–112. [CrossRef]
22. Le Guen, L.; Huchet, F.; Tamagny, P. Drying and heating modelling of granular flow: Application to the mix-asphalt process. *J. Appl. Fluid Mech.* **2011**, *4*, 71–80.
23. Peinado, D.; de Vega, M.; Garcia-Hernando, N.; Marugan-Cruz, C. Energy and exergy analysis in an asphalt's rotary dryer. *Appl. Therm. Eng.* **2011**, *31*, 1039–1049. [CrossRef]
24. Wang, H.; Hao, P.; Xue, L. Laboratory evaluation of microwave heating method for hot in-place recycling. *J. Test. Eval.* **2011**, *39*, 1070–1077.
25. Carpenter, S.H.; Wolosick, J.R. Modifier influence in the characterization of hot-mix recycled material. *Transp. Res. Rec. J. Transp. Res. Board.* **1980**, *777*, 15–22.
26. Nahar, S.N.; Mohajeri, M.; Schmets, A.J.M.; Scarpas, A.; van de Ven, M.F.C.; Schitter, G. First observation of blending-zone morphology at interface of reclaimed asphalt binder and virgin bitumen. *Transp. Res. Rec. J. Transp. Res. Board.* **2013**, *2370*, 1–9. [CrossRef]
27. Willis, J.R.; Turner, O.; de Goes Padula, F.; Tran, N.; Julian, G. Effects of changing virgin binder grade and content on high reclaimed asphalt pavement mixture properties. *Transp. Res. Rec. J. Transp. Res. Board.* **2013**, *2371*, 66–73. [CrossRef]
28. Dony, A.; Colin, J.; Bruneau, D.; Drouadainc, I.; Navaro, J. Reclaimed asphalt concretes with high recycling rates: Changes in reclaimed binder properties according to rejuvenating agent. *Constr. Build. Mater.* **2013**, *41*, 175–181. [CrossRef]
29. Lin, J.; Guo, P.; Xie, J.; Wu, S.; Chen, M. Effect of rejuvenator scaler materials on the properties of aged asphalt binder. *J. Mater. Civ. Eng.* **2013**, *25*, 829–835. [CrossRef]
30. Čygas, D.; Mučinis, D.; Sivilevičius, H.; Abukauskas, N. Dependence of the recycled asphalt mixture physical and mechanical properties on the grade and amount of rejuvenating bitumen. *Balt. J. Road Bridge Eng.* **2011**, *6*, 124–134. [CrossRef]
31. Huang, S.-C.; Qin, Q.; Grimes, W.R.; Pauli, A.T.; Galer, R. Influence of Rejuvenators on the Physical Properties of RAP Binders. *J. Test. Eval.* **2015**, *43*, 594–603. [CrossRef]
32. Shirodkar, P.; Mehta, Y.; Nolan, A.; Sonpal, K.; Norton, A.; Tomlison, C.; Dubois, E.; Sullivan, P.; Sauber, R. A study to determine the degree of partial blending of reclaimed asphalt pavement (RAP) binder for high RAP hot mix asphalt. *Constr. Build. Mater.* **2011**, *25*, 150–155. [CrossRef]
33. Karlsson, R.; Isacsson, U. Material-related aspects of asphalt recycling—State-of-art. *J. Mater. Civ. Eng.* **2006**, *18*, 81–92. [CrossRef]
34. Wang, F.; Wang, Z.; Li, C.; Xiao, Y.; Wu, S.; Pan, P. The rejuvenating effect in hot asphalt recycling by mortar transfer ratio and image analysis. *Materials* **2017**, *10*, 574. [CrossRef] [PubMed]
35. Poulikakos, L.D.; dos Santos, S.; Bueno, M.; Kuentzel, S.; Hugener, M.; Partl, M.N. Influence of short and long-term aging on chemical, microstructural and macro-mechanical properties of recycled asphalt mixtures. *Constr. Build. Mater.* **2014**, *51*, 414–423. [CrossRef]
36. Huang, B.; Li, G.; Vukosavljevic, D.; Shu, X.; Egan, B. Laboratory investigation of mixing hot-mix asphalt with reclaimed asphalt pavement. *Transp. Res. Rec. J. Transp. Res. Board.* **2005**, *1929*, 37–45. [CrossRef]
37. Bražiūnas, J.; Sivilevičius, H.; Virbickas, R. Dependences of SMA mixture and its bituminous binder properties on bitumen batching system, mixing time and temperature on asphalt mixing plant. *J. Civ. Eng. Manag.* **2013**, *19*, 862–872. [CrossRef]
38. Vislavičius, K.; Sivilevičius, H. Effect of reclaimed asphalt pavement gradation variation on the homogeneity of recycled hot-mix asphalt. *Arch. Civ. Mech. Eng.* **2013**, *13*, 345–353. [CrossRef]
39. Aurangzeb, Q.; Al-Qadi, I.L.; Abuawad, I.M.; Pine, W.J.; Trepanier, J.S. Achieving desired volumetrics and performance for mixtures with high percentage of reclaimed asphalt pavement. *Transp. Res. Rec. J. Transp. Res. Board.* **2012**, *2294*, 34–42. [CrossRef]
40. Sivilevičius, H.; Vislavičius, K.; Bražiūnas, J. Technological and economic design of asphalt mixture composition based on optimization methods. *Technol. Econ. Dev. Econ.* **2017** *23*, 627–648. [CrossRef]

41. Valdés, G.; Pérez-Jiménez, F.; Miró, R.; Martínez, A.; Botella, R. Experimental study of recycled asphalt mixtures with high percentages of reclaimed asphalt pavement (RAP). *Constr. Build. Mater.* **2011**, *25*, 1289–1297. [CrossRef]
42. Mučinis, D.; Sivilevičius, H.; Oginskas, R. Factors determining the inhomogeneity of reclaimed asphalt pavement and estimation of its components content variation parameters. *Balt. J. Road Bridge Eng.* **2009**, *4*, 69–79. [CrossRef]
43. Mogawer, W.S.; Austerman, A.J.; Bonaquist, R. Determining the influence of plant type and production parameters on performance of plant-produced reclaimed asphalt pavement mixtures. *Transp. Res. Rec. J. Transp. Res. Board.* **2012**, *2268*, 71–81. [CrossRef]
44. Silva, H.M.R.D.; Oliveira, J.R.M.; Jesus, C.M.G. Are totally recycled hot mix asphalts a sustainable alternative for road paving? *Resour. Conserv. Recycl.* **2012**, *60*, 38–48. [CrossRef]
45. Zhao, S.; Huang, B.; Shu, X.; Woods, M.E. Quantitative Characterization of Binder Blending. *Transp. Res. Rec. J. Transp. Res. Board.* **2015**, *2506*, 72–80. [CrossRef]
46. Babtista, A.M.; Picado-Santos, L.G.; Capitão, S.D. Design of hot-mix recycled asphalt concrete produced in plant without preheating the reclaimed material. *Int. J. Pavement Eng.* **2013**, *14*, 95–102. [CrossRef]
47. DeDene, C.D.; Voller, V.R.; Marasteanu, M.O.; Dave, E.V. Calculation of particle heating times of reclaimed asphalt pavement material. *Road Mater. Pavement Des.* **2014**, *15*, 721–732. [CrossRef]
48. National Cooperative Highway Research Program (NCHRP). *A Manual for Design of Hot Mix Asphalt with Commentary. National Cooperative Highway Research Program. Transportation Research Board of the National Academics*; NCHRP Report 673; NCHRP: Washington, DC, USA, 2011; p. 273.
49. Ma, T.; Huang, X.; Zhao, Y.; Zhang, Y.; Wang, H. Influences of preheating temperature of RAP on properties of hot-mix recycled asphalt mixture. *J. Test. Eval.* **2016**, *44*, 762–769. [CrossRef]
50. Baroux, R. Recyclage des enrobés bitumineux en centrale d'enrobage. Les problémes de matériel. *Bulletin de Liaison des Ratoires des Ponts et Chaussées* **1980**, *105*, 98–102.

7

Permeability and Stiffness Assessment of Paved and Unpaved Roads with Geocomposite Drainage Layers

Cheng Li [1,*], Jeramy Ashlock [1], David White [2] and Pavana Vennapusa [2]

1 Department of Civil Construction and Environmental Engineering, Iowa State University, Ames, IA 50011, USA; jashlock@iastate.edu

2 Ingios Geotechnics, Inc., Northfield, MN 55057, USA; david.white@ingios.com (D.W.); pavana.vennapusa@ingios.com (P.V.)

* Correspondence: cheng@iastate.edu

Academic Editor: Zhanping You

Abstract: Poor subsurface drainage is frequently identified as a factor leading to the accelerated damage of roadway systems. Geocomposite drainage layers offer an alternative to traditional methods but have not been widely evaluated, especially in terms of the impact of changes on both drainage capacity and stiffness. In this study, both paved and unpaved test sections with and without an embedded geocomposite drainage layer were constructed and tested. The geocomposite layers were installed directly beneath the roadway surface layers to help the rapid drainage of any infiltrated water and thus prevent water entering the underlying foundation materials. The laboratory, field, and numerical analysis results showed that the geocomposite layers increased the permeability of roadway systems by two to three orders of magnitude and that it can effectively prevent the surface and foundation materials from becoming saturated during heavy rainfall events. For the stiffness of the sections, the paved sections with and without a geocomposite layer showed that the composite modulus values measured at the surface were more reflective of the foundation layer support conditions beneath the geocomposite layer than the geocomposite layer itself. The unpaved road section with the geocomposite layer yielded lower composite modulus values than the control section but showed overall better road surface conditions after a rain event due to the improved subsurface drainage condition.

Keywords: geocomposite; permeability; drainage; stiffness; elastic modulus; concrete pavement; asphalt pavement; unpaved road; water infiltration model

1. Introduction

Poor subsurface drainage is frequently identified as a factor leading to the accelerated damage of both paved and unpaved roadway systems. Typically, subsurface drainage is controlled with ditches, edge drains, drainable aggregate layers, and/or roadway crowns. These methods typically require maintenance, and the drainage capacity degrades with time. Geocomposite drainage layers offer an alternative to traditional methods but have not been widely field tested, especially in terms of the impact of changes on both drainage capacity and pavement foundation stiffness.

Geocomposites usually consist of two geotextile outer layers and an internal drainage layer (i.e., geonet) and are typically designed to provide three-dimensional subsurface drainage, soil separation, and filtration [1,2]. Previous lab and field studies have shown the potential for using geocomposite drainage layers to reduce drainage-related damage for both paved and unpaved roads [3–7]. However, the influence of geocomposite layers on the composite stiffness of pavement systems and the drainage performance under transient water flow conditions are not well documented. A numerical analysis showed that geotextiles on either side of a geocomposite layer can decrease plastic deformation through combined mechanistic and hydraulic actions for both paved and unpaved roads, but increasing the surface course or base course thickness reduces the reinforcement benefits [3]. The effects of the location of a geocomposite drainage layer in an asphalt pavement system were studied by Christopher et al. [4], in which it was found that geocomposite drainage layers were quickest at removing water during spring thaws when placed on or within the subgrade. Falling weight deflectometer (FWD) test results from the same study showed that a section with geocomposite in the subgrade had a higher stiffness than sections with geocomposites embedded at higher positions in the pavement foundation.

In this study, paved (with either concrete or asphalt surfaces) and unpaved roadway test sections with and without an embedded geocomposite drainage layer were constructed and tested in Iowa, USA. The geocomposite drainage layers were placed directly beneath the roadway surface layers to help the rapid drainage of any infiltrated water through the pavement joints, cracks, or unpaved road surface materials and thus prevent water entering into the underlying foundation layers. However, since the locations of the relatively soft geocomposite layers are close to the roadway surfaces, the impact of changes on both the drainage capacity and stiffness of the systems need to be evaluated. A laboratory large-scale horizontal permeameter test (HPT) device was developed to measure the horizontal saturated hydraulic conductivities of the system with and without an embedded geocomposite drainage layer. Core-hole permeameter (CHP) and air permeameter test (APT) devices were also used to evaluate the in situ drainage conditions of the test sections. To assess the influence of the geocomposite layers on the composite stiffness of the roadway systems, falling weight deflectometer (FWD) tests were used to measure the composite elastic modulus of the test sections. A two-dimensional (2D) water infiltration model was also developed based on Richard's equation and Haverkamp's soil water retention characteristic (WRC) model to simulate a heavy rain event (i.e., transient flow condition) and to compare the effectiveness of the geocomposite drainage layer with the current practice of improving the drainage conditions of unpaved roads by building and maintaining a 4% crown to enable lateral water flow.

2. Descriptions of Materials and Test Sections

In this study, the two different geocomposite materials (GC-1 and GC-2) consisting of two layers of non-woven geotextile with synthetic polymer geonet cores were evaluated. The properties of the two materials provided by the manufactures are shown in Table 1. Geocomposite GC-1 (Figure 1a,b) possessed a thicker non-woven geotextile and a stiffer geonet core than GC-2 (Figure 1c,d). The stiffness values of the geocomposite materials were not provided by the manufacturers.

Table 1. Properties of the GC-1 and GC-2 geocomposite materials.

Property	GC-1	GC-2	Testing Method
Thickness (mm)	11	9	NA (used Calipers)
Strength (kN)	Exceeds Class 2	17	AASHTO [2] M 288 [8] for GC-1 ASTM [3] D5035 [9] for GC-2
Water Flow Rate (L/min/m^2)	4481	4100	ASTM D 4491 [10]
AOS [1] of the Geotextile (mm)	0.212	0.212	ASTM D 4751 [11]

[1] AOS stands for apparent opening size; [2] AASHTO is the abbreviation of American Association of State Highway and Transportation Officials; [3] ASTM is the abbreviation of American Society for Testing and Materials.

Figure 1. Photos of the (**a**,**b**) Roadrain RD-5 (GC-1) and (**c**,**d**) Macdrain W 1091 (GC-2) geocomposite materials placed beneath paved and unpaved surface courses, respectively.

A total of six test sections were constructed at two different sites, with the Portland cement concrete (PCC) and warm-mixed asphalt (WMA) sections located at the same site. The nominal cross-section profiles of the test sections are shown in Figure 2. For the paved test sections, GC-1 was installed at the top of the base layer (directly beneath the pavement layer) prior to paving. For the unpaved test sections, GC-2 was placed at the interface of the subgrade and surface aggregate layers. The soil index properties of the geomaterials used in this study are summarized in Table 2.

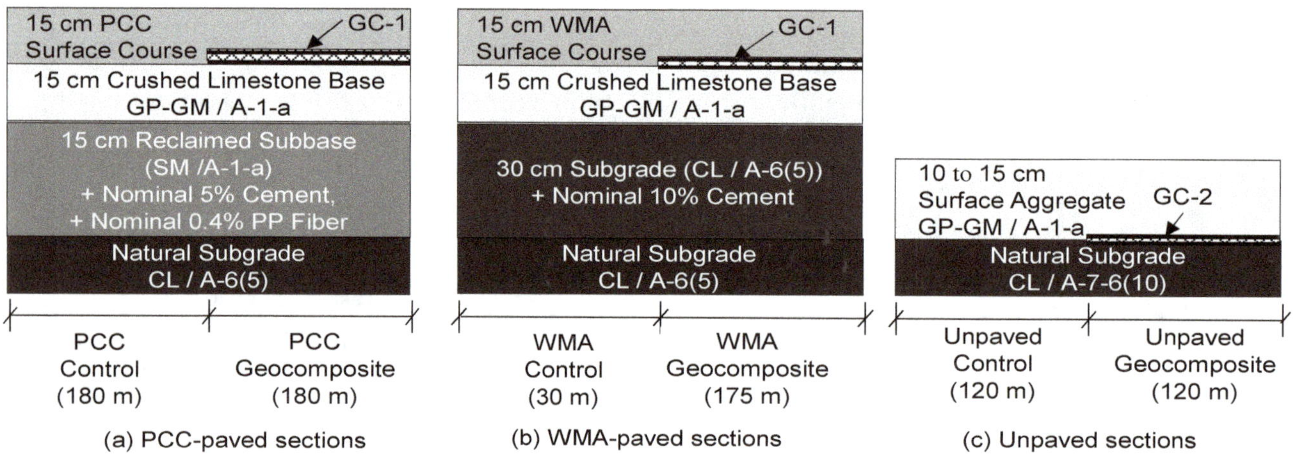

Figure 2. Nominal cross-section profiles (not to scale) of the (**a**) Portland cement concrete (PCC) paved; (**b**) warm-mixed asphalt (WMA) paved and (**c**) unpaved test sections.

Table 2. Summary of the gradation, plasticity, and classifications of the geomaterials used in this study.

Parameter	Crushed Limestone Base	Subgrade of Paved Sections	Unpaved Road Surface Aggregate	Subgrade of Unpaved Sections
Gravel content (%) (>4.75 mm)	65.2	5.3	57.9	0.9
Sand content (%) (4.75–0.075 mm)	27.7	39.7	30.3	39.8
Silt content (%) (4.75–0.005 mm)	3.9	29.3	9.2	30.6
Clay content (%) (<0.005 mm)	3.2	25.7	2.6	28.7
D_{10} (mm)	0.3	–	0.049	–
D_{30} (mm)	3.6	0.08	1.731	0.003
D_{60} (mm)	10.1	0.12	9.720	0.081
Coefficient of uniformity, c_u	33.7	–	198.96	–
Coefficient of curvature, c_c	4.36	–	6.31	–
Liquid limit (%)	Non-plastic	33	Non-plastic	43
Plastic limit (%)		15		22
AASHTO classification	A-1-a	A-6(5)	A-1-a	A-7-6(10)
Unified soil classification system (USCS) group symbol	GP-GM	CL	GP-GM	CL
USCS group name	Poorly graded gravel with silt and sand	Sandy lean clay	Poorly graded gravel with silt and sand	Sandy lean clay

For the PCC-paved sections, discrete fibrillated polypropylene (PP) fibers with 5% Portland cement (PC) were used to stabilize the reclaimed subbase layer. Two WMA sections were constructed at the same site as the PCC sections (Figure 2). The top 30 cm of natural subgrade of the WMA sections were stabilized with 10% Portland cement. The detailed test section designs, construction processes, material properties, and costs are beyond the scope of this paper but are reported in [12,13].

The unpaved road test sections were constructed at a different site. The final thickness of the surface aggregate layer after construction ranged between 10 and 15 cm, and the subgrade had a slightly higher fines content than the subgrade of the paved test sections. The GC-2 geocomposite was placed at the interface of the surface aggregate layer and subgrade. The construction processes and costs of the test sections are described in detail in [14].

3. Test Methods

To compare the drainage conditions of the roadway systems with and without a geocomposite drainage layer and determine inputs for the numerical simulations, a large-scale laboratory horizontal permeameter test (HPT) was conducted to evaluate the saturated horizontal hydraulic conductivity of the systems. For the field-test sections, two innovative devices, core-hole permeameter (CHP) and air permeameter test (APT) devices, designed at Iowa State University, were used to measure the in situ hydraulic conductivities of the different test sections. The influences of the geocomposite layers on the composite stiffness of the roadway systems were assessed using the falling weight deflectometer (FWD) test. The descriptions, theories, and testing procedures of the different testing methods are described in the following sections.

3.1. Laboratory Large-Scale Horizontal Permeameter Test (HPT)

The large-scale horizontal permeameter test (HPT) was developed to measure the saturated hydraulic conductivity of the aggregate materials under horizontal flow conditions because most permeability tests employ vertical flow, which does not accurately represent how water typically drains or flows through the pavement base layers horizontally in the field. The device can simulate direct horizontal flow situations under different pressure heads and is large enough (soil tank dimensions are 1 m × 0.46 m × 0.33 m) to effectively test multiple material layers, as shown in Figure 3. The inside wall of the HPT soil tank has several ribs (1 cm in height) installed perpendicular to the flow direction to prevent water flowing through the interface of the soil specimen and the inside wall of the soil tank.

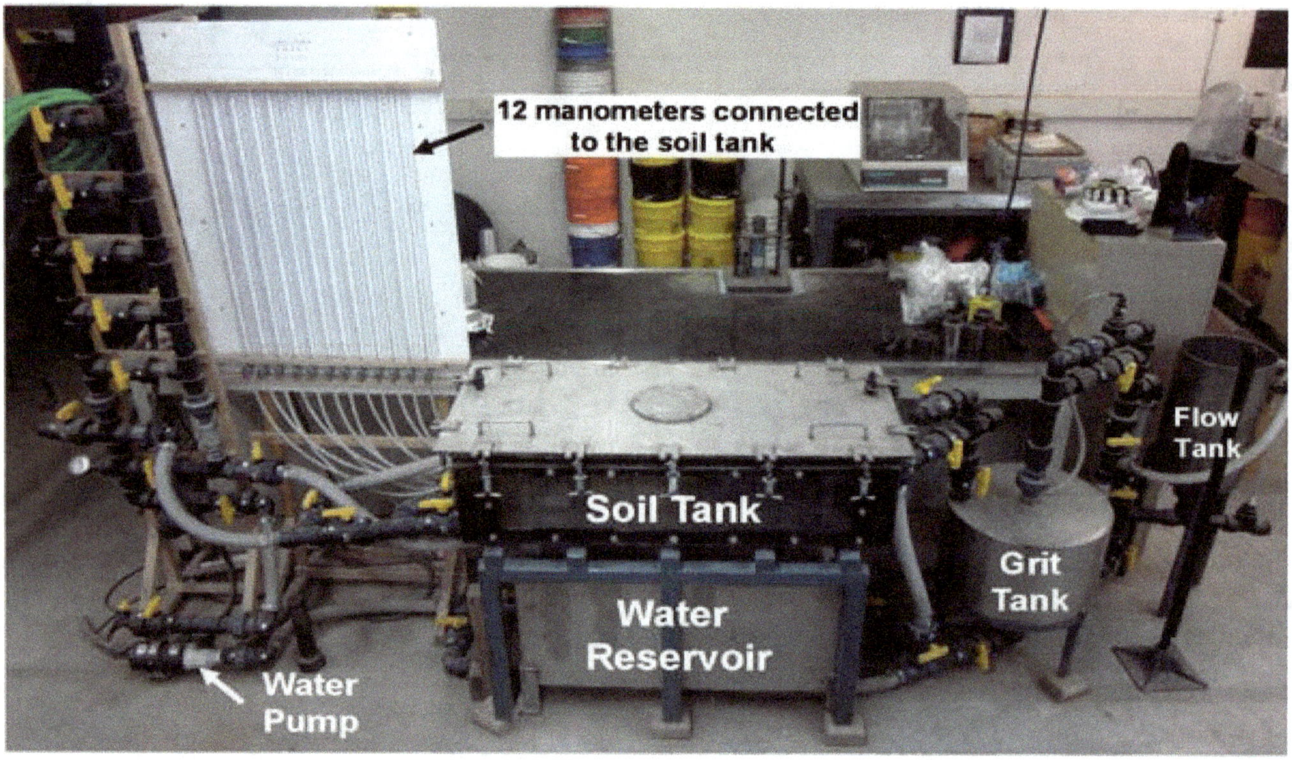

Figure 3. Photo of the large-scale horizontal permeameter test (HPT) device used in this study.

To quantify the improvement in horizontal drainage offered by the geocomposite layer, representative base material (Unified soil classification system (USCS) group symbol: GP-GM) were collected from the field and compacted in the HPT soil tank with and without a layer of geocomposite. Three HPTs were conducted; one specimen consisted of the base material only, and the other two had one of the two types of geocomposite drainage materials embedded (i.e., GC-1 and GC-2). The testing specimens were compacted to the standard Proctor maximum dry unit weight of 22.0 kN/m^3, determined in accordance with ASTM D698 [15]. The saturated horizontal hydraulic conductivities of the three specimens were measured under four different constant water heads of 50, 100, 150 and 200 mm. In the field, the geocomposite materials are placed at the interface of material layers. However, for the HPT test, the geocomposite layer was embedded at the middle of the test specimen to prevent any boundary issues such as the water flowing through the gap between the geocomposite layer and the bottom of the soil tank.

3.2. Core-Hole Permeameter (CHP) Test

The CHP device (Figure 4a) was developed to assess the in situ drainage capacities of pavement base systems. The test method follows the procedure in ASTM D6391 [16]. In this study, the CHP test results were used to compare the relative drainage capacities of the foundation system with and without a geocomposite layer placed at the pavement/base layer interface. The condition this simulates is the ability to drain any infiltrated water through the pavement surface via joints/cracks. Vennapusa et al. [17] showed that by using the American Association of State Highway and Transportation Officials (AASHTO) 1993 design guide PCC pavement design procedure, an increase in the drainage coefficient (C_d) value of the base material from 1.0 to 1.2 can reduce the required PCC pavement thickness by up to 10%. The CHP test uses the falling head method to measure the in situ hydraulic conductivity of the drainable base and geocomposite drainage layers of the paved test sections after drilling cores, as shown in Figure 4b,c. The test involves drilling a 15 cm diameter core hole through the pavement surface to the underlying base or geocomposite layer, inserting the device and sealing it against the bottom foam ring and interior of the core hole using a rubber tube inflated to 135–175 kPa, and recording the rate of water head loss from the device over a period of 30 to 60 min. The hydraulic conductivity (K) of the tested layer was calculated using Equation (1). The testing procedure is described in detail by Zhang et al. [18].

$$K = R_t G_1 \frac{\ln(H_1/H_2)}{(t_2 - t_1)} \quad (1)$$

$$R_t = 2.2902\left(0.9842^T\right)/T^{0.1702} \quad (2)$$

$$G_1 = \left(\pi d^2/11D_1\right)\left[1 + a(D_1/4b_1)\right] \quad (3)$$

where H_1 and H_2 are the effective heads (cm) at time t_1 and t_2 (s), respectively; R_t is the ratio of the kinematic viscosity of the permeant at the temperature of the test during time increment t_1 to t_2 to that of water at 20 °C; T is the temperature of the test permeant (°C); d is the inside diameter of the standpipe (3.6 cm for the top standpipe and 33 cm for the middle standpipe); D_1 is the inside diameter of bottom casing (12.7 cm); b_1 is the thickness of tested layer (cm); and a is equal to +1 for an impermeable base at b_1, 0 for the infinite depth of the tested layer (i.e., $b_1 > 20D_1$), and −1 for a permeable base at b_1.

Figure 4. Photos of (**a**) core-hole permeameter (CHP) device; (**b**) CHP test conducted on the paved test sections and (**c**) GC-1 material after the CHP test.

3.3. Air Permeameter Test (APT)

The APT device was developed for rapid in situ saturated hydraulic conductivity determination for granular materials (Figure 5a). The APT device consists of a contact ring, differential pressure gauges, precision orifices, and a programmable digital display (Figure 5b).

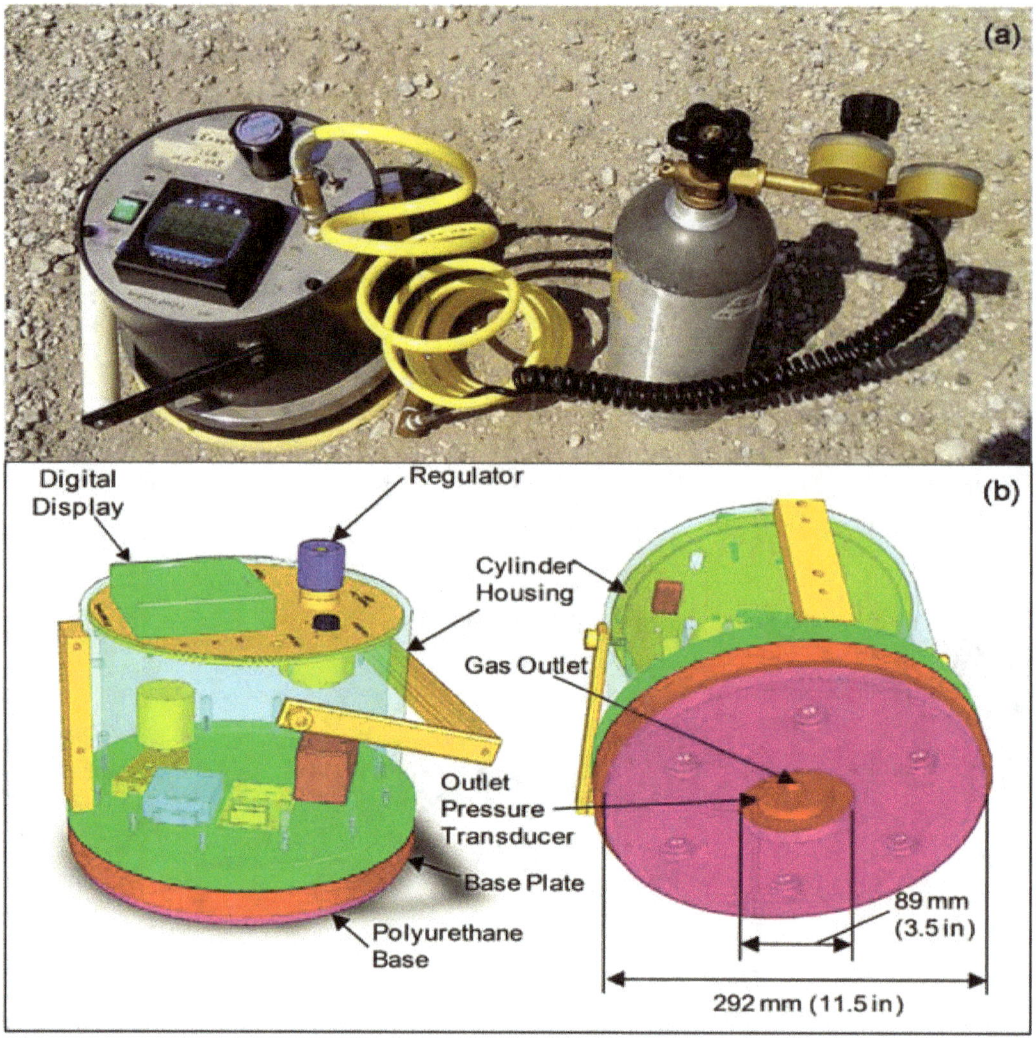

Figure 5. (**a**) Air permeameter test (APT) test conducted on unpaved test section and (**b**) schematic showing the major components of the APT device [13].

The APT device measures the gas pressure on the inlet and outlet sides of the precision orifice and calculates the gas flow rate. From these measurements and the material parameters, the gas permeability of the material being tested can be derived and converted to water saturated hydraulic conductivity (K_{sat}) by taking into account the effect of partial saturation (Equation (4)).

$$K_{sat} = \left[\frac{2\mu_{gas} Q P_{2a}}{r G_0 ({P_{2a}}^2 - {P_{atm}}^2)}\right] \times \frac{\rho g}{\mu_{water}(1 - S_e)^2 (1 - {S_e}^{((2+\lambda)/\lambda)})} \tag{4}$$

where K_{sat} is the saturated hydraulic conductivity (cm/s), μ_{gas} is the kinematic viscosity of the gas (Pa-s), Q is the flow rate (cm^3/s), P_{2a} is the absolute gas pressure on the surface (Pa), r is the radius at the outlet (cm), G_0 is the geometric factor determined from test layer thickness, P_{atm} is the atmospheric pressure (Pa), ρ is the density of water (g/cm^3), g is the acceleration due to gravity (cm/s^2), μ_{water} is the kinematic viscosity of water, S_e is the effective saturation, and λ is the Brooks-Corey pore size distribution index.

The derivation of Equation (4) involves an expansion of Darcy's Law considering the compressibility and viscosity of the gas, the gas flow under partially saturated conditions, and the Brooks-Corey pore size distribution index. The development of the testing method, description of the device, and derivations of the theoretical relationship to calculate the hydraulic conductivity from the gas flow are detailed by White et al. [19]. The repeatability of the APT measurements, expressed as a coefficient of variation (COV), is ≤1% [20]. In this study, the APT test was conducted to measure the in situ saturated hydraulic conductivity as a function of depth for the unpaved road test sections.

3.4. Falling Weight Deflectometer (FWD) Test

FWD tests were conducted using a Kuab Model 150 2m device with a 300-mm diameter segmented loading plate to apply a relatively uniform stress distribution. A 40 kN impact load was applied on the roadway surface, which is the AASHTO standard axle load [21]. The induced roadway surface deflection at the center of the loading plate was recorded by a seismometer. For each test location, a single equivalent composite elastic modulus of the roadway system was calculated based on Boussinesq's solution using Equation (5);

$$E_{\text{Composite}} = \frac{(1-\nu^2)\sigma_0 A}{d_0} \times f \tag{5}$$

where $E_{\text{Composite}}$ is the composite elastic modulus (MPa), d_0 is the measured deflection under the center of the loading plate (mm), σ is the Poisson's ratio (assumed to be 0.4), σ_0 is the normalized applied peak stress (MPa), A is the radius of the plate (mm), and f is the shape factor, assumed to be 2 because of the assumed uniform stress distribution.

4. Laboratory Horizontal Permeameter Test (HPT) Results

The saturated horizontal hydraulic conductivities of the specimens with and without an embedded geocomposite layer were measured under four different constant water heads (i.e., 50, 100, 150 and 200 mm). The relationships between the hydraulic conductivity (K) and hydraulic gradient (i) of the three specimens are shown in Figure 6. The horizontal saturated hydraulic conductivity of the specimen without a geocomposite layer at 20 °C ($K_{20\,°C}$) was approximately 0.8 m/day and remained relatively constant as the hydraulic gradient increased. Compared to the specimen without a geocomposite layer, the geocomposite drainage layer increased the saturated hydraulic conductivity by two to three orders of magnitude.

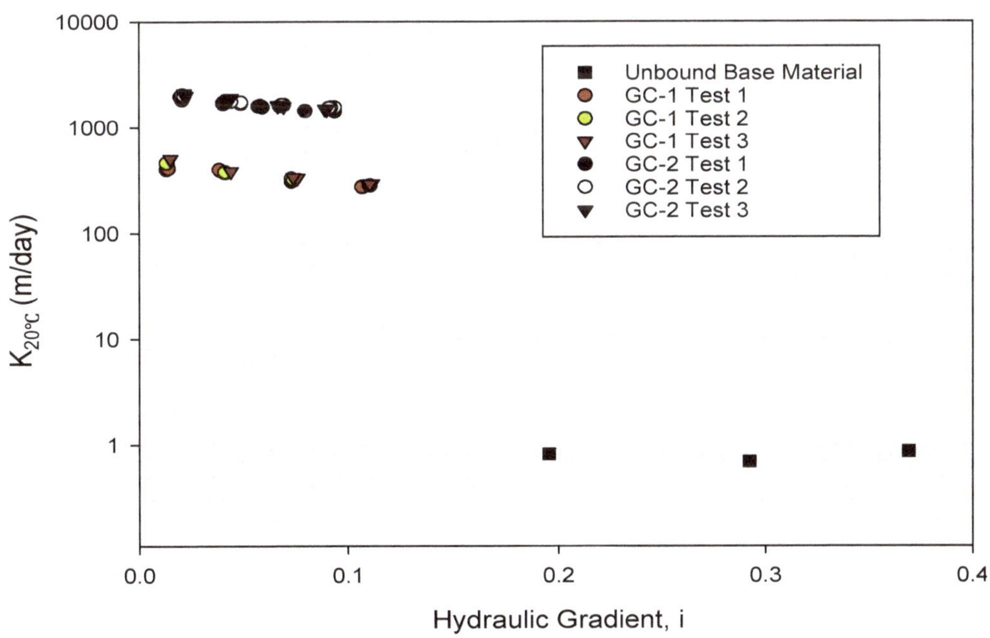

Figure 6. Large-scale horizontal permeameter test (HPT) results of the specimens with and without an embedded geocomposite drainage layer.

Figure 6 also shows that the GC-2 geocomposite had a higher horizontal hydraulic conductivity (~1500 m/day) than the GC-1 geocomposite (~350 m/day). The total thickness of GC-1 is greater than that of GC-2 due to the thicker geotextile on each side of the geonet core for GC-1 However, the geonet core of GC-2 is more porous and has the approximately the same thickness as GC-1.

5. In Situ Permeability and Stiffness of The Test Sections

5.1. In Situ Permeability Assessment

The CHP test was conducted on the PCC and WMA sections to compare the subsurface drainage conditions of the GC-1 geocomposite layer and conventional base materials. The CHP test results showed that the in situ hydraulic conductivity of the GC-1 layer varied between 637 and 695 m/day with a coefficient of variation (COV) of 5.8%, whereas that of the crushed limestone base layer was between 0.1 and 34.2 m/day with a COV of 110%. The hydraulic conductivity of the geocomposite varied within a much smaller range compared to the base material, which may indicate that the segregation of the traditional base material can cause significant variation in the drainage performance of roadway systems.

The APT device was used to quantify the improvement in drainage offered by the geocomposite layer (GC-2) for the unpaved road test sections. For each test location, APTs were performed at different depths within the overlying aggregate layer to determine the variation of saturated hydraulic conductivity (K_{sat}) with depth. The test results are summarized in Figure 7. For the control section, the K_{sat} values at the three test locations were similar (~19 m/day) and remained relatively constant with depth. However, for the geocomposite section, K_{sat} in the surface aggregate layer increased consistently with depth from the roadway surface. The K_{sat} of the GC-2 drainage layer is approximately 1920 m/day, which is more than two orders of magnitude greater than that of the control section. The field-measured K_{sat} of the GC-2 agreed very well with the HPT test results shown in Figure 6.

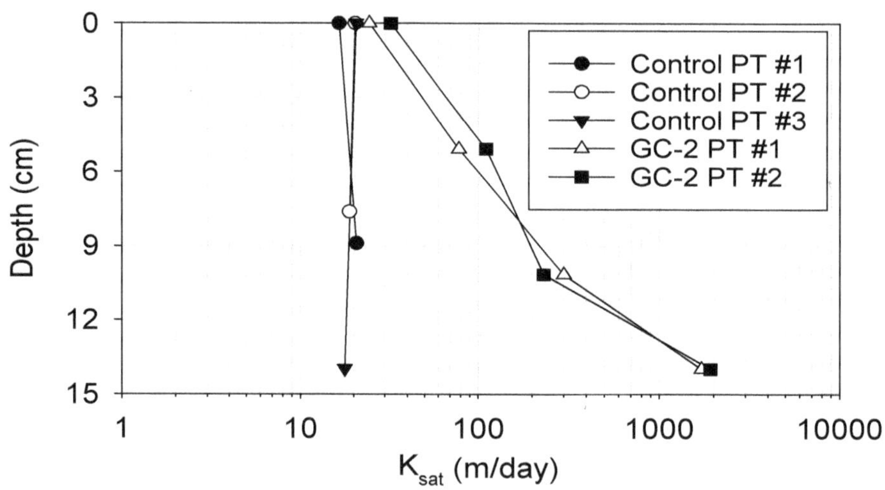

Figure 7. APT-measured saturated hydraulic conductivity versus depth for the surface aggregate layers of the unpaved geocomposite and control sections.

5.2. In Situ Stiffness Assessment

To assess the influence of the geocomposite layer on the composite stiffness of the PCC- and WMA-paved sections, two groups of FWD tests were conducted; (1) on the base layer prior to placing the geocomposite layers on the base layer and (2) on the pavement surfaces after paving, with the geocomposite layer installed at the pavement/base layer interface. The FWD tests were conducted at the same test locations on both the base and the pavement surface layers. A comparison of the two groups of test results shows the influence of the geocomposite on the composite stiffness (Figure 8).

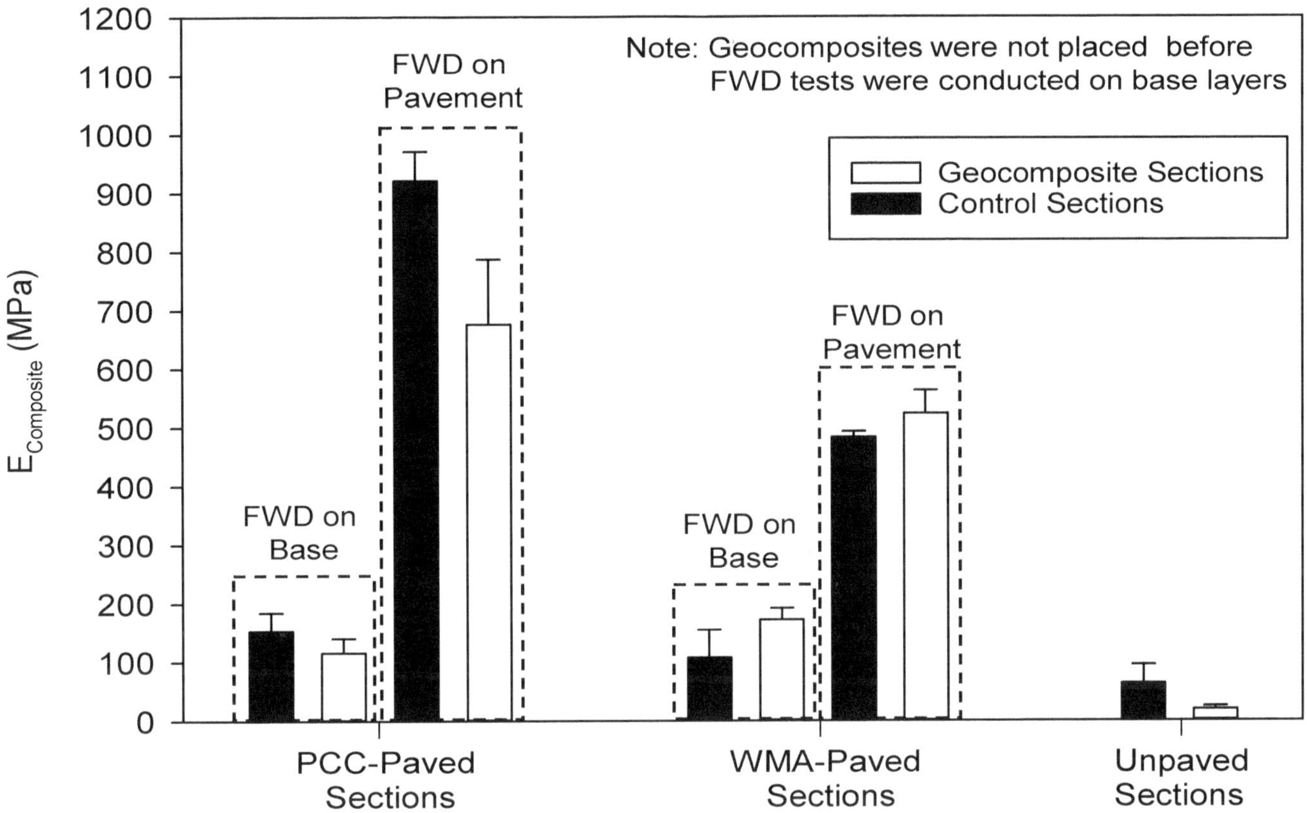

Figure 8. Summary of the falling weight deflectometer (FWD) test results of the PCC-paved, WMA-paved, and unpaved sections.

For the PCC sections prior to paving, the foundation layer in the geocomposite section had a lower average $E_{\text{Composite}}$ value than the control section. After placing the geocomposite layer and paving, the FWD tests conducted on the pavement surface showed similar trends, wherein the control section had a higher average $E_{\text{Composite}}$ value than the geocomposite section. For the WMA sections, the composite modulus values in the geocomposite section were higher than in the control section for tests conducted both on the foundation layer and the pavement layer. These FWD results from the paved road test sections, although limited, suggest that the composite stiffness measurements obtained at the pavement surface were more reflective of the foundation layer support conditions beneath the geocomposite layer than of the stiffness of the geocomposite layer itself.

FWD tests were also conducted on the unpaved sections after the spring thaw in 2015 because the spring thaw is when unpaved roads are most susceptible to moisture-related damage. Figure 8 shows that the average $E_{\text{Composite}}$ of the geocomposite section is less than half that of the control section. However, the roadway surface performance of the geocomposite section was much better than that of the control section based on observations after a heavy rainfall and subsequent traffic, as can be clearly seen in Figure 9. The control section suffered much greater rutting and moisture retention than the geocomposite section. These field observations after the rain event indicate that the improved subsurface drainage conditions offered by the geocomposite layer can effectively prevent the surface material from becoming saturated during a heavy rain event, reduce the amount of water infiltrating into the subgrade, and thus reduce the surface damage. To further examine this hypothesis, a numerical analysis is conducted in the following section to compare the drainage performance of the geocomposite and control sections after a simulated heavy rain event.

Figure 9. Surface conditions of the unpaved road sections after a heavy rainfall (photos taken on 2 September 2015).

6. Numerical Simulation of Drainage Performance under Heavy Rainfall

To improve the drainage conditions of unpaved roads, most agencies choose to build and maintain a 4% roadway crown to enable lateral water flow to drain excess water from the surface course into side drains or ditches. However, this practice requires frequent maintenance, and the effectiveness of the method needs to be quantified. In this study, a numerical simulation was performed to compare the drainage capacity offered by the geocomposite drainage layer with that of a 4% roadway surface crown for unpaved roads. A 2D water infiltration model was programmed in Matlab using Richard's equation and Haverkamp's soil water retention characteristic (WRC) model [22,23].

Richards' equation (Equation (6)) is used to predict volumetric water content (θ) and matric potential (h) changes of the unpaved road surface material during transient water flow;

$$\frac{\partial \theta}{\partial t} = \frac{\partial}{\partial z}\left[K(h)\left(\frac{\partial h}{\partial z} + 1\right)\right] \quad (6)$$

Liquid water flow in the vertical and horizontal directions is expressed using the well-known Darcy-Buckingham flux equations:

$$q = -K(h)\left(\frac{dh}{dz} - \cos\alpha\right) \text{ (for vertical direction)} \quad (7)$$

$$q = -K(h)\left(\frac{dh}{dx} - \sin\alpha\right) \text{ (for horizontal direction)} \quad (8)$$

where q is the water flux in cm/s, $K(h)$ is the unsaturated hydraulic conductivity in cm/s, h is the matric potential in cm, $\frac{dh}{dz}$ or $\frac{dh}{dx}$ is the matric potential gradient, α is the slope of the soil layer in degrees, and the material is assumed to be isotropic and homogenous.

Haverkamp's water retention characteristic (WRC) model was used to estimate relationships between unsaturated hydraulic conductivity, matric potential, and the water content of the unpaved surface material during transient flow, as shown in Equations (9) and (10):

$$\theta(h) = \theta_r + \frac{\alpha(\theta_s - \theta_r)}{\alpha + |h|^{\beta}} \quad (9)$$

where $\theta(h)$ is the volumetric water content at the corresponding matric potential (h), θ_r is the residual volumetric water content, θ_s is the saturated volumetric water content, α is equal to 1.611×10^6, and β is equal to 3.96.

$$K(h) = K_s \frac{A}{A + |h|^{\beta}} \quad (10)$$

where $K(h)$ is the unsaturated hydraulic conductivity at the corresponding matric potential (h) in cm/s, K_s is the saturated hydraulic conductivity in cm/s, A is equal to 1.175×10^6, and, β is equal to 4.74.

In the above two equations, the coefficient values (α, β, and A) were those suggested by Haverkamp et al. for granular materials based on infiltration experiments [22]. The residual and saturated volumetric water contents (0.034 and 0.27 cm^3/cm^3) and the saturated hydraulic conductivity (34.56 cm/day) of the base material were measured in the laboratory using the HPT tests. The saturated hydraulic conductivity of the subgrade of the unpaved road sections was measured using the laboratory rigid-wall compaction mold permeameter test in accordance with ASTM D5856 [24]. The measured saturated hydraulic conductivity of the subgrade is 1.8×10^{-8} cm/day, which can be considered impermeable relative to the surface aggregate material.

In the present numerical analysis, the rate of the simulated rainfall event was taken as 12.7 cm/h, which can be considered a violent rain event for the local weather conditions. This rainfall rate is less than the saturated hydraulic conductivity (K_{sat}) of the surface aggregate so surface runoff cannot occur. The rainfall duration was 30 min, and the total simulation time was 90 min. The rainfall direction is assumed to be perpendicular to the ground surface. The left side boundary condition is fixed at zero flux due to symmetry about the centerline of the road section; therefore water can only drain laterally to the bottom and the ditch side, which is open to the air. Two different bottom boundary conditions were prescribed to simulate the test sections with and without a geocomposite layer. For the section without the geocomposite, the natural subgrade (Table 2) was assumed to be impermeable due to the extremely low saturated hydraulic conductivity. For the test section with a geocomposite drainage layer, the hydraulic conductivity of the bottom boundary was updated after each time step of 0.25 s and set equal to the unsaturated hydraulic conductivity of the bottom layer of the surface material at the previous time step because the flow capacity of the geocomposite is much greater than that of the overlying surface aggregate. Therefore, the hydraulic conductivity of the bottom boundary (the geocomposite layer) is controlled by that of the surface course material.

The numerical analysis results in terms of the 2D volumetric water content distributions of half of the roadway cross-section with and without the geocomposite layer are shown in Figure 10. At the end of the simulated rainfall event, the volumetric water content of the control section surface aggregate is close to the lab-measured saturated volumetric water content of 0.27 cm^3/cm^3 and increases with depth (Figure 9a). One hour after the rain stopped, the water content of the bottom layer remained relatively unchanged, as shown in Figure 10a,b, and the shoulder material was not significantly drier than the centerline material.

Compared to the control section with the 4% crown, the system with a geocomposite layer shows a slightly lower water content at the end of the rain event as shown in Figure 10c, but the water content is greatly reduced one hour after the rain stops (Figure 10d). For the particular rainfall event simulated, these comparisons clearly demonstrate that the 4% road crown cannot remove water from the unpaved road surface material as effectively as the geocomposite drainage layer, which significantly improves the subsurface drainage conditions and effectively prevents the unpaved road surface material from becoming saturated.

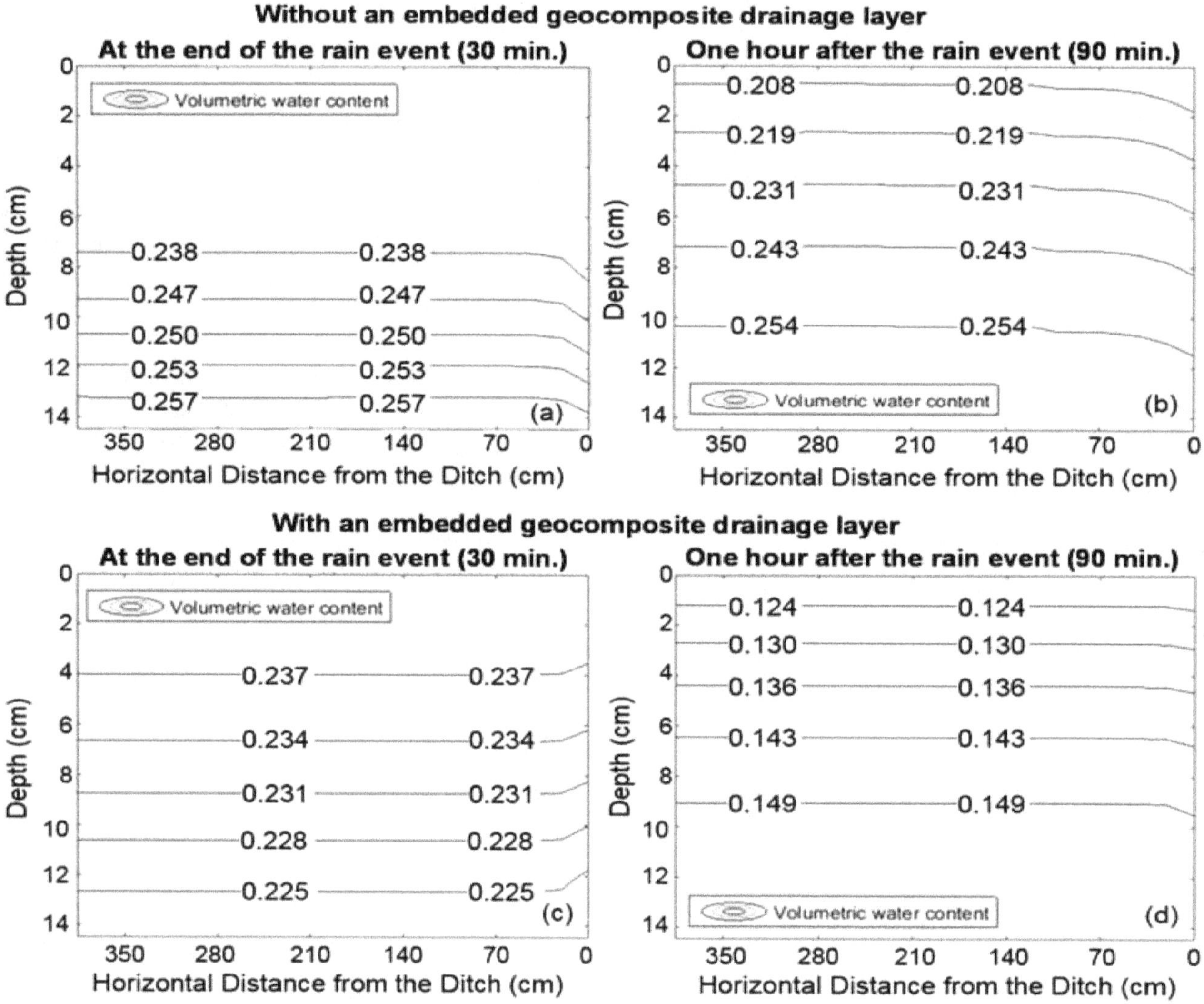

Figure 10. Comparisons of volumetric water content distributions of unpaved road system (**a**,**b**) without and(**c**,**d**) with an embedded geocomposite drainage layer, (**a**,**c**) immediately and (**b**,**d**) one hour after the simulated rainfall event.

7. Summary and Conclusions

This study evaluated the influence of an embedded geocomposite drainage layer on the permeability and stiffness of both paved and unpaved roadway systems using laboratory tests, field tests, and numerical simulations. Both the paved and unpaved test sections constructed using two geocomposite materials were tested in Iowa, USA. A two-dimensional water infiltration model was developed to compare the drainage performance of the geocomposite drainage layer with the 4% roadway surface crown after a simulated heavy rain event. Based on the experimental and numerical analysis data, although limited, some conclusions can be drawn from this study.

For the PCC and WMA sections, the GC-1 geocomposite layer increased the horizontal and vertical saturated hydraulic conductivities by two to three orders of magnitude without significant changes in composite stiffness of the pavement systems. For the unpaved test section, the GC-2 geocomposite layer yielded a lower composite stiffness than the control section due to the relatively softer geonet core of the GC-2. However, the visual observations, field permeability test, and numerical analysis results indicate that the unpaved road section with a geocomposite drainage layer performs better than the control section after a heavy rainfall event and traffic loading, which is attributed to the geocomposite drainage layer rapidly draining water out of the system and effectively preventing the overlying surface material and top layer of subgrade from becoming saturated.

The field permeameter test results indicate that the drainage performance can be significantly influenced by the segregation of the traditional base material, but the geocomposite drainage layer can provide a more uniform subsurface drainage condition. The numerical analysis also showed that the current practice of using the 4% surface crown cannot remove water from the unpaved road surface material as effectively as the geocomposite drainage layer.

The initial construction costs of using the geocomposite for pavement systems might be considered high, but a breakeven cost analysis conducted for the unpaved road project showed that the geocomposite will begin to provide cost savings after 11 years relative to continuation of the current maintenance practices [25]. In addition, the better performance of the roadway system with a geocomposite layer can provide benefits beyond economic ones such as significantly improved ride quality.

Acknowledgments: The research projects were sponsored by the Iowa Department of Transportation and the Federal Highway Administration. The authors would like to thank the Hamilton County Secondary Roads Departments for assistance with constructing the test sections. The assistance of Yang Zhang, Jinhui Hu, and Ji Lu in performing the field tests is also greatly appreciated.

Author Contributions: David White, Pavana Vennapusa, Jeramy Ashlock, and Cheng Li conceived and designed the experiments; David White, Jeramy Ashlock, Pavana Vennapusa, and Cheng Li designed and supervised the constructions of the test sections. Cheng Li, Pavana Vennapusa and Jeramy Ashlock performed the experiments; Cheng Li, Pavana Vennapusa, and Jeramy Ashlock analyzed the data; David White and Pavana Vennapusa contributed materials and analysis tools; Cheng Li, Jeramy Ashlock, David White, and Pavana Vennapusa wrote the paper.

References

1. Han, J. Design of Planar Geosynthetic-Improved Unpaved and Paved Roads. In *First International Symposium on Pavement and Geotechnical Engineering for Transportation Infrastructure*; Huang, B., Bowers, B.F., Mei, G.-X., Luo, S.-H., Zhang, Z., Eds.; American Society of Civil Engineers (ASCE): Nanchang, China, 2011; pp. 31–41.
2. Holtz, R.D.; Christopher, B.R.; Berg, R.R. Geosynthetic design and construction guidelines. In *Highway Capacity Manual: A Guide for Multimodal Mobility Analysis*, 6th ed.; Federal Highway Administration, U.S. Department of Transportation: Washington, DC, USA, 2008; p. 592.
3. Bahador, M.; Evans, T.M.; Gabr, M.A. Modeling effect of geocomposite drainage layers on moisture distribution and plastic deformation of road sections. *J. Geotech. Geoenviron. Eng.* **2013**, *139*, 1407–1418. [CrossRef]
4. Christopher, B.R.; Hayden, S.A.; Zhao, A. Roadway base and subgrade geocomposite drainage layers. In *The Symposium of 'Testing and Performance of Geosynthetics in Subsurface Drainage*; ASTM: Seattle, WA, USA, 2000; pp. 35–51.
5. Henry, K.S.; Holtz, R.D. Geocomposite Capillary Barriers to Reduce Frost Heave in Soils. *Can. Geotech. J.* **2001**, *38*, 678–694. [CrossRef]
6. Stormont, J.C.; Ramos, R.; Henry, K.S. Geocomposite Capillary Barrier Drain Systems with Fiberglass Transport Layer. *Transp. Res. Rec. J. Transp. Res. Board* **2001**, *1772*, 131–136. [CrossRef]
7. Sweet, J.G. Vertical Stiffness Characterization of a Geocomposite Drainage Layer for PCC Highway Pavements. Master's Thesis, West Virginia University, Morgantown, WV, USA, 2005.
8. AASHTO. Geotextile Specification for Highway Applications. In *Standard Specifications for Transportation Materials and Methods of Sampling and Testing, AASHTO M 288*; American Association of State Highway and Transportation Officials: Washington, DC, USA, 2014.
9. ASTM. Standard Test Method for Breaking Force and Elongation of Textile Fabrics (Strip Method). In *Annual Book of ASTM Standards ASTM D5035-11*; ASTM International: West Conshohocken, PA, USA, 2013.
10. ASTM. Standard Test Methods for Water Permeability of Geotextiles by Permittivity. In *Annual Book of ASTM Standards ASTM D4491-99*; ASTM International: West Conshohocken, PA, USA, 2013.
11. ASTM. Standard Test Methods for Determining Apparent Opening Size of a Geotextile. In *Annual Book of ASTM Standards ASTM D4751-12*; ASTM International: West Conshohocken, PA, USA, 2013.

12. White, D.J.; Becker, P.; Vennapusa, P.K.; Dunn, M.J.; White, C.I. Assessing Soil Stiffness of Stabilized Pavement Foundations. *Transp. Res. Rec. J. Transp. Res.* **2013**, *2335*, 99–109. [CrossRef]
13. White, D.J.; Vennapusa, P.K.; Becker, P.J.; Zhang, Y.; Dunn, M.J. Performance Assessment of Cement Stabilized, Polymer Fiber-Reinforced Pavement Foundation Layers. In Proceedings of the Geosynthetics Conference, Portland, OR, USA, 15–18 February 2015.
14. Li, C.; Ashlock, J.C.; White, D.J.; Vennapusa, P. Mechanistic-based comparisons of stabilised base and granular surface layers of low-volume roads. *Int. J. Pavement Eng.* **2017**. [CrossRef]
15. ASTM. Standard Test Methods for Laboratory Compaction Characteristics of Soil Using Standard Effort (12,400 ft-lbf/ft_3 (600 kN-m/m_3)). In *Annual Book of ASTM Standards ASTM D698-07*; ASTM International: West Conshohocken, PA, USA, 2013.
16. ASTM. Standard Test Method for Field Measurement of Hydraulic Conductivity Using Borehole Infiltration. In *Annual Book of ASTM Standards ASTM D6391-11*; ASTM International: West Conshohocken, PA, USA, 2013.
17. Vennapusa, P.; White, D.J.; Jahren, C.T. In-Situ Permeability of Unbound Granular Bases Using the Air Permeameter Test. In Proceedings of the 85th Annual Transportation Research Board Conference, Washington, DC, USA, 22–26 January 2006.
18. Zhang, J.; White, D.; Taylor, P.C.; Shi, C. A case study of evaluating joint performance in relation with subsurface permeability in cold weather region. *Cold Reg. Sci. Technol.* **2015**, *110*, 19–25. [CrossRef]
19. White, D.J.; Vennapusa, P.K.R.; Suleiman, M.T.; Dahren, C.T. An in-situ device for rapid determination of permeability for granular bases. *Geotech. Test. J.* **2007**, *30*, 282–291.
20. White, D.J.; Vennapusa, P.; Zhao, L. Verification and repeatability analysis for the in situ air permeameter test. *Geotech. Test. J.* **2014**, *37*, 365–376. [CrossRef]
21. AASHTO. *AASHTO Guide for Design of Pavement Structures*; American Association of State Highway and Transportation Officials: Washington, DC, USA, 1993.
22. Haverkamp, R.; Vauclin, M.; Touma, J.; Wierenga, P.J.; Vachaud, G. A Comparison of Numerical Simulation Models for One-Dimensional Infiltration. *Soil Sci. Soc. Am. J.* **1977**, *41*, 285–294. [CrossRef]
23. Jury, W.A.; Horton, R. *Soil Physics*, 6th ed.; John Wiley & Sons: Hoboken, NJ, USA, 2004.
24. ASTM. Standard Test Method for Measurement of Hydraulic Conductivity of Porous Material Using a Rigid-Wall, Compaction-Mold Permeameter. In *Annual Book of ASTM Standards ASTM D5856-07*; ASTM International: West Conshohocken, PA, USA, 2013.
25. Li, C.; Ashlock, J.C.; White, D.J.; Vennapusa, P. *Low-Cost Rural Surface Alternatives: Demonstration Project*; Iowa Highway Research Board (IHRB) Project TR-664; Iowa State University: Ames, IA, USA, 2015.

8

Effect of Fibers on Mixture Design of Stone Matrix Asphalt

Yanping Sheng [1,*], Haibin Li [2], Ping Guo [3], Guijuan Zhao [2], Huaxin Chen [1] and Rui Xiong [1]

1 School of Materials Science and Engineering, Chang'an University, Xi'an 710064, China; chx92070@163.com (H.C.); xiongr61@126.com (R.X.)
2 School of Architecture and Civil Engineering, Xi'an University of Science and Technology, Xi'an 710054, China; lihaibin1212@126.com (H.L.); guijuanzhao@126.com (G.Z.)
3 Xi'an Highway Research Institute, Xi'an 710054, Shaanxi, China; guoping8088@163.com
* Correspondence: syp@chd.edu.cn

Academic Editors: Zhanping You, Qingli (Barbara) Dai and Feipeng Xiao

Abstract: Lignin fibers typically influence the mixture performance of stone matrix asphalt (SMA), such as strength, stability, durability, noise level, rutting resistance, fatigue life, and water sensitivity. However, limited studies were conducted to analyze the influence of fibers on the percent voids in mineral aggregate in bituminous mixture (VMA) during the mixture design. This study analyzed the effect of different fibers and fiber contents on the VMA in SMA mixture design. A surface-dry condition method test and Marshall Stability test were applied on the SMA mixture with four different fibers (i.e., flocculent lignin fiber, mineral fiber, polyester fiber, blended fiber). The test results indicated that the bulk specific gravity of SMA mixtures and asphalt saturation decreased with the increasing fiber content, whilst the percent air voids in bituminous mixtures (VV), Marshall Stability and VMA increased. Mineral fiber had the most obvious impact on the bulk specific gravity of bituminous mixtures, while flocculent lignin fiber had a minimal impact. The mixture with mineral fiber and polyester fiber had significant effects on the volumetric properties, and, consequently, exhibited better VMA over the conventional SMA mixture with lignin fiber. Modified fiber content range was also provided, which will widen the utilization of mineral fiber and polyester fiber in the applications of SMA mixtures. The mixture evaluation suggested no statistically significant difference between lignin fiber and polyester fiber on the stability. The mineral fiber required a much larger fiber content to improve the mixture performance than other fibers. Overall, the results can be a reference to guide SMA mixture design.

Keywords: stone matrix asphalt; volume parameters; Marshall Stability; flocculent lignin fiber; polyester fiber; mineral fiber; fiber content

---

1. Introduction

Stone matrix asphalt (SMA) is a hot asphalt mixture in which coarse aggregate interlocks to form a stone skeleton that resists permanent deformation. SMA was first used in Europe as a mixture that would resist the wear of studded tires. Then it was used successfully in the United States in 1990, and is now widely used in China. The advantages of SMA include high resistance to rutting, excellent low-temperature performance, improved macrotexture, long service life, low tire noise, less water spray from tires, and weak light reflection on rainy nights [1–3]. However, the coarse texture of an SMA mixture may result in more internal air voids that are related to performance degradation, even when the volume of air voids is the same as that of common asphalt mixtures [4]. The coarse surface texture makes it more difficult to differentiate between mixture air voids and surface texture.

Fiber additive is important for SMA due to its oil absorptive characteristics. A certain quantity of fiber should be added into the SMA mixture in order to prevent asphalt from flowing out due to the high asphalt content. The outflow of the asphalt can result in fat spots on the pavement surface [5]. The mineral skeleton of coarse aggregate supplies the mixture with a strong particle interlock increasing the mixture resistance, and the mastic supplies the mixture with better durability. Since the fiber occupies some space, the gap between aggregates will be increased if it blocks the contact of the aggregates, and then the mixture performance will be reduced through the influence of volumetric parameters. The volumetric parameters are the direct controlling indicators in the design and preparation of the SMA mixtures.

In the early stage of hot-mix-asphalt (HMA) mix design, percent voids in mineral aggregate in bituminous mixtures (VMA) were determined and maintained throughout the mix-design procedure. VMA includes the air voids and the volume occupied by the effective asphalt content. This volumetric property is correlated to mechanical properties [6–8], e.g., small percent air voids in bituminous mixtures (VV) will cause bleeding and high VV may lead to water damage or instability in asphalt pavement. In addition to the size gradation, VMA is one of the most important HMA design criteria to obtain durable pavement, and it significantly affects the permanent deformation and fatigue performance of a compacted mix [9,10]. The use of VMA criteria for mix design is a time-honored and fairly successful tool. The VMA requirements for HMA mixtures were initially developed in the 1950s and were considered one of the most important volumetric parameters for HMA and SMA mixtures [11,12]. Then other influence factors of VMA, such as aggregate factors and volumetric basis, were pointed out, and VMA specifications were strongly emphasized during the process of asphalt mixture design and analysis [12–15].

In order to determine VMA, the bulk density, percent air voids in bituminous mixtures (VV), and percent voids in mineral aggregate that are filled with asphalt in bituminous mixtures (VFA) have to be obtained first because they are critical parameters to obtain proper VMA in design and practice. Studies have reported the difficulty of meeting the minimum VMA requirement in an efficient manner [10,16,17]. It indicated that the minimum VMA should be based on the minimum asphalt film thickness rather than the minimum asphalt content [18]. Although both Bailey's method and the NCHRP 9-33 manual have provided suggestions for adjusting the mix design to achieve the target VMA, the determination of VMA still requires a large amount of experimental testing [15,19]. As another point of view, VMA was to incorporate at least the minimum permissible asphalt content into the mixture to ensure its durability. VMA and the shape of aggregate particles influence workability, shear resistance, fatigue, and durability of the mixture [20–24].

The most commonly adopted fibers in SMA mixtures are lignin fibers. The success in SMA mixtures spurred the adoption of the fiber for many major highway projects. Then, lignin fiber, glass fiber, and mineral fiber have been studied in asphalt mixtures [25,26]. From then on, other types of fibers, such as carpet fiber, polyester fiber, waste tires, cellulose oil palm fiber, waste glass fiber, and coconut fiber, were used to study the service properties of the HMA mixture and SMA mixture [26–30]. The studies focused on the mixture to obtain a better performance, such as strength, stability, durability, reduction of noise, rutting resistance, fatigue life, and water sensitivity. However, fiber types may influence oil absorption and fiber content will affect the VV. Then it can affect the VMA directly. Limited studies were conducted to analyze the influence of fibers on the VMA during the mixture design.

Therefore, this study investigated the effects of four different fibers on the mixture volume parameter during the SMA mixture design, with the goal of identifying the adaptability of polyester fiber and mineral fiber for satisfactory binder performance. The mineral fiber has a similar density with aggregates and smaller oil absorption and specific surface area, which means it cannot absorb much asphalt binder to fill the mineral outside space, and is less sensitive to the content change. Polyester fiber has better asphalt absorption and higher ductility; therefore, it can form much more space in the SMA mixture [31]. High content of polyester fiber means low asphalt content, which potentially reduces the adhesion between the aggregate and asphalt binder.

Empirical binder tests were conducted to identify volume parameters and appropriate contents of different fibers. The bulk specific gravity of bituminous mixtures, VV, VFA, and VMA, were studied with a surface-dry condition method test. Then, the Marshall Stability of the SMA mixtures with optimized mixing procedures was evaluated to check the effect of the fibers on the mechanical performance. The flocculent lignin fiber was used as a control fiber. The suggested fiber content of this study for different fibers could provide better performance of the SMA mixture. The results provide effective references for the SMA mixture design.

2. Materials and Methods

2.1. Materials

2.1.1. Asphalt Binder

A modified asphalt binder, i.e., styrene-butadiene-styrene (SBS) (I-C), which has been regularly used in pavement engineering, was selected in this study. Table 1 shows the measured technical indicators of the SBS asphalt binder. Basic binder tests, such as the penetration, softening point, and ductility were conducted to evaluate the fundamental characteristics of SBS asphalt binder which may influence the SMA mixture.

Table 1. Technical indicators of the asphalt binder.

Test Properties		Unit	Test Results	Specification Requirements
Penetration (25 °C, 100 g, 5 s)		0.1 mm	71.7	60~80
Softening point		°C	97	≥55
Ductility (5 °C, 5 cm/min)		cm	33.4	≥30
Penetration index		-	0.21	≥−0.4
Density (15 °C, g/cm^3)		-	1.032	-
Viscosity (135 °C)		Pa·s	1.83	≤3
Flash point		°C	328	≥230
Solubility (Trichloroethylene)		%	99.37	≥99
Segregation, 48 h D-value of Softening point		°C	2.1	≤2.5
Elastic recovery (25 °C)		%	98	≥65
Short-term oven aging, 163 °C, 75 min	Mass change	%	0.006	≤±1.0
	Penetration ratio, 25 °C	%	76.4	≥60
	Ductility, 5 °C	cm	29.3	≥20

The asphalt penetration test is used to evaluate the asphalt's soft and hard levels and its shear resistance. The test reflects the asphalt's relative viscosity. The softening point test is used to determine the temperature at which the asphalt becomes soft and achieves a certain viscosity. Ductility is mainly about deformability of asphalt and indirectly reflects low-temperature anti-cracking property. It is an important index that can be used to evaluate asphalt plasticity such that the larger the ductility value, the better plasticity of the asphalt. All of these are part of the basic performance index to evaluate the asphalt binder.

2.1.2. Aggregate

The diabase gravel and the limestone sand were chosen as coarse aggregate and fine aggregate in this study. Some important technical indicators are listed in Tables 2 and 3, respectively. To create a better adhesion between the aggregate and the asphalt binder during the mixing procedure, the aggregates were first cleaned and then dried well. The required amounts of aggregates and fillers were placed into an oven at 105 °C for 5 h, and then the temperature rose to 180 °C for mixing.

Table 2. Technical indicators of coarse aggregate.

Test Properties		Unit	Test Results	Specification Requirements
Apparent relative density		-	2.927	≥2.60
Crushing value		%	8.3	≤26
Sturdiness		%	9.8	≤12
LA abrasion value		%	10.2	≤28
Water absorption		%	0.48	≤2.0
Adhesion with asphalt		Grade	5	5
<0.075 mm Grain content		%	0.3	≤1
Soft stone content		%	1.1	≤3
Needle and plate particle content	Mixture	%	6.2	≤15
	>9.5 mm	%	5.3	≤12
	<9.5 mm	%	9.3	≤18

Table 3. Technical indicators of fine aggregate.

Test Properties	Unit	Test Results	Specification Requirements
Sturdiness (>0.3 mm)	%	14	≥12
Apparent relative density	-	2.745	≥2.50
Methylene blue value (g/Kg)	%	10.6	≤25
Angularity (flow time)	s	43.1	≥30

2.1.3. Mineral Filler

The mineral filler was produced by limestone. Some important technical indicators of mineral filler were shown in Table 4.

Table 4. Technical indicators of mineral filler.

Test Properties	Unit	Test Results	Specification Requirements
Apparent density	t/m^3	2.726	≥2.50
Hydrophilic coefficient	-	0.7	<1
Plasticity index	%	2.3	<4
Water content	%	-	≤1.0

2.1.4. Fiber

Four different fibers, i.e., flocculent lignin fiber, mineral fiber, polyester fiber, and blended fiber, were selected in order to analyze the effect on VMA at different fiber contents. The blended fiber was made up of flocculent lignin fiber and polyester fiber with mass ratio of 2:1. Some important technical indicators of these fibers are listed in Table 5.

Table 5. Technical indicators of different fibers.

Test Properties	Unit	Flocculent Lignin Fiber	Polyester Fiber	Mineral Fiber
Relative density	g/cm^3	1.813	1.390	2.720
Length	mm	5	6	4
Thickness	mm	0.047	-	-
Diameter	μm	-	20	5
Ash content (by weight)	%	16	-	8.8
PH value	-	6.9	-	-
Water content rate (by weight)	%	3	2.43	-
Oil absorption rate	times	6.5	4.1	-
Melting points	°C	-	260	>1000
Tensile strength	MPa	-	570	935

2.2. Test Methods

2.2.1. Mixing Proportion Determination

A typical SMA mixture, i.e., SMA-13 with a nominal maximum aggregate size of 13.2 mm, which has been regularly used in asphalt pavement construction, was selected to study volumetric parameters and mixture stability. Table 6 and Figure 1 show the gradation of the SMA-13. As an additive, the fiber was added to make mixture specimens.

Table 6. Gradation of SMA-13 asphalt mixture.

Composite	Percentage	Mesh Size (mm/%)									
		16.0	13.2	9.5	4.75	2.36	1.18	0.6	0.3	0.15	0.075
9.5~16 mm	44%	100	85.1	12.3	0.40	0.40	0.40	0.40	0.40	0.40	0.40
4.75~9.5 mm	33%	100	100	97.9	8.6	0.5	0.50	0.50	0.50	0.50	0.50
0~2.36 mm	13%	100	100	100	100	92.6	62.0	36.1	19.9	13.6	9.9
Mineral filler	10%	100	100	100	100	100	100	100	99.3	95.4	85.3

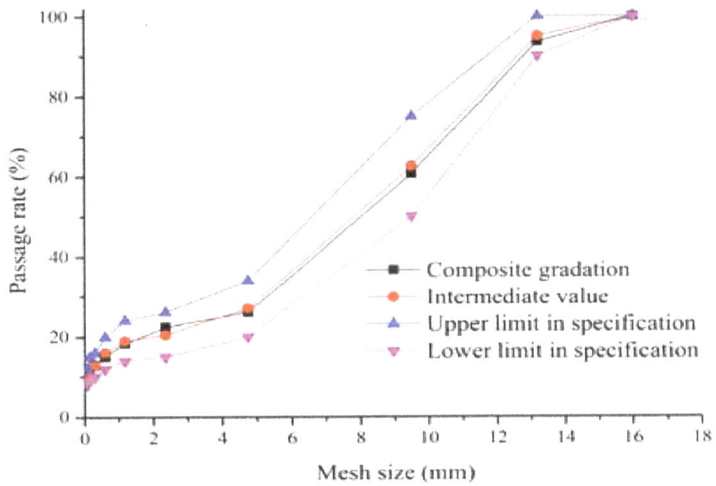

Figure 1. Gradation of the SMA-13 mixture.

Using the materials and the aggregate gradation described above, the SMA mixtures with different fibers and 0.3% content (by the weight of total mix) was prepared for laboratory testing. The optimal asphalt aggregate ratio of SMA mixtures with different fibers shown in Table 7 were determined according to the Marshall mixture design method of the Chinese technical specification for construction of highway asphalt pavements (JTG F40-2004).

Table 7. Asphalt aggregate ratio of SMA mixtures with different fibers.

Different Fibers	Flocculent Lignin Fiber	Mineral Fiber	Polyester Fiber	Blended Fiber
Asphalt aggregate ratio (%)	5.9	5.5	5.7	5.9

2.2.2. SMA Sample Preparation

In this study, SMA specimens were prepared using the compaction method. The dimensions were 101.6 mm × 63.5 mm. According to the standard requirement and field construction experience, the fiber content (by the weight of the total mix) used in the SMA mixture was selected, which was 0.1%, 0.2%, 0.3%, 0.4%, 0.5%, and 0.6%, respectively. The fiber types and contents are shown in Table 8.

Table 8. Different fiber types and fiber content in SMA mixtures.

Type	**Flocculent Lignin Fiber**					**Mineral Fiber**					
Content (%)	0.1	0.2	0.3	0.4	0.5	0.1	0.2	0.3	0.4	0.5	0.6
Type	Blended fiber					Polyester fiber					
Content (%)	0.1	0.2	0.3	0.4	0.5	0.1	0.2	0.3	0.4	0.5	-

The density was determined with the surface-dry condition method (T0705-2011/JTG E20-2011), which was very similar to ASTM D2726-14. There were mainly two differences. First, ASTM D2726-14 explained if the temperature of the specimen differs from the temperature of the water bath by more than 2 °C (3.6 °F), the specimen should be immersed in the water bath for 10 to 15 min; instead of 3 to 5 min. The immersed time in the water bath was only 3 to 5 min in the T0705-2011. Secondly, ASTM D2726-14 required, after determining the mass in water and in a saturated-surface dry condition, thoroughly drying the specimen to a constant mass at 110 ± 5 °C (230 ± 9°F). While it only required making the specimens thoroughly dry, it did not illustrate the temperature in the T0705-2011.

The loose fibers (by the weight of total mix) were first blended with the hot aggregates to prevent the asphalt binder from draining during the mixing procedure. The heated asphalt binder was added after mixing 1 to 1.5 min. Then the heated filler was added until all of the aggregate was completely covered. The total mixing time was 3 min. During the mixing process, the temperature was kept between 170 °C and 180 °C. Finally, the Marshall specimens were made with dimensions of 101.6 mm × 63.5 mm using the compaction method. For note, the final temperature was greater than 145 °C.

2.2.3. Marshall Stability and Flow Tests

The Marshall stability and flow tests were conducted to evaluate the resistance of asphalt mixtures to distortion, displacement, rutting, and shearing stresses. The stability test measures the maximum load sustained by the specimen at a loading rate of 50.8 mm/min. Basically, the applied testing load increases until the specimen splits into two pieces, then the loading is finished and the maximum load is recorded as the Marshall Stability.

3. Results and Discussion

3.1. Effect of Fiber on Bulk Specific Gravity, VV, and VFA

Figure 2 displayed the bulk specific gravity of SMA mixtures with different fibers and fiber contents. The data show an inverse correlation between bulk specific gravity of SMA mixtures and fiber content for fiber types. The SMA mixture with mineral fiber had the largest bulk specific gravity values, followed by the mixture with polyester fiber. The mixture with flocculent lignin fiber had the smallest values. All of the bulk specific gravity values were between 2.465 and 2.523. The bulk specific gravity of SMA mixtures with polyester fiber and blended fiber decreased with fiber content increasing from 0.1% to 0.4%, and then it maintained a slight decrease with fiber content from 0.4% to 0.5%. The mixture with mineral fiber had a similar variety, with the only difference being the relatively smaller reduction in fiber content from 0.5% to 0.6%.

These results indicated that it was not necessarily true that the larger bulk specific gravity of the SMA mixture results from a higher fiber content. The higher fiber content results in lower bulk specific gravity. In terms of the four different fibers, the mineral fiber has a density very close to that of the aggregate, and it was much easier to combine with the asphalt binder than other fibers. Under the same compaction effort, the SMA mixture with mineral fiber can reach a larger dry mass per unit volume. Therefore, it appeared to have a larger bulk specific gravity value.

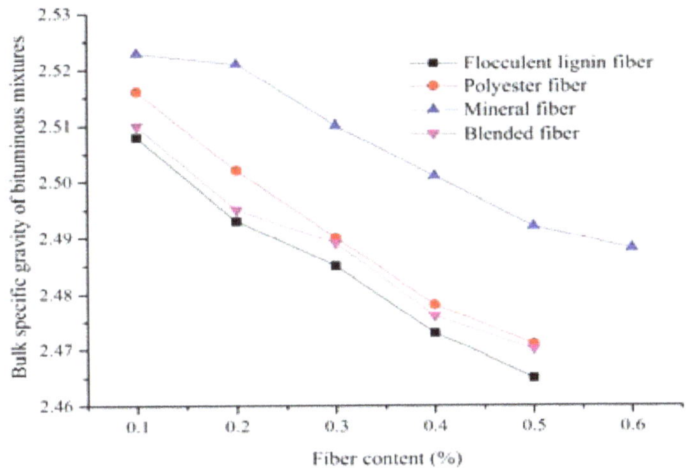

Figure 2. Bulk specific gravity of SMA mixtures with four different fibers.

Figure 3 displayed the percent air voids in SMA mixtures with different fibers and fiber contents. The data showed a positive correlation between VV and fiber content. Higher fiber content resulted in better asphalt absorption and adsorption in the SMA mixture. The measured VV values of different fiber types and contents were between 2.9% to 4.4%.

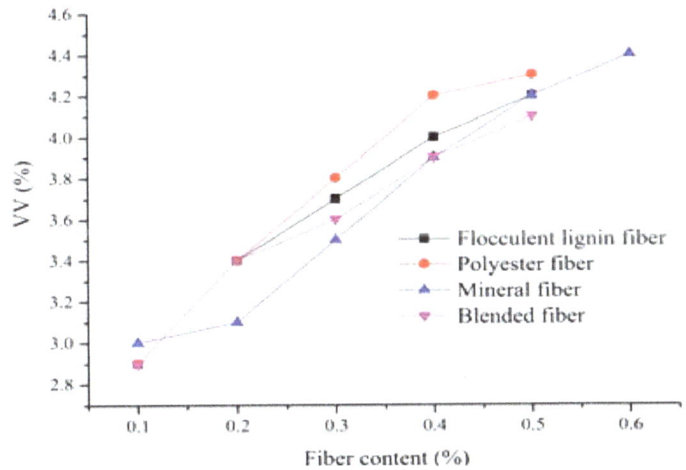

Figure 3. Percent air voids in SMA mixtures with four different fiber types.

It was found from the results that the polyester fiber had an important impact on the VV values when the content was up to 0.2%. It indicated that polyester fiber prevented the aggregates contacting each other and formed much more space in the mixture due to its higher ductility. The VV with polyester fiber, flocculent lignin fiber, and blended fiber increased with the fiber content increasing from 0.1% to 0.4%, and then it maintained a slight increase with the fiber content from 0.4% to 0.5%, and the increments of flocculent lignin fiber, polyester fiber, and blended fiber were 4.5%, 1.9%, and 4.9%, respectively. However, mineral fiber had little impact on VV values when the content was less than 0.2% due to the negligible impact on the asphalt absorption. However, when the content was more than 0.2%, mineral fiber began to show its effect on absorptive action and the mixture's adsorption, for which the VV showed a significant increase. At the same content, the mixture with polyester fiber had the largest VV value.

In practice, the optimum fiber type and content should be selected to achieve satisfactory performance of the SMA mixture in the production process. This study suggests that the minimum content is above 0.3% for mineral fiber and above 0.2% for the other fibers. However, the specific value may depend more on other factors, such as cost considerations, availability, and ease of field construction application because of the difference between lab tests and field construction.

Figure 4 showed the percentage of voids in mineral aggregate (VFA) that are filled with asphalt in SMA mixtures with different fibers and fiber contents. The data from Figure 4 indicated that fiber types and contents had an obvious impact on VFA, which decreased as fiber content increased. All of the VFA values were between 73% and 82%. The percentage of voids in mineral aggregate that are filled with asphalt in SMA mixtures with blended fiber, flocculent lignin fiber, and polyester fiber decreased with the fiber content increasing from 0.1% to 0.4%, and then it maintained a slight decrease with fiber content from 0.4% to 0.5%. When the additive was mineral fiber, it had similar variety, with the only difference being the relatively smaller reduction in the content from 0.1% to 0.2% and 0.5% to 0.6%.

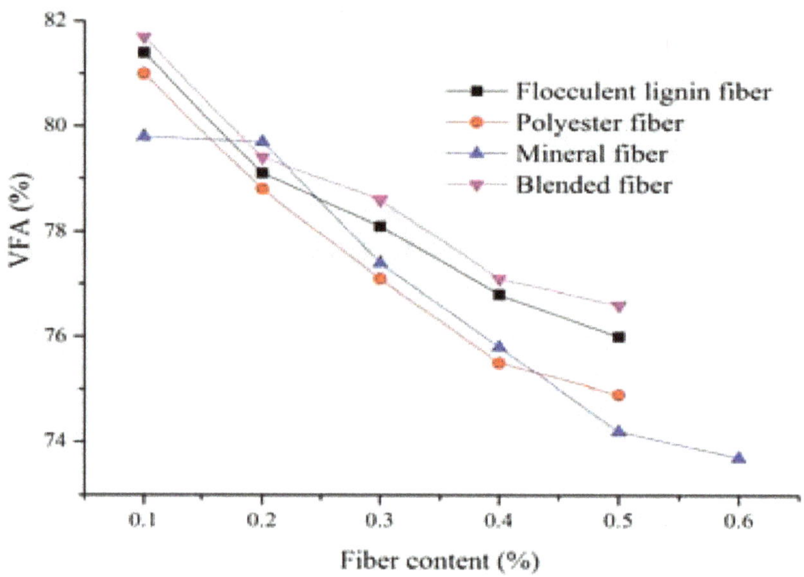

Figure 4. VFA of SMA specimens with four different fiber types.

The content change of mineral fiber had a negligible impact on the VFA values due to its large density and low oil absorption rate when the content was less than 0.2%. Then VFA showed a significant decrease after the content was more than 0.2%. However, when the mineral fiber content was up to 0.4%, it began to show its absorptive action and the mixture's adsorption, which made the VFA much smaller than that with polyester fiber. The three other fibers had obvious impacts on the VFA value, and the SMA mixture with polyester fiber had a much smaller VFA than that with blended fiber and flocculent lignin fiber.

The results indicated that the mineral fiber and polyester fiber had better asphalt absorption and adsorption in the SMA mixture. To obtain the same VFA value, lower content is needed for mineral fiber and polyester fiber. Therefore, the fiber selection should comprehensively consider both absorptive action and the mixture's adsorption. Higher contents of mineral fiber and polyester fiber were not the best choice. Fiber with higher adsorption, but lower absorption, will improve the SMA mixture volume index.

3.2. *Effect of Fiber on Percent Voids in Mineral Aggregate in Bituminous Mixtures (VMA)*

Figure 5 displayed the percent voids in mineral aggregate in SMA mixtures. The VMA increased with fiber content increasing. The measured VMA values were between 15% and 18%. At the same fiber content, the VMA value had the minimum gap between lignin fiber and blended fiber. It was only 16.4% even when the mineral fiber content was up to 0.5%. The values were more than 17.3% for the other fibers. This indicated that the mineral fiber had the minimum impact on VMA. The same trend was found for Marshall Stability. Compared with the previous literature, the mix designer was able to judge the proper VMA requirement for each kind of fiber.

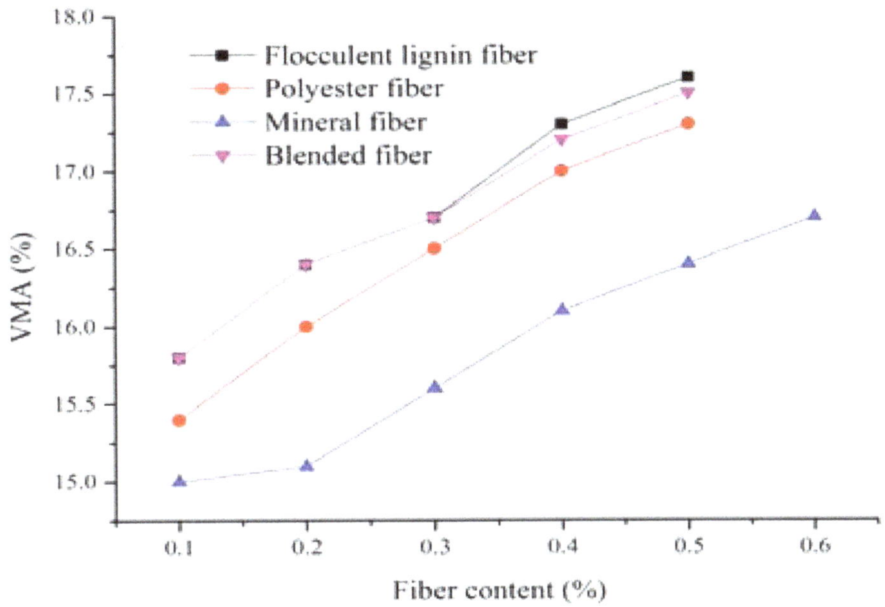

Figure 5. VMA of SMA specimens with four different fiber types.

The VMA with polyester fiber, flocculent lignin fiber and blended fiber increased with fiber content increasing from 0.1% to 0.4%, and then the increment gradually reduced with fiber content from 0.4% to 0.5%. However, the content change of mineral fiber had a negligible impact on VMA when the content was less than 0.2%. The mixture with mineral fiber had a similar variety with the only difference being the relatively lower VMA values with fiber contents from 0.2% to 0.6%. When the content of flocculent lignin fiber, blend fiber, and polyester fiber was 0.3%, and mineral fiber content was 0.4%, the VMA values of SMA mixtures were 16.7%, 16.7%, 16.5%, and 16.1%, respectively, which explained much more mineral fiber content was needed to achieve a target VMA value in the SMA mixture design.

Furthermore, mineral fiber had a similar density with aggregate and smaller oil absorption and smaller specific surface area; it cannot absorb much asphalt binder to fill the outside space of the mineral fiber. Therefore, effective asphalt content was not sensitive to the content change. Therefore, the SMA mixture with mineral fiber can obtain a lower VMA under the same compaction effort. These test findings can be a reference and used to promote the use of fibers in SMA mixtures.

3.3. Effect of Fiber on Marshall Stability

Figure 6 displayed the Marshall Stability of SMA mixtures. It had the similar trend with VV. All of the Marshall Stability values were between 7 and 14 kN. Compared with the previous literature, this was larger than with cellulose oil palm fibers. The data from Figure 6 indicated that fiber types and contents had evident correlations between fiber content and Marshall Stability, which increased with fiber content increasing. When the fiber content was less than 0.3%, the SMA mixture with polyester fiber showed better stability. When the fiber content was larger than 0.3%, the SMA mixture with blended fiber showed better stability than that with the other three fibers.

The SMA mixture with the mineral fiber had the lowest Marshall Stability values within the content ranges. The amount of polyester fiber was the smallest at the same Marshall Stability when the fiber content was larger than 0.2%, followed by blended fiber, flocculent lignin fiber, and mineral fiber. Meanwhile, the Marshall Stability of SMA mixtures with polyester fiber and flocculent lignin fiber increased with the fiber content increasing from 0.1% to 0.3%. The Marshall Stability then maintained a slight increase with the fiber content from 0.3% to 0.5%, and 0.4% to 0.6% for the blended fiber and the mineral fiber.

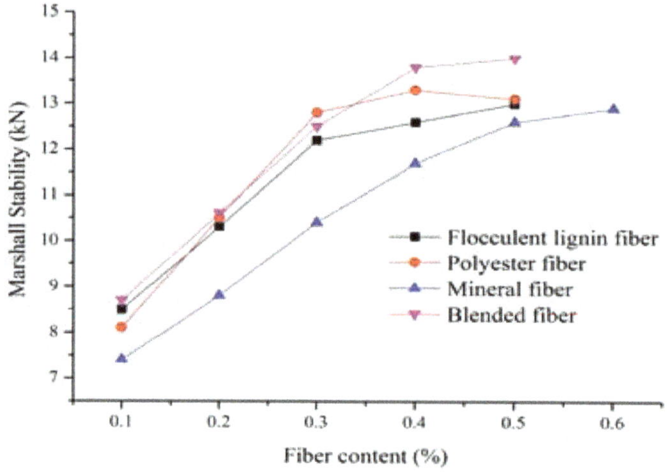

Figure 6. Marshall Stability of the SMA mixtures with four different fibers.

4. Conclusions

HMA and SMA are commonly used pavement materials which are composed of aggregates, fillers, binders, and fibers. These loose materials are mixed together by controlling the volumetric properties to obtain an optimum mechanical performance. Thus, the volumetric properties are the direct indicators in design and construction practice. This paper analyzed the effects of four fibers on the volumetric parameters and Marshall Stability (mechanical performance) of SMA mixtures. As a control fiber, the flocculent lignin fiber was compared with the other three fibers.

This paper found that both the polyester fiber and mineral fiber had an obvious impact on VV, VFA, VMA, and Marshall Stability of the SMA mixture compared to the regular flocculent lignin fiber. The experiment data indicated that increasing the content of polyester fiber and mineral fiber benefitted the mechanical performance of SMA mixtures, which potentially enlarged the fibers' application range. This study suggests that the optimum fiber content is to be larger than 0.3% for flocculent lignin fiber, blended fiber, and polyester fiber, and larger than 0.4% for mineral fiber.

When fiber content was up to 0.5%, the VMA of the sample with the mineral fiber and the polyester fiber was both larger than 16.4%. However, it was more than 17.5% for lignin fiber and blended fiber. Therefore, SMA mixtures with mineral fiber and polyester fiber will be better in heavy traffic sections or hot areas with larger high-temperature stability. They can provide more options in SMA mixture design, which indicates the significant potential for their application. Fiber with higher adsorption, but lower absorption, is a better choice as a potential alternative for lignin fiber in SMA mixture design and construction.

The study proved the adaptability of polyester fiber and mineral fiber in SMA mixture design and provided proper content ranges of the fibers. However, other factors (e.g., cost, availability, and ease of field construction application) should be considered in practice. The fiber type and fiber content should be selected according to field construction experience. The results obtained from the lab tests are a reference to bring convenience to field projects.

In the future, microstructure analysis will be the research emphasis. The dispersion and dissolution of mineral fiber or polyester fiber in the asphalt binder will be characterized with micro-analytical instruments. Then, the chemical reactions between polyester fiber and the asphalt binder can be evaluated. Finally, the high-temperature stability and cracking performance (load-induced, fatigue, top-down, etc.,) of SMA mixtures with mineral fiber and polyester fiber will be evaluated in different areas. Through synthetic consideration of microscopic analysis and field construction, the application range of mineral fiber and polyester fiber will be further expanded in pavement industries and create economic benefits.

Acknowledgments: The project was supported by the National Natural Science Foundation of China (No. 51608046), Qinghai Transportation Science and Technology Project (No. 2014-GX-A2A), the Special Fund for Basic Scientific Research of Central Colleges Chang'an University (No. 310831171009 and 310831171001) and China Postdoctoral Science Foundation (No. 2015T81000 and 2015M582590). Y. Sheng also appreciates the funding from the Chinese Scholarship Council (CSC) of the Ministry of Education, P.R. China.

Author Contributions: Yanping Sheng and Haibin Li conceived and designed the experiments; Ping Guo and Guijuan Zhao performed the experiments; Huaxin Chen and Rui Xiong analyzed the data; Yanping Sheng and Haibin Li wrote the paper.

References

1. Cooley, L.A., Jr.; Hurley, G.C. Potential of using stone matrix asphalt (SMA) in Mississippi. Available online: http://mdot.ms.gov/documents/research/Reports/Interim%20and%20Final%20Reports/Using%20Stone%20Matrix%20Asphalt%20(SMA)%20In%20Mississippi.pdf (accessed on 11 March 2017).
2. Dong, Y.; Tan, Y. Mix design and performance of crumb rubber modified asphalt SMA. *J. Mater. Civil Eng.* **2011**, *10*, 78–86.
3. Xue, Y.J.; Hou, H.B.; Zhu, S.J.; Zha, J. Utilization of municipal solid waste incineration ash in stone mastic asphalt mixture: Pavement performance and environmental impact. *Constr. Build. Mater.* **2009**, *23*, 989–996. [CrossRef]
4. Xie, H.B.; Watson, D. Determining air voids content of compacted stone matrix asphalt mixtures. *Transp. Res. Record: J. Transp. Res. Board* **2014**. [CrossRef]
5. Micheal, L.; Burke, G.; Schwartz, C.W. Performance of stone matrix asphalt pavements in Maryland. *Asph. Paving Technol.: Assoc. Asph. Paving Technol.* **2003**, *72*, 287–314.
6. Christensen, D.W.; Pellinen, T.; Bonaquist, R.F. Hirsch model for estimating the modulus of asphalt concrete. *J. Assoc. Asph. Paving Technol.* **2003**, *72*, 97–121.
7. You, Z.P.; Dai, Q.L. Dynamic complex modulus predictions of hot-mix asphalt using a micromechanical-based finite element model. *Can. J. Civil Eng.* **2007**, *34*, 1519–1528. [CrossRef]
8. Goh, S.W.; You, Z.P.; Williams, R.; Li, X.J. Preliminary dynamic modulus criteria of HMA for field rutting of asphalt pavements: Michigan's experience. *J. Transp. Eng.* **2011**, *137*, 37–45. [CrossRef]
9. Harvey, J.; Eriksen, K.; Sousa, J.; Monismith, C. Effects of laboratory specimen preparation on aggregate–asphalt structure, air-voids contents measurement, and repetitive simple shear test results. *Transp. Res. Board* **1994**, 113–122.
10. Hinrichsen, J.; Heggen, J. Minimum VMA in HMA based on gradation and volumetric properties. In Proceedings of the 75th Transportation Research Board Annual Meeting, Washington, DC, USA, 7–11 January 1996; Volume 1545, pp. 75–79.
11. McLeod, N.W. *Void Requirements for Dense-Graded Bituminous Paving Mixtures*; American Society for Testing and Materials (ASTM): Philadelphia, PA, USA, 1959. [CrossRef]
12. Roberts, F.L.; Kandhal, P.S.; Brown, E.R.; Lee, D.-Y.; Kennedy, T.W. *Hot Mix-Asphalt Materials, Mixture Design, and Construction*, 2nd ed.; NAPA Research and Education Foundation: Lanham, MD, USA, 1991.
13. Hislop, W.; Corree, B. A Laboratory Investigation Into the Effects of Aggregate-Related Factors of Critical VMA in Asphalt Paving Mixtures. Available online: http://publications.iowa.gov/11672/1/tr415.pdf (accessed on 11 March 2017).
14. Sengoz, B.; Topal, A. Minimum voids in mineral aggregate in hot-mix asphalt based on asphalt film thickness. *Build. Environ.* **2007**, *42*, 3629–3632. [CrossRef]
15. Christensen, D.W. Draft final report to the National Cooperative Highway Research Program (NCHRP) on Project NCHRP 9-33: A mix design manual for hot mix asphalt. Advanced Asphalt Technologies: Sterling, VA, USA, 2009; Available online: http://www.trb.org/Main/Blurbs/161248.aspx (accessed on 17 March 2017).
16. Anderson, R.M.; Bahia, H.U. Evaluation and selection of aggregate gradations for asphalt mixtures using Superpave. *Transp. Res. Rec.: Transp. Res. Board* **1997**, *1583*, 91–97. [CrossRef]
17. Mallick, R.B.; Buchanan, M.S.; Kandhal, P.S.; Bradbury, R.L.; McClay, W. Rational approach of specifying the voids in the mineral aggregate for dense-graded hot-mix asphalt. In Proceedings of the 79th Annual Meeting of the Transportation Research Board, Washington, DC, USA, 9 January 2000.

18. Kandhal, P.S.; Foo, K.; Mallick, R. Critical review of voids in mineral aggregate requirements in Superpave. *Transp. Res. Rec.: Transp. Res. Board* **1998**, *1609*, 28–35. [CrossRef]
19. Vavrik, W.R.; Huber, G.; Pine, W.J.; Carpenter, S.H.; Bailey, R. *Bailey Method for Gradation Selection in Hot-Mix-Asphalt Mixture Design*; Transportation Research E-Circular, E-C044; Transportation Research Board: Washington, DC, USA, 2002; Available online: www.trb.org/publications/circulars/ec044.pdf (accessed on 17 March 2017).
20. Asphalt Institute Superpave Mix Design Series No.2 (SP-2). Lexington, Kentucky, 1996. Available online: http://www.asphaltinstitute.org/superpave-documents (accessed on 17 March 2017).
21. Coree, B.J.; Hislop, W.P. A laboratory investigation into the effect of aggregate related factors of critical VMA in asphalt paving mixture. *J. Assoc. Asph. Paving Technologists* **2001**, *70*, 70–131.
22. Hargett, E.R. Effect of size, surface texture, and shape of aggregate particles on the properties of bituminous mixtures. *Highw. Res. Board Special Rep.* **1970**, 25–26.
23. Barksdale, R.D.; Kemp, M.A.; Sheffield, W.J.; Hubbard, J.L. Measurement of aggregate shape, surface area, and roughness. *Transp. Res. Rec.* **1991**, 107–116.
24. Adhikari, S.; You, Z.P. Investigating the sensitivity of aggregate size within sand mastic by modeling the microstructure of an asphalt mixture. *J. Mater. Civil Eng.* **2011**, *23*, 580–586. [CrossRef]
25. Brown, E.R.; Manglorkar, H. *Evaluation of Laboratory Properties of SMA Mixtures*; NCAT Rep. No. 93–5; Auburn Univ.: Auburn, AL, USA, 1993.
26. Hassan, J.H.; Israa, J.Y. The effect of using glass power filler on hot asphalt concrete mixture properties. *J. Eng. Technol.* **2010**, *29*, 44–57.
27. Fu, Z.; Dang, Y.N.; Guo, B.; Huang, Y. Laboratory investigation on the properties of asphalt mixtures modified with double-adding admixtures and sensitivity analysis. *J. Traffic Transp. Eng. (Engl. Ed.)* **2016**, *3*, 412–426. [CrossRef]
28. Putman, B.J.; Amirkhanian, S.N. Utilization of waste fiber in stone matrix asphalt mixtures. *Resour. Conserv. Recycl.* **2004**, *42*, 265–274. [CrossRef]
29. Muniandy, R.; Huat, B.B.K. Laboratory diameteral fatigue performance of stone matrix asphalt with cellulose oil palm fiber. *Am. J. Appl. Sci.* **2006**, *3*, 2005–2010. [CrossRef]
30. Al-Hadidy, A.I.; Tan, Y.Q. Mechanistic analysis of ST and SBS-modified flexible pavements. *Constr. Build. Mater.* **2009**, *23*, 2941–2950. [CrossRef]
31. Chen, H.X.; Xu, Q.W. Experimental study of fibers in stabilizing and reinforcing asphalt binder. *Fuel* **2010**, *89*, 1616–1622. [CrossRef]

9

Towards an Alternate Evaluation of Moisture-Induced Damage of Bituminous Materials

Aboelkasim Diab [1], Zhanping You [2,*], Xu Yang [3] and Mohd Rosli Mohd Hasan [4]

1. Department of Civil Engineering, Aswan University, Aswan 81542, Egypt; adiab@aswu.edu.eg or daali@mtu.edu
2. Department of Civil and Environmental Engineering, Michigan Technological University, Houghton, MI 49931, USA
3. Department of Civil Engineering, Monash University, Clayton, VIC 3800, Australia; xu.yang@monash.edu
4. School of Civil Engineering, Universiti Sains Malaysia, Engineering Campus, Nibong Tebal Penang 14300, Malaysia; cerosli@usm.my

* Correspondence: zyou@mtu.edu

Abstract: Moisture-induced damage is widely known to cause multiple distresses that affect the durability of constructed pavements and eventually lead to the costly maintenance of pavement structures. The reliability and practicality of the assessment protocol to evaluate moisture susceptibility of flexible pavements presents a dilemma within the asphalt community that arises from the complexity and interrelation of moisture mechanisms in the asphalt–aggregate system. Researchers worldwide are continuously trying to develop suitable evaluation methods to simulate the combined destructive field-induced effects of moisture in the laboratory to help practitioners identify and alleviate this complex problem. The main objective of this article is to provide insights and highlight the challenges and opportunities of this important topic in order to extend and share knowledge towards finding a realistic assessment protocol of moisture damage in the laboratory. Two scenarios are proposed in this article: (1) a damage rate concept that accounts for the change of mechanical property (e.g., indirect tensile strength) with respect to the conditioning time, and (2) the establishment of a database using a surface free energy concept to help stakeholders select appropriate asphalt–aggregate combinations without the need to run additional moisture susceptibility tests.

Keywords: asphalt pavement; moisture damage; damage rate; surface free energy

---

1. Outlook

The distresses related to moisture-induced damage still remain one of the most common but complex issues of bituminous pavements [1,2]. Mixtures that are not properly designed or evaluated against ever increasing traffic loading alongside the expected exposure to moisture in the field could potentially affect the pavement durability that can lead to progressive degradation in the form of several distresses, such as raveling, rutting, or cracking [3–5]. In service life, the bituminous pavement is subjected to different environmental conditions combined with traffic loading, which both affect the durability and life-cycle cost of the constructed structure. Displacement, detachment, spontaneous emulsification, pore-pressure-induced damage, hydraulic scour, pH instability, and climatic conditions are considered contributing mechanisms associated with moisture damage in the field [6,7]. So far, the moisture damage of asphalt pavements is not well understood, as it is a complex phenomenon that is affected by the physicochemical properties of the constituents forming the asphalt mixture. In the design stage, the asphalt materials are usually characterized for the moisture effect at the macro scale using compacted asphalt samples. Although the evaluation of compacted asphalt mixtures is commonly used in the pavement engineering community, there are several problems still unsolved, and different perspectives should be considered towards this issue. During the mixture design, the asphalt mixture must pass the tensile strength ratio (TSR) test to ensure the moisture susceptibility of the designed mixture. As a supporting step, in turn, this perspective article is intended to highlight the current issue of moisture damage evaluation and provide an overview of the opportunities to be considered for future researches.

2. Challenges

Various, predominantly empirical, mechanical test methods to evaluate the moisture susceptibility of asphalt mixtures have been developed and are discussed elsewhere [8]. The AASHTO T 283 test method is frequently used by highway agencies worldwide for the sake of evaluation of moisture damage of compacted asphalt concrete mixtures. Basically, in the T 283 test, the TSR, which is calculated as the ratio of the indirect tensile strength (ITS) of conditioned specimens to that for dry specimens, is used to determine the resistance of the asphalt mixture to moisture damage. A minimum TSR of 80% is usually used as a threshold to ensure that the mixture is moisture-resistant, while the mixture is considered moisture-prone if the TSR is any lower. However, several concerns of this ratio have been discussed [9,10]:

- Does this value guarantee a satisfactory resistance to moisture damage in the long-term? The current practice of the TSR concept is based on short-term aged samples, but the moisture susceptibility is a concern throughout the service life of asphalt pavement.
- What is the proper conditioning method to mimic the moisture damage in real life? Freeze–thaw is used in AASHTO T 283, and the moisture induced stress tester (MIST) was introduced in recent years. Can these methods, or some other method, better mimic real moisture damage?
- As the unconditioned or dry compacted specimens are different from the conditioned or wet specimens, does specimen-to-specimen variability affect the results? Even for the same mixture, the prepared compacted samples would have different distributions of aggregate particles and/or air voids, which affects the measured property. Therefore, the test still lacks repeatability, and its accuracy is questionable.
- Does the saturation level while conditioning affect the repeatability of the results?
- Based on previous studies on the AASHTO T 283 test, a number of successful cases that pass in the laboratory would fail in the field.

3. Opportunities

As for suggestions, two scenarios for the evaluation of moisture damage are discussed in this article, the future studies may evaluate.

3.1. First Scenario

The moisture-induced damage of asphalt mixtures can be considered as a time-dependent damage due to the combined effects of moisture and traffic loading. The mechanical property $MP(t)$ at time t is affected by the initial mechanical property $MP(0)$ (i.e., dry specimen) and the induced damage $D(t)$ during conditioning. The progressive damage due to moisture and/or traffic loading can be represented as

$$MP(t) = MP(0) - D(t). \tag{1}$$

Based on Equation (1), Figure 1 depicts a schematic of the mechanical property (such as ITS) with respect to the time of conditioning. As examples, the figure compares possible trends of two different mixes.

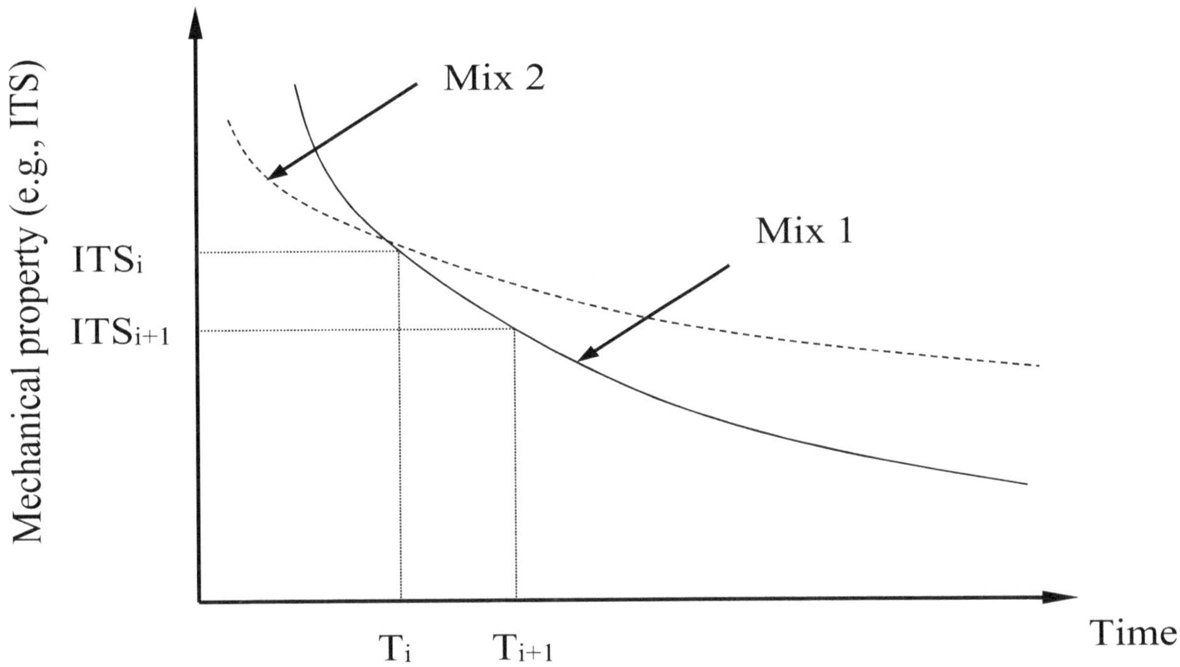

Figure 1. Mechanical property as a function of conditioning time due to the effect of environmental conditions and traffic loading.

This method should involve the conditioning and evaluation steps. In the conditioning step, realistic or field-like conditions should be adopted to induce reasonable effects of moisture and traffic. In a recently published work [9], the authors suggested a conditioning procedure by exposing compacted asphalt specimens to a temperature as high as the highest performance grading (PG) of the used asphalt binder, while the traffic loading was selected based on realistic effects from the literature to simulate the pressure of vehicle wheel on the wet surface. These effects were integrated into the proposed conditioning system in the laboratory. In regard to the evaluation, rather than using a single criterion (i.e., TSR), a model was developed to identify the time-dependent damage of the mixture using the damage rate D_r concept:

$$D_r = \frac{ITS_i - ITS_{i+1}}{T_{i+1} - T_i}. \quad (2)$$

In this model, one can quantify the moisture damage over time and determine how much the mixture degrades with respect to time of conditioning. Using the concept of damage rate, the moisture effects can be evaluated at different times, which is more promising as the mixture might have different behaviors over time. Although its usefulness to discriminate between insusceptible and moisture susceptible mixtures in a rational manner, the concept necessitates the development of a threshold to help quantitatively perform the judgment.

3.2. Second Scenario

The second suggested scenario is beneficial to pavement contractors and can save more time. Under the effects of moisture and traffic, the susceptibility to moisture damage of asphalt–aggregate systems is related to cohesive failure within the matrix and/or adhesive failure at the interface between the asphalt binder and aggregate (Figure 2).

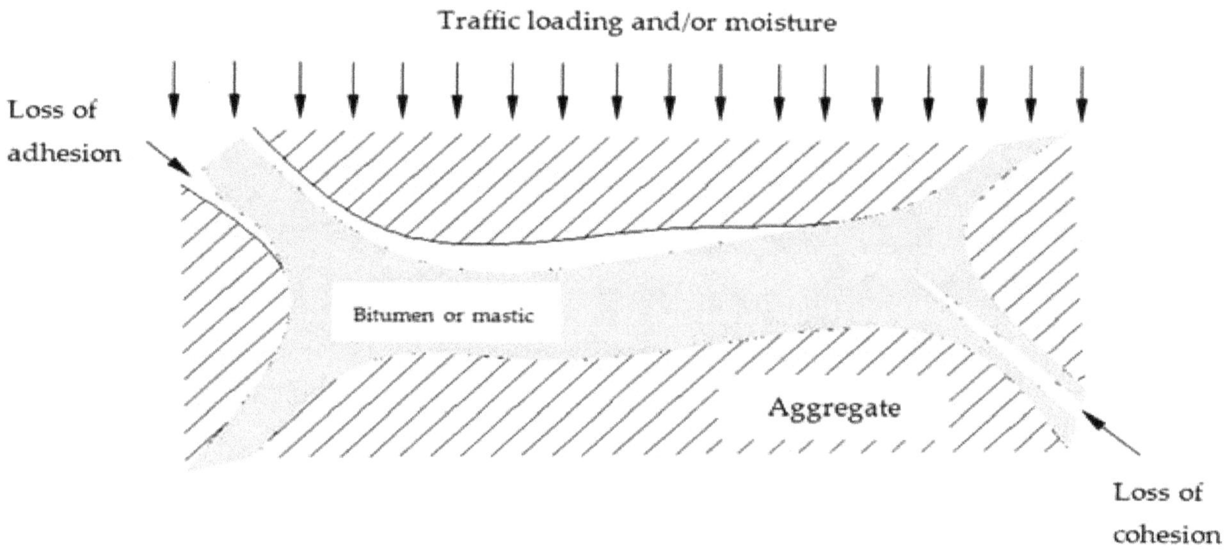

Figure 2. Possible failures due to moisture damage and traffic loading.

The moisture damage of asphalt mixtures can be studied through fundamental approaches to understand the mechanisms causing the susceptibility to moisture damage. Based on the theory of thermodynamics, the surface free energy (SFE) can be used to evaluate the adhesive bond between asphalt binder and aggregate and the cohesive bond between binder molecules. The SFE concept can be applied to select compatible asphalt–aggregate combination(s) based on a satisfactory resistance to moisture damage without the need to performing additional tests (e.g., TSR). It is recommended that highway agencies establish a database of the compatible combinations of the available aggregate and asphalt binders and provide an appropriate strategy to minimize the moisture susceptibility of the asphalt pavement to a satisfactory level by using appropriate adhesion promoters (e.g., hydrated lime or liquid anti-stripping agents) for the incompatible combinations. The existing successful pavements with minimal or no stripping can be used as a reference to establish this database. The minimum cohesive and adhesive bonding between asphalt binder and aggregate can be used to provide a criterion to distinguish between compatible and incompatible combinations based on this concept; hereby, a threshold can be developed based on successfully performing pavements to evaluate different combinations of aggregates and asphalt binders. The importance of this strategy is related to the fact that the practitioners can access the developed database for further implementation and help save cost and time because in this case there is no need to run additional tests to assess the moisture susceptibility of the mixtures for different combinations.

Author Contributions: Aboelkasim Diab wrote the manuscript, while Zhanping You, Xu Yang, and Mohd Rosli Mohd Hasan took care of the revision.

References

1. Tarefder, R.A.; Zaman, A. Characterization of Asphalt Materials for Moisture Damage Using Atomic Force Microscopy and Nanoindentation. In *Nanotechnology in Civil Infrastructure: A Paradigm Shift*; Gopalakrishnan, K., Birgisson, B., Taylor, P., Attoh-Okine, N.O., Eds.; Springer: Berlin/Heidelberg, Germany, 2011; pp. 237–256.
2. Gubler, R.; Partl, M.N.; Canestrari, F.; Grilli, A. Influence of water and temperature on mechanical properties of selected asphalt pavements. *Mater. Struct.* **2005**, *38*, 523–532. [CrossRef]
3. Alam, M.M.; Vemuri, N.; Tandon, V.; Nazarian, S.; Picornell, M. *A Test Method for Identifying Moisture Susceptible Asphalt Concrete Mixes. Research Project 0–1455: Evaluation of Environmental Conditioning System (ECS) for Predicting Moisture Damage Susceptibility of HMAC*; The University of Texas at El Paso: El Paso, TX, USA, 1998.

4. Cho, D.-W.; Kim, K. The mechanisms of moisture damage in asphalt pavement by applying chemistry aspects. *KSCE J. Civ. Eng.* **2010**, *14*, 333–341. [CrossRef]
5. Mohd Hasan, M.R.; You, Z.; Porter, D.; Goh, S.W. Laboratory moisture susceptibility evaluation of WMA under possible field conditions. *Constr. Build. Mater.* **2015**, *101*, 57–64. [CrossRef]
6. Kiggundu, B.M.; Roberts, F.L. *Stripping in HMA Mixtures: State-of-the-Art and Critical Review of Test Methods*; National Center for Asphalt Technology: Auburn, AL, USA, 1988.
7. Hamzah, M.; Hasan, M.; van de Ven, M.; Voskuilen, J. Development of Dynamic Asphalt Stripping Machine for Better Prediction of Moisture Damage on Porous Asphalt in the Field. In *7th RILEM International Conference on Cracking in Pavements*; Springer: New York, NY, USA; pp. 71–81. [CrossRef]
8. Solaimanian, M.; Fedor, D.; Bonaquist, R.; Soltani, A.; Tandon, V. Simple performance test for moisture damage prediction in asphalt concrete. *Assoc. Asph. Paving Technol. AAPT* **2006**, *75*, 345–380.
9. Diab, A.; Sangiorgi, C.; Enieb, M.; You, Z. A conditioning method to evaluate moisture influence on the durability of asphalt mixture materials. *Can. J. Civ. Eng.* **2016**, *43*, 943–948. [CrossRef]
10. Epps, J.; Sebaaly, P.E.; Penaranda, J.; Maher, M.R.; McCann, M.B.; Hand, A.J. *Compatibility of a Test for Moisture-Induced Damage with Superpave Volumetric Design*; National Academy Press: Washington, DC, USA, 2000.

10

Study of the Diffusion of Rejuvenators and its Effect on Aged Bitumen Binder

Yue Xiao [1,*], Chao Li [1], Miao Wan [1], Xinxing Zhou [2], Yefei Wang [3] and Shaopeng Wu [1,*]

1 State Key Laboratory of Silicate Materials for Architectures, Wuhan University of Technology, Wuhan 430070, China; lic@whut.edu.cn (C.L.); alexwanm@163.com (M.W.)
2 Key Laboratory of Highway Construction and Maintenance in Loess Region, Shanxi Transportation Research Institute, Taiyuan 030006, China; zxx09432338@whut.edu.cn
3 Jinan Urban Construction Group Co., Ltd., Jinan 250031, China; wangyefei0803@163.com
* Correspondence: xiaoy@whut.edu.cn (Y.X.); wusp@whut.edu.cn (S.W.)

Academic Editors: Zhanping You, Qingli (Barbara) Dai and Feipeng Xiao

Abstract: Aged asphalt mixture is heavily involved in pavement maintenance and renewed construction because of the development of recycling techniques. The aged bitumen binder has partially lost its viscous behavior. Rejuvenators are therefore designed and used in this recycling procedure to enhance the behavior of such aged reclaimed bitumen. However, tests have not yet been clearly specified to understand the diffusion characteristics of rejuvenators in aged bitumen. In this research, molecular dynamic simulation is proposed and conducted with Materials Studio software to study the diffusion behavior of rejuvenators in aged bitumen at the molecular level. Two rejuvenators, named R-1 and R-2, were included. The difference between these two rejuvenators is their chemical composition of C=O. The diffusion coefficient is determined by studying the molecular movement. Results illustrate that the proposed models can be used to study the diffusion of rejuvenators in aged bitumen sufficiently. In the meantime, a dynamic shear rheometer (DSR) is used to evaluate the recovery influence on aged bitumen resulting from rejuvenators. The experimental results strengthen the model simulations and indicate that the aging index of bitumen has a significant influence on the rejuvenating effect. Research results indicate that rejuvenators have a sufficient rejuvenating effect on the long-term aged bitumen and a limited effect on short-term aged bitumen.

Keywords: bitumen binder; molecular simulation; diffusion of rejuvenator; aged bitumen; rheological properties

1. Introduction

Reclaimed asphalt [1], which is of great importance for conserving resources and protecting the environment, is one of the largest fractions of recycled raw materials used in road construction [2–4]. Rejuvenators are employed in many cases to improve the flexibility, which means better stress relaxation and softer properties of such aged reclaimed asphalt. The performance of these reclaimed asphalts depends on the mixing degree of rejuvenators and aged bitumen [5]. Two of the main factors that assist in the mixing efficiency are the blending condition and the diffusion process [6,7]. This study therefore focused on the diffusion process of rejuvenators in aged bitumen during hot recycling.

Molecular diffusion plays an important role in bitumen recovery process as well as in numerous engineering applications [8]. Mazzotta used FTIR (Fourier transform infrared spectroscopy) to carry out chemical analysis of molecular changes during the bitumen aging process [9]. Indexes of $I_{C=O}$ and $I_{S=O}$ were defined and successfully used to characterize the changes of functional groups. Many studies on diffusion in bituminous materials have been successfully carried out recently. A method based on FTIR-ATR (Fourier Transform Infrared Spectroscopy using Attenuated Total Reflectance) was developed and described by Karlsson and Isacsson [10,11]. Using this method, diffusion rates are influenced by the temperature, size, and shape of the diffusion molecules. A positive outcome of that study was that aging showed very little influence on diffusion [12]. Karlsson and Isacsson [13] also did some laboratory studies on diffusion in bitumen using markers. They selected diphenyl-silane and t-butyldiphenylsilyl (DPS) as markers. They monitored the diffusion through the binder, and the importance of diffusant size and polarity on diffusion rate was demonstrated. Research on the diffusion of rejuvenators in aged bitumen has been further investigated using molecular dynamic simulation to calculate the diffusion coefficient.

Molecular simulation technology can be used to calculate the molecular conformations and diffusion process between different materials [14]. The establishment of the model mainly relies on system parameters such as the sizes of scale, the degree of light and shade, the colors, and the different structures of the microscopic particles themselves. The molecular simulation technique is an experimental technology based on computer analysis. This technology can simulate the static characteristics and motor behavior of molecules by setting up a reasonable molecular model to define the chemical and physical properties of the micro system [15]. Dong and Wang successfully used molecular dynamic simulation method to study the asphalt–aggregate interface [16,17]. Wang also used molecular dynamics simulation to characterize the oxidative aging effect in asphalt binder. Mean square displacement and diffusion coefficient were analyzed, and aged binder was found to have a higher activation energy barrier that decreased its viscoelastic behavior [18,19]. Huang applied molecular dynamic simulation to investigate the diffusion between virgin binder and aged binder, which was then verified with gel permeation chromatography [20,21]. They reported that the diffusion ability is a function of the diffusion ability itself and the diffusion environment. All of this research leads to a powerful method of studying the diffusion of rejuvenators in aged bitumen binder.

In this research, the chemical bond morphology of the investigated rejuvenators was first characterized to define structures for molecular dynamic simulation. Then, the molecular dynamic method was used, and the COMPASS force field was adopted to discuss the diffusion properties of the modeled rejuvenators in aged bitumen. Einstein's model was employed to analyze the results obtained from a computer and to calculate the diffusion coefficient [22]. At the same time, rheological properties of rejuvenated bitumen treated by different rejuvenators were characterized. These laboratory tests were conducted to further understand the findings from the molecular simulation on the diffusion and rejuvenating effect.

2. Materials and Methods

2.1. Materials

The 80/100 penetration grade bitumen provided by the Panjin asphalt company (Sichuan, China) was used in this research. The short-term aged bitumen used in this research was aged using the standard rolling thin-film oven test (RTFOT) at a temperature of 163 °C for 85 min. Long-term aged bitumen was obtained by means of a pressure aging vessel (PAV) and an ultraviolet aging oven (UV). The PAV was conducted at 95 °C with pressure of 2.1 MPa for 20 h. The ultraviolet aging oven was set with an ultraviolet intensity of 79.5 W/m^2 and aged for 168 h. Both short-term aged and long-term aged binders were then treated with rejuvenators. The rejuvenating contribution to short-term aged binder was studied to understand the recovering ability during hot mixing.

The rejuvenator-treated aged bitumen was prepared in two steps. Firstly, the aged bitumen was blended with 8% of R-1 or R-2, separately, using the propeller mixer (GS-1, Tianjin, China) for 30 min at 120 °C with a constant speed of 200 rpm. Secondly, the rejuvenator-treated bitumen was then allowed to sit at room temperature for 3 days before being used for further rheological analysis to ensure completed diffusion.

The rejuvenators, named R-1 and R-2, used throughout this research were provided by CRAFCO Company (Wuxi, P.R. China). Table 1 displays the basic technical data of R-1 and R-2. They are liquid and made from emulsified bitumen, containing about 61.8% to 76.5% of residue after fully cured.

Table 1. Technical data of Rejuvenators R-1 and R-2.

Parameter	R-1	R-2	Test Methods
Appearance	Viscose liquid	Viscose liquid	–
Color	tawny	brown	–
Viscosity (25 °C), SFS	40	15	ASTM D-244
Residue, wt %	65	60	ASTM D-244
Weight ratio for rejuvenate, %	8	8	ASTM D-2006-70
Asphaltenes, wt %	0.4	0.75	–

2.2. Molecular Dynamic Simulation

Molecular dynamic simulation was performed using Materials Studio software (Materials Studio 8.0, BIOVIA, Beijing, China). The compass force field was used only for describing atomic level interactions in molecular models. According to the literature, the saturates, aromatics, and asphaltenes in aged bitumen were modeled as a straight chain alkane $C_{22}H_{46}$, 1,7-dimethyl-naphthalene [23,24], and Groenzin's and Mullin's model $C_{72}H_{98}S$ [25], respectively.

Components of rejuvenators were first determined, and R-1 and R-2 were then defined as simplified monomer that consists of functional groups as Table 2 indicates. For simplified calculations, $C_{13}H_{15}NO_3$ and $C_{11}H_{13}NO_2$ were defined for R-1 and R-2, respectively.

Table 2. Functional groups in the rejuvenators used.

Rejuvenator	Functional Groups					
R-1	HC−	−CH2	−CH3	−NH2	−C(=O)−	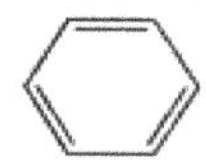
R-2	HC−	−CH2	−CH3	−NH2	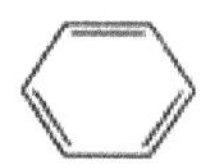	-

The simulation process of diffusion coefficients mainly includes two parts: energy minimization and a dynamic simulation process. After energy optimization, a canonical ensemble (NVT) and an isobaric–isothermal ensemble (NPT) were simultaneously applied for the first 50 picoseconds (ps) to ensure a density stable system [26]. NVT is a statistical ensemble that represents the possible states of a mechanical system in thermal equilibrium with a heat bath at a fixed temperature [27]. The system can exchange energy with the heat bath, so the states of diffusion system will differ in total energy. NPT is a statistical mechanical ensemble that maintains a constant temperature and a constant applied pressure [28].

The molecular dynamic simulation was running at a simulated temperature of 25 °C. Then, NVT dynamic simulation was used in the following 50 ps. The movement locus of the molecule was recorded to study the diffusion coefficient. Figure 1 presents the diffusion models. R-1 and R-2 are located in the boundary of the RTFOT aged asphalt at the initial stage, and are diffused into the center at a stable stage after full diffusion.

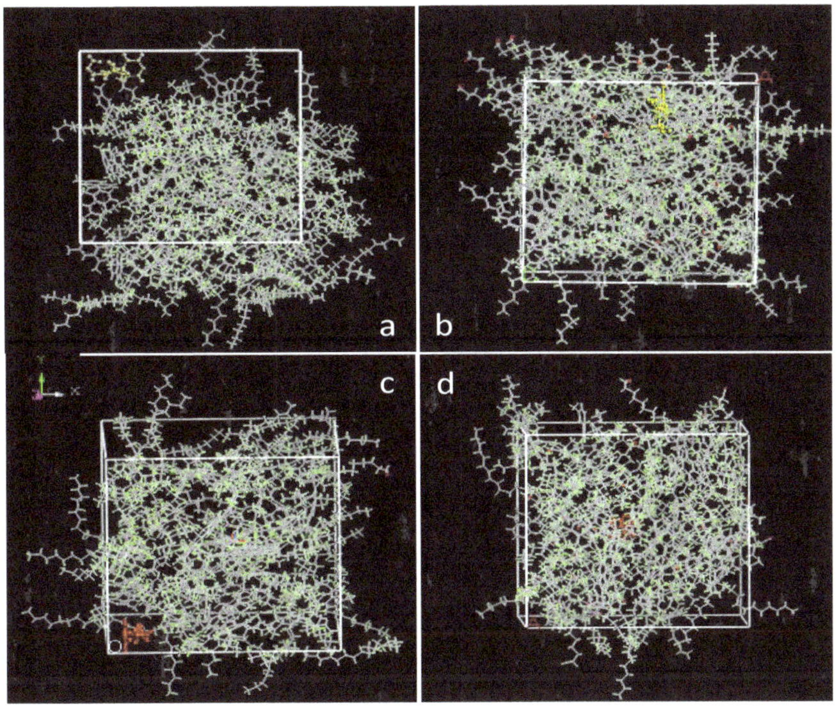

Figure 1. Diffusion models: the initial stage (**a**) and the stable stage (**b**) of R-1 in the rolling thin-film oven test (RTFOT) aged asphalt models; the initial stage (**c**) and the stable stage (**d**) of R-2 in the RTFOT aged asphalt models.

2.3. Rheological Properties Test

The dynamic shear rheometer (DSR) (Anton Paar, Graz, Austria) is usually used to characterize both viscous and elastic behaviors of bitumen by measuring the complex shear modulus and phase angle. A temperature sweep test was conducted. A sinusoidal signal of stress was applied. The complex shear modulus represents the sample's total resistance to the applied deformation during sinusoidal stress conditions. Phase angle is described as the lag between applied shear stress and the resulting shear strain. The higher the complex shear modulus value, the stiffer the bitumen binder is. The larger the measured phase angle, the more viscous the binder is [29]. The temperature sweep test was performed at a fixed frequency of 10 rad/s and in temperature ranges from −10 to 60 °C with temperature increments of 2 °C per minute.

In this research, DSR tests were conducted on virgin bitumen, aged bitumen (RTFOT, PAV, UV), and rejuvenated bitumen with 8% R-1 or 8% R-2. According to the test specification [30], different geometries of specimens are specified for different temperature conditions. The diameter of the plate that contains the bitumen binder is 8 mm, and the thickness is 2 mm when the temperature is lower than 20 °C. When the temperature is higher than 20 °C, the plate diameter should be 25 mm, and the thickness of bitumen binder should be 1 mm. Strain sweep tests with increasing applied strain were first conducted to study the viscoelastic range of virgin, aged, and rejuvenated bitumen binder. Based on strain sweep tests, strains were selected based on viscoelastic ranges of every type of binder for temperature sweep analysis, ensuring that all of the performed DSR moduli and phase angle evaluations are conducted in the viscoelastic range of the tested bitumen binder. Table 3 summarizes the applied strains.

Table 3. Applied shear stains during dynamic shear rheometer (DSR) analysis.

Temperature Ranges	Virgin Binder, Rejuvenated Binders (%)	RTFOT, PAV and UV Binders (%)
−10–30 °C	0.03	0.02
30–60 °C	1.5	1

3. Results and Discussion

3.1. Molecular Calculation Results

3.1.1. Mean Squared Displacement

The diffusion coefficients and diffusion mechanisms in R-1- and R-2-treated RTFOT, PAV, and UV aged asphalt were modeled and studied. The molecular dynamic simulation of the diffusion process can be investigated by recording the movement locus of the molecular particles. The mean squared displacement (MSD) of the analyzed particles can therefore be finalized by means of this movement.

Figure 2 compares MSD values along with diffusion time for rejuvenators in aged bitumen binder. In both R-1-modified and R-2-modified bitumen binder, the MSD values decrease in order of UV aged, PAV aged, and RTFOT aged binder. Rejuvenator R-1 performs higher MSD values than R-2 in bitumen binder with the same aging index, which means R-1 diffuses faster than R-2 in aged bitumen.

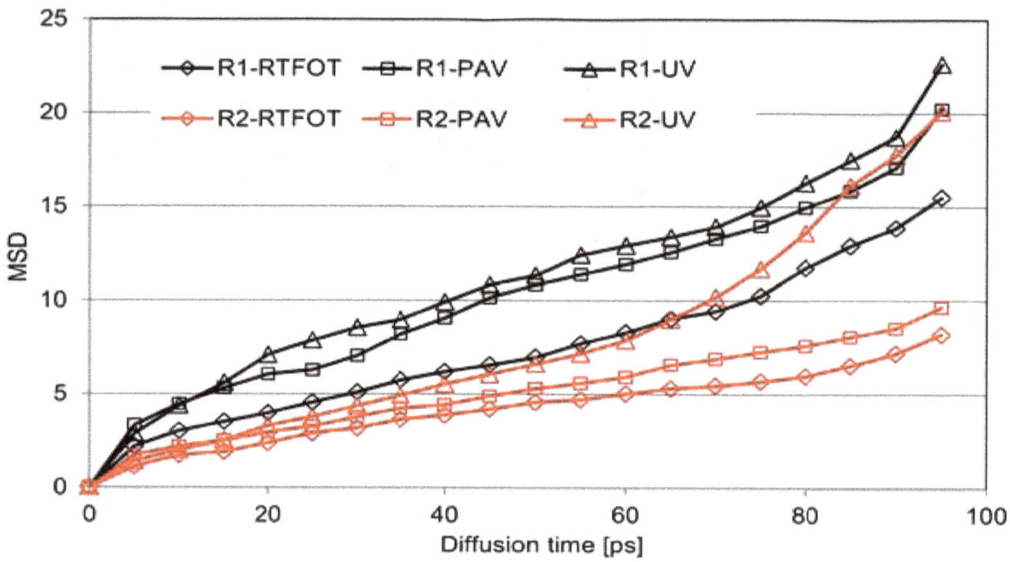

Figure 2. Mean squared displacement (MSD) curves of rejuvenators in aged bitumen at 150 °C.

3.1.2. Diffusion Coefficient

MSD values are then used to calculate the diffusion coefficient (D) via Einstein's formula as follows:

$$MSD = \lim_{t \to \infty} \frac{d}{dt} \sum_{i=1}^{N\alpha} \left\langle [r_i(t) - r_i(0)]^2 \right\rangle \tag{1}$$

where $r_i(t)$ is the displacement of particle i at time t, and $r_i(0)$ is the displacement at the starting time. In this formula, N represents the diffused particles of the system. When the MSD curve presents a linear trend, the equation can be simplified as follows:

$$D = \frac{1}{6T} MSD \tag{2}$$

where T is the total time of atoms or molecule movements. D is the $1/6$ slope of the simplified linear curve.

In the R-1-treated bitumen, the calculated D of the rejuvenator in RTFOT, PAV, and UV aged bitumen is 2.26×10^{-10}, 2.82×10^{-10}, and 3.09×10^{-10}, respectively. The calculated D of Rejuvenator R-2 in RTFOT, PAV, and UV aged bitumen is 1.18×10^{-10}, 1.35×10^{-10}, and 3.05×10^{-10}. Firstly, this illustrates that the severity level of aging has a significant influence on the diffusion of rejuvenators. Investigated rejuvenators diffuse easier and faster in the long-term aged binder, such as the UV aged binder, but diffuse slower in the short-term aged bitumen binder. Secondly, R-1 presents a higher diffusion coefficient in aged bitumen than does R-2, which indicates that R-1 has a better diffusion coefficient in the aged bitumen.

3.2. The Effect of Rejuvenators on the Rheological Properties

Figure 3 presents the complex shear modulus and phase angle of virgin bitumen, RTFOT aged bitumen, and rejuvenated bitumen with 8% R-1 or 8% R-2 in the higher temperature range. It is shown that the complex shear modulus of RTFOT aged bitumen is much higher, and the phase angle is much lower than that of the virgin bitumen. This indicates that aged bitumen is more elastic and stiffer. The introduction of 8% R-1 or 8% R-2 into aged bitumen decreases the complex shear modulus and obviously increases the phase angle. This means that the addition of R-1 or R-2 can reduce the aging degree of bitumen. Furthermore, the complex shear modulus of RTFOT aged bitumen treated with 8% R-1 is lower than that of the aged bitumen treated with 8% R-2, while the phase angle of R-1-treated bitumen is higher and the curve is closer to the virgin bitumen, compared to that of R-2-treated specimens. These results indicate that, although both R-1 and R-2 can recover the rheological properties of RTFOT aged bitumen, R-1 obviously has a stronger recovery capability of rheological properties than does R-2.

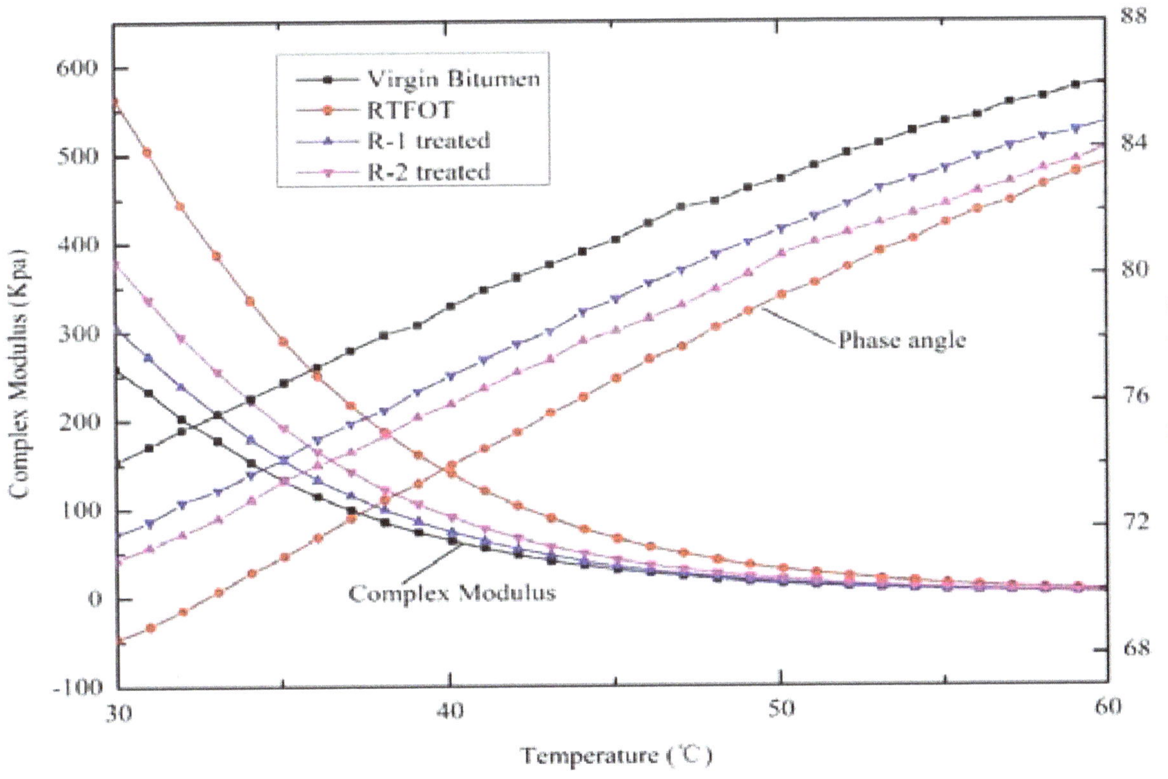

Figure 3. Effect of rejuvenators on RTFOT aged bitumen at higher temperatures.

Figures 4 and 5 reveal the complex shear modulus and phase angle of PAV aged bitumen and UV aged bitumen in the higher temperature range, respectively. The rejuvenating effect of the investigated rejuvenators on PAV aged bitumen is similar to the effect on RTFOT aged bitumen. While for the UV aged bitumen, the impact of R-2 is limited. Figure 5 shows that the phase angle curves of UV aged bitumen and R-2-treated bitumen overlap. The previous molecular calculation shows a higher diffusion coefficient of R-1 in aged binder than that of R-2. Figures 3–5 provide strong evidence that R-1 exhibits a much greater rejuvenating influence on the viscoelastic behavior at higher temperature conditions than does R-2.

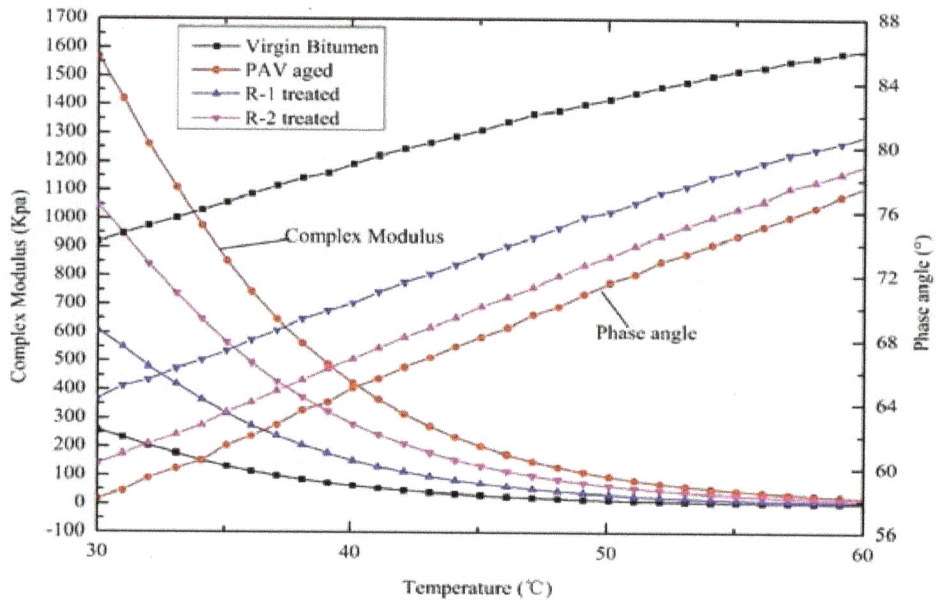

Figure 4. Effect of rejuvenators on pressure aging vessel (PAV) aged bitumen at higher temperatures.

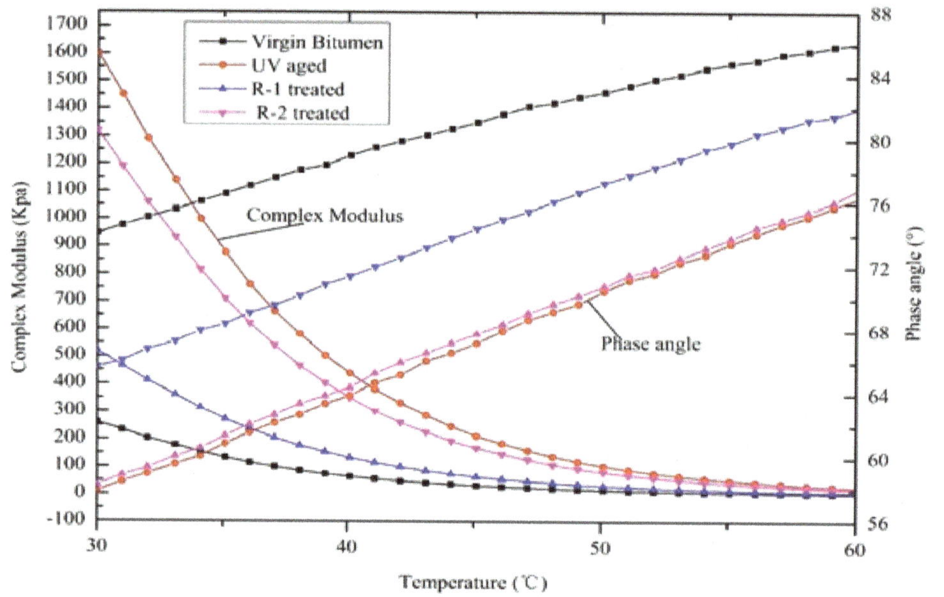

Figure 5. Effect of rejuvenators on UV aged bitumen at higher temperatures.

Figures 6–8 present the complex shear modulus and phase angle at lower temperatures of virgin bitumen, RTFOT (or PAV, or UV) aged bitumen, and rejuvenator-treated bitumen. The complex shear modulus of aged bitumen is higher than that of virgin bitumen. We all know that aged bitumen is very brittle and easily cracks under loading at low temperature. The addition of R-1 and R-2 can also decrease the complex shear modulus and increase the phase angle at lower temperatures. This does mean, however, that both R-1 and R-2 can significantly improve low-temperature properties of RTFOT aged bitumen and extend the service life of bitumen in a lower temperature condition. The modulus curve of aged bitumen treated with R-1 is closer to the curve of virgin bitumen, while the curve of aged bitumen treated with R-2 is closer to the curve of aged bitumen. This phenomenon indicates once again that R-1 has a stronger recovery capability of lower temperature behavior of aged bitumen than does R-2.

Furthermore, Figures 3–8 illustrate that the curve of aged bitumen treated with R-1 in the lower temperature range is closer to the curve of virgin bitumen than that in the higher temperature range. This indicates that the recovery capacity of R-1 on the lower temperature behavior is more significant than that on the higher temperature behavior. Figures 3–8 also show that the distinction between regenerative capacity of R-1 and R-2 is more obvious on the lower temperature properties than that on the higher temperature properties.

Practically speaking, RTFOT, which represents short-term aging, results in a minor property deterioration of bituminous materials. Therefore, in the long-term aged bitumen, such as UV aged binder, the rejuvenating effect resulting from R-1 and R-2 presents a significant difference from that in the case of short-term aged binder. In Figures 3 and 6, both R-1 and R-2 present an obvious rejuvenating contribution. Nevertheless, in Figures 5 and 8, R-1 shows a nice rejuvenating effect, while R-2 has a very limited influence on the properties of modulus and phase angle. This illustrates that the key to designing a good rejuvenator is the improvement of its effect on long-term aged binder, instead of focusing on short-term aged binder. An improved rejuvenator should have a sufficient effect on rejuvenating long-term aged bitumen.

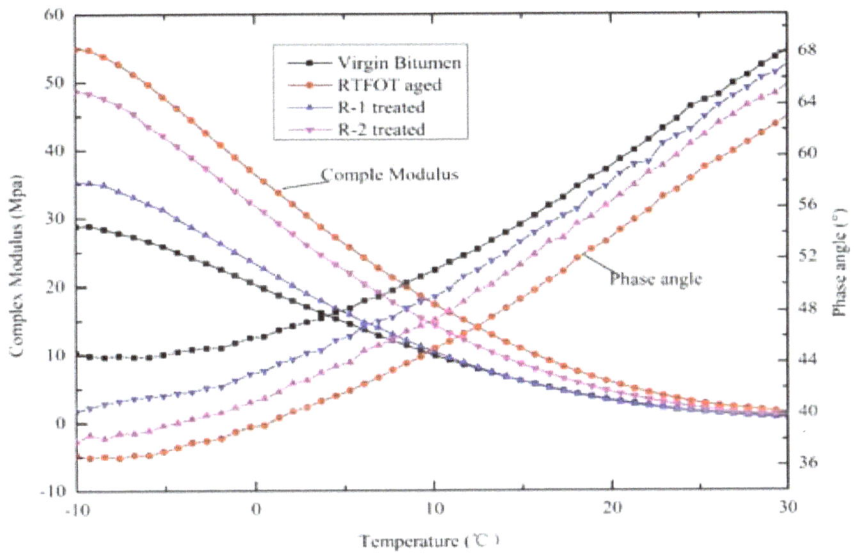

Figure 6. Effect of rejuvenators on RTFOT aged bitumen at lower temperatures.

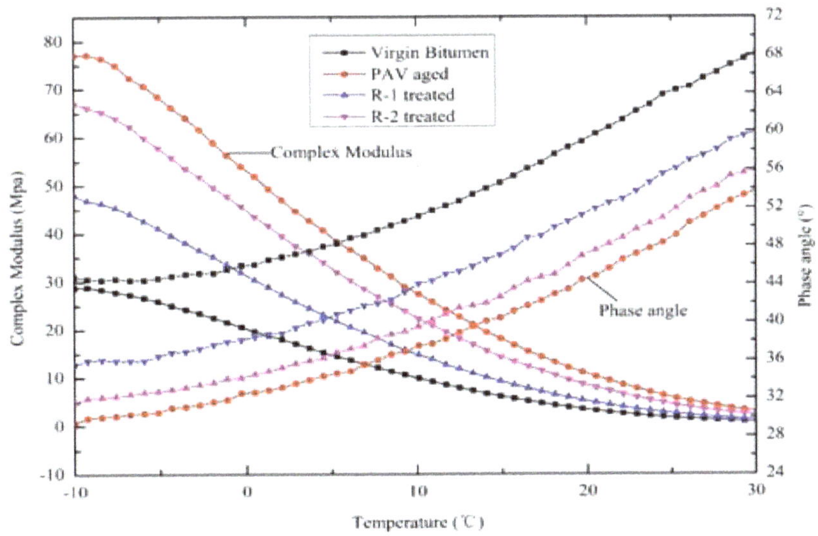

Figure 7. Effect of rejuvenators on PAV aged bitumen at lower temperatures.

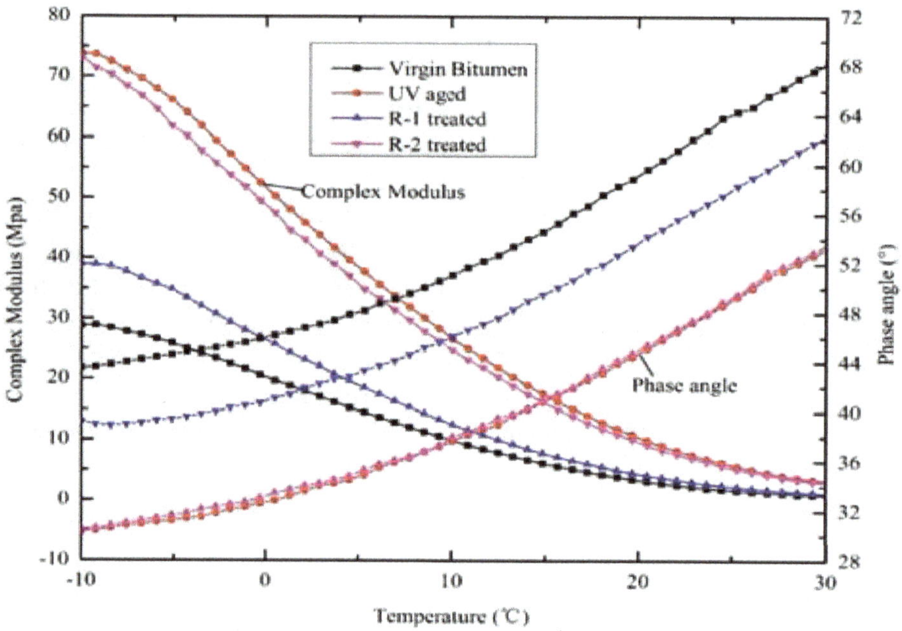

Figure 8. Effect of rejuvenators on UV aged bitumen at lower temperatures.

4. Conclusions

Molecular dynamic simulation was established in this research to simulate the actual movement of rejuvenators in aged bitumen and to characterize their diffusion coefficient. Then, DSR tests were carried out to further investigate the rejuvenating effect. Based on the discussed results, the following can be concluded:

1. Molecular dynamic simulation and rheological analysis results indicate that the severity level of aging has a significant influence on the diffusion coefficient and rejuvenating effect of rejuvenators. The investigated rejuvenators diffuse faster and more easily in the long-term aged binder, such as UV aged binder, but diffuse slower in the short-term aged bitumen binder. Consequently, there is a sufficient rejuvenating effect on the long-term aged bitumen and a limited effect on the short-term aged bitumen.
2. Both R-1 and R-2 can reduce the complex modulus and increase the phase angle of the long-term aged bitumen, but R-1 has a greater recovery capacity than R-2. The rejuvenators presented greater rejuvenate ability at lower temperatures than at higher temperatures.
3. Molecular simulation indicates that R-1 has a higher diffusion coefficient in aged bitumen binder than does R-2, while DSR proves that R-1 has greater recovery capacity. Molecular simulation and experiment results from the DSR test match quite well. This means that the molecular simulation proposed in this research can be successfully used to characterize the diffusion coefficient of rejuvenators in aged bitumen.

Acknowledgments: This study was supported by the National Natural Science Foundation of China (No. 51408447) and the National Key Scientific Apparatus Development Program from the Ministry of Science and Technology of China (No. 2013YQ160501).

Author Contributions: Yue Xiao, Chao Li and Yefei Wang conceived and designed the experiments; Chao Li, Miao Wan, and Yefei Wang performed the experiments; Yue Xiao, Chao Li, Miao Wan, and Yefei Wang analyzed the data; Xinxing Zhou and Shaopeng Wu contributed analysis tools; Yue Xiao and Chao Li wrote the paper.

References

1. Wu, S.; Qiu, J.; Mo, L.; Yu, J.; Zhang, Y.; Li, B. Investigation of temperature characteristics of recycled hot mix asphalt mixtures. *Resour. Conserv. Recycl.* **2007**, *51*, 610–620. [CrossRef]

2. Cui, P.; Wu, S.; Xiao, Y.; Wan, M.; Cui, P. Inhibiting effect of layered double hydroxides on the emissions of volatile organic compounds from bituminous materials. *J. Clean. Prod.* **2015**, *108*, 987–991. [CrossRef]
3. Li, X.-G.; Lv, Y.; Ma, B.-G.; Chen, Q.-B.; Yin, X.-B.; Jian, S.-W. Utilization of municipal solid waste incineration bottom ash in blended cement. *J. Clean. Prod.* **2012**, *32*, 96–100. [CrossRef]
4. Chen, Z.; Wu, S.; Xiao, Y.; Zeng, W.; Yi, M.; Wan, J. Effect of hydration and silicone resin on basic oxygen furnace slag and its asphalt mixture. *J. Clean. Prod.* **2016**, *112*, 392–400. [CrossRef]
5. Chen, M.; Xiao, F.; Putman, B.; Leng, B.; Wu, S. High temperature properties of rejuvenating recovered binder with rejuvenator, waste cooking and cotton seed oils. *Constr. Build. Mater.* **2014**, *59*, 10–16. [CrossRef]
6. Ma, T.; Wang, H.; Zhao, Y.; Huang, X.; Pi, Y. Strength mechanism and influence factors for cold recycled asphalt mixture. *Adv. Mater. Sci. Eng.* **2015**, *2015*, 181853. [CrossRef]
7. Ma, T.; Huang, X.; Zhao, Y.; Zhang, Y. Evaluation of the diffusion and distribution of the rejuvenator for hot asphalt recycling. *Constr. Build. Mater.* **2015**, *98*, 530–536. [CrossRef]
8. Sheikha, H.; Mehrotra, A.K.; Pooladi-Darvish, M. An inverse solution methodology for estimating the diffusion coefficient of gases in athabasca bitumen from pressure-decay data. *J. Pet. Sci. Eng.* **2006**, *53*, 189–202. [CrossRef]
9. Dondi, G.; Mazzotta, F.; Simone, A.; Vignali, V.; Sangiorgi, C.; Lantieri, C. Evaluation of different short term aging procedures with neat, warm and modified binders. *Constr. Build. Mater.* **2016**, *106*, 282–289. [CrossRef]
10. Karlsson, R.; Isacsson, U.; Ekblad, J. Rheological characterisation of bitumen diffusion. *J. Mater. Sci.* **2007**, *42*, 101–108. [CrossRef]
11. Karlsson, R.; Isacsson, U. Application of ftir-atr to characterization of bitumen rejuvenator diffusion. *J. Mater. Civ. Eng.* **2003**, *15*, 157–165. [CrossRef]
12. Karlsson, R.; Isacsson, U. Investigations on bitumen rejuvenator diffusion and structural stability (with discussion). *J. Assoc. Asph. Paving Technol.* **2003**, *72*, 463–501.
13. Karlsson, R.; Isacsson, U. Laboratory studies of diffusion in bitumen using markers. *J. Mater. Sci.* **2003**, *38*, 2835–2844. [CrossRef]
14. Mueller-Plathe, F.; Rogers, S.C.; Van Gunsteren, W.F. Diffusion coefficients of penetrant gases in polyisobutylene can be calculated correctly by molecular-dynamics simulations. *Macromolecules* **1992**, *25*, 6722–6724. [CrossRef]
15. Hofmann, D.; Fritz, L.; Ulbrich, J.; Paul, D. Molecular simulation of small molecule diffusion and solution in dense amorphous polysiloxanes and polyimides. *Comput. Theor. Polym. Sci.* **2000**, *10*, 419–436. [CrossRef]
16. Dong, Z.; Liu, Z.; Wang, P.; Gong, X. Nanostructure characterization of asphalt-aggregate interface through molecular dynamics simulation and atomic force microscopy. *Fuel* **2017**, *189*, 155–163. [CrossRef]
17. Xu, G.; Wang, H. Molecular dynamics study of interfacial mechanical behavior between asphalt binder and mineral aggregate. *Constr. Build. Mater.* **2016**, *121*, 246–254. [CrossRef]
18. Xu, G.; Wang, H. Molecular dynamics study of oxidative aging effect on asphalt binder properties. *Fuel* **2017**, *188*, 1–10. [CrossRef]
19. Pan, P.; Wu, S.; Xiao, Y.; Liu, G. A review on hydronic asphalt pavement for energy harvesting and snow melting. *Renew. Sustain. Energy Rev.* **2015**, *48*, 624–634. [CrossRef]
20. Ding, Y.; Huang, B.; Shu, X.; Zhang, Y.; Woods, M.E. Use of molecular dynamics to investigate diffusion between virgin and aged asphalt binders. *Fuel* **2016**, *174*, 267–273. [CrossRef]
21. Pan, P.; Wu, S.; Hu, X.; Liu, G.; Li, B. Effect of material composition and environmental condition on thermal characteristics of conductive asphalt concrete. *Materials* **2017**, *10*, 218. [CrossRef]
22. Sonnenburg, J.; Gao, J.; Weiner, J. Molecular dynamics simulations of gas diffusion through polymer networks. *Macromolecules* **1990**, *23*, 4653–4657. [CrossRef]
23. Zhang, L.; Greenfield, M.L. Analyzing properties of model asphalts using molecular simulation. *Energy Fuels* **2007**, *21*, 1712–1716. [CrossRef]
24. Bhasin, A.; Bommavaram, R.; Greenfield, M.L.; Little, D.N. Use of molecular dynamics to investigate self-healing mechanisms in asphalt binders. *J. Mater. Civ. Eng.* **2010**, *23*, 485–492. [CrossRef]
25. Groenzin, H.; Mullins, O.C. Molecular size and structure of asphaltenes from various sources. *Energy Fuels* **2000**, *14*, 677–684. [CrossRef]
26. Zhou, X.; Wu, S.; Liu, G.; Pan, P. Molecular simulations and experimental evaluation on the curing of epoxy bitumen. *Mater. Struct.* **2016**, *49*, 241–247. [CrossRef]
27. Sadus, R.J. *Molecular Simulation of Fluids*; Elsevier Science Ltd.: New York, NY, USA, 2002.

28. Bosko, J.T.; Todd, B.; Sadus, R.J. Molecular simulation of dendrimers and their mixtures under shear: Comparison of isothermal-isobaric (NPT) and isothermal-isochoric (NVT) ensemble systems. *J. Chem. Phys.* **2005**, *123*, 034905. [CrossRef] [PubMed]
29. Xiao, Y.; van de Ven, M.; Molenaar, A.; Su, Z.; Zandvoort, F. Characteristics of two-component epoxy modified bitumen. *Mater. Struct.* **2011**, *44*, 611–622. [CrossRef]
30. CEN-European Committee for Standardization. *Nen-en 14770 Bitumen and Bituminous Binders-Determination of Complex Shear Modulus and Phase Angle-Dynamic Shear Rheometer (DSR)*; BSI: London, UK, 2012.

11

Numerical Study on the Asphalt Concrete Structure for Blast and Impact Load using the Karagozian and Case Concrete Model

Jun Wu [1,2], Liang Li [1,*], Xiuli Du [1] and Xuemei Liu [3]

1 The Key Laboratory of Urban Security and Disaster Engineering, Ministry of Education, Beijing University of Technology, Beijing 100124, China; cvewujun@sues.edu.cn (J.W.); duxiuli@bjut.edu.cn (X.D.)

2 School of Urban Railway Transportation, Shanghai University of Engineering Science, Shanghai 201620, China

3 School of Civil Engineering and Built Environment, Queensland University of Technology, Brisbane 4001, Australia; x51.liu@qut.edu.au

* Correspondence: liliang@bjut.edu.cn

Academic Editor: Zhanping You

Abstract: The behaviour of an asphalt concrete structure subjected to severe loading, such as blast and impact loadings, is becoming critical for safety and anti-terrorist reasons. With the development of high-speed computational capabilities, it is possible to carry out the numerical simulation of an asphalt concrete structure subjected to blast or impact loading. In the simulation, the constitutive model plays a key role as the model defines the essential physical mechanisms of the material under different stress and loading conditions. In this paper, the key features of the Karagozian and Case concrete model (KCC) adopted in LSDYNA are evaluated and discussed. The formulations of the strength surfaces and the damage factor in the KCC model are verified. Both static and dynamic tests are used to determine the parameters of asphalt concrete in the KCC model. The modified damage factor is proposed to represent the higher failure strain that can improve the simulation of the behaviour of AC material. Furthermore, a series test of the asphalt concrete structure subjected to blast and impact loadings is conducted and simulated by using the KCC model. The simulation results are then compared with those from both field and laboratory tests. The results show that the use of the KCC model to simulate asphalt concrete structures can reproduce similar results as the field and laboratory test.

Keywords: asphalt concrete; constitutive model; impact loading; blast loading; numerical simulation

1. Introduction

The behaviour of structures or infrastructures under extreme loadings has become a hot topic in the area of civil, mechanical and material engineering. Critical infrastructures, such as runway pavement designed for normal aircraft landing and taking off, are expected to have adequate resistance when subjected to extreme loadings, such as impact or blast loadings (e.g., heavy airplane landing or taking off, air plane crash or terrorist attack). Asphalt concrete (AC) is made of bitumen binder and coarse aggregate. It is usually used as the surface course for both highway and runway flexible pavement [1]. The dynamic load in the daily application for AC pavement normally corresponds to a strain rate less than 10^{-1} s^{-1}. Tashman et al. [2] conducted experiments of AC under triaxial compressive loading at the strain rate from 10^{-6} s^{-1} to 10^{-3} s^{-1}. The results showed that the failure stress increased with the increase of the applied strain rate. Seibi et al. [3] studied AC subjected to uniaxial compressive loading with strain rate from 0.064 s^{-1} to 0.28 s^{-1}. It was found that the yield stress was significantly dependent on the strain rate. Park et al. [4] carried out tests on AC under uniaxial and triaxial compression with the strain rate changing from 10^{-4} s^{-1} to 0.07 s^{-1}. The results showed that with the increase of the applied strain rates, the yield stress and failure stress increased, and the strain rate dependency was clearly showing up at the higher strains. It was also found that the viscous behaviour of AC decreased with the increase of the strain rate. However, when the pavement structure is subjected to the impact loading from further heavy traffic loading or aircraft loading, the corresponding strain rate exceeds 10^{-1} s^{-1}, especially for heavy aircraft loading, the related strain rate reaches around 100 s^{-1}. Based on previous research [5,6], the compressive strength of AC material can be enhanced with the increase of strain rate, and AC material exhibits high plastic behaviour at the high strain rate. However, it is very expensive to conduct field test to investigate the actual behaviour of AC under severe dynamic loading, the numerical simulation is an effective alternative solution.

There are many factors that influence the reliability of numerical simulation. Among these factors, the material model plays a key role because it should reproduce the essential physical mechanisms of the material under various loading conditions. Seibi et al. [3] and Park et al. [4] used the Drucker–Prager yield function to simulate the compressive behaviour of AC under dynamic strain loading (strain rate from 0.0001 s^{-1} to 0.0701 s^{-1}). Tashman et al. [2] developed a microstructure-based viscoplastic continuum model to take into account the effect of temperature in AC material with the strain rate ranging from 10^{-6} s^{-1} to 10^{-3} s^{-1}. Since the late 1990s, microstructural-based discrete element models (DEM) have been used for better understanding of asphalt pavement concrete (e.g., [7–10]). In DEM models, the assumption of elastic or viscoelastic behaviour was employed for the simulation of the static or creep behaviour of AC under normal traffic loading. However, it was found that when the pavement structure was under blast or high impact load, plastic deformation and severe damage would occur. Thus, a robust material model should be developed to consider the strain rate effect, strain hardening, strain softening and damage of the AC material under the severe dynamic loading. Recently, several concrete-like material models subjected to dynamic loadings have been developed, such as the Riedel-Hiermaier-Thoma (RHT) model [11], the Advanced Fundamental Concrete (AFC) model [12,13], the Karagozian and Case concrete model (KCC) [14] and the Holmquist-Johnson-Cook (HJC) concrete model [15,16]. These robust material models are capable of capturing varying concrete-like materials' behaviour under different loading conditions. When subjected to dynamic loading, such as blast loading or high impact loading, concrete-like materials show a highly non-linear response. Besides, due to the general complexity of the constitutive models, the determination of the parameters (i.e., residual strength, failure strain and failure criteria in model) also plays an important role in achieving the actual performance of the concrete-like materials. This requires sufficient understanding of the modelling formulation and the associated considerations. In the current study, the KCC model [14] is used to simulate the AC material. This model is capable of capturing the varied concrete-like material behaviours under different loading conditions. However, it should be noticed that this model cannot consider the temperature effect. In this study, the dynamic behaviour of AC under a high strain rate is investigated, and the temperature effect is not considered. Hence, the KCC can be used; otherwise, the temperature issue should come up.

In this paper, several key features of the KCC model are firstly discussed, and then, the determination of the parameters in the KCC model for AC is provided. An application example on the AC structure under blast and impact loading is also illustrated and validated based on a field blast test and a laboratory drop weight impact test.

2. Review on KCC Model

When subjected to blast loading or high impact loading, concrete or other concrete-like materials shows a highly non-linear response. They usually exhibit pressure hardening and strain hardening under static loading and strain rate hardening in tension and compression under dynamic loading. A number of material models has been developed to model concrete-like materials recently [11–16]. Among them, the KCC model is widely used to analyse concrete-like materials' response to blast and impact loading due to its simple implementation. In addition, the KCC model can capture the non-linear behaviour of the material under dynamic loading [14]. The key features of the model are discussed briefly in the following section.

2.1. Strength Surface in KCC Model

The KCC model decouples stress into the hydrostatic pressure and deviatoric stress as shown in Equation (1):

$$\sigma_{ij} = s_{ij} + \frac{1}{3}\sigma_{ii}\delta_{ij} \tag{1}$$

where σ_{ij} is the stress tensor, s_{ij} is the deviatoric stress tensor and σ_{ii} is the hydrostatic pressure tensor. It should be noted that stress is positive in tension, and pressure is positive in compression. The hydrostatic pressure is related to the volumetric change of material, while the deviatoric stress is related to shear resistance of the material and is usually expressed by the second invariant of the deviatoric stress tensor, J_2:

$$J_2 = \frac{1}{2}s_{ij}s_{ji} = \frac{s_1^2 + s_2^2 + s_3^2}{2} \tag{2}$$

The KCC model has three independent strength surfaces: maximum strength surface, yield surface and residual strength surface, which are shown in Figure 1. The general formation of strength surfaces can be written as:

$$\Delta\sigma = \sqrt{3J_2} = f(p, J_2) \tag{3}$$

in which $\Delta\sigma$ is the principal stress difference and p is the hydrostatic pressure. Usually, the above Equation (3) refers to the compressive meridian. The whole failure curve can be obtained through rotation of the compressive meridian around the hydrostatic pressure axis by multiplying $r_3(\theta_L)$, which has the formation:

$$\Delta\sigma = r_3(\theta_L) \cdot \sqrt{3J_2} = f(p, J_2, J_3) \tag{4}$$

$$r_3(\theta_L) = \frac{r}{r_c} = \frac{2(1-\psi^2)\cos\theta_L + (2\psi - 1)\sqrt{4(1-\psi^2)\cos^2\theta_L + 5\psi^2 - 4\psi}}{4(1-\psi^2)\cos^2\theta_L + (1-2\psi)^2} \tag{5}$$

where $\psi = r_t/r_c$ and r_t and r_c are the radius of tensile and compressive meridian, respectively. According to Equation (5), it can be found that the $r_3(\theta_L)$ depended on ψ and θ_L. The parameter ψ in turn relies on the hydrostatic pressure. For the concrete-like material, the value of ψ varies from $\frac{1}{2}$ at negative (tensile) pressures to unity at high compressive pressures [14]. The value of Lode angle θ_L can be obtained from:

$$\cos\theta_L = \frac{\sqrt{3}}{2}\frac{s_1}{\sqrt{J_2}} \quad \text{or} \quad \cos 3\theta_L = \frac{3\sqrt{3}}{2}\frac{J_3}{J_2^{3/2}} \tag{6}$$

During the initial increase of hydrostatic pressure p, the deviatoric stresses $\Delta\sigma$ remains in the elastic region until the yield surface is reached. Deviatoric stress can be further developed until the maximum strength surface is reached, and the material will subsequently start to fail. After failure is initiated, the material will gradually lose its load-carrying capacity and reaches its residual strength surface. The formations of these three surfaces are also given in Equations (7)–(9).

$$\text{Yield surface } \Delta\sigma_y = a_{0y} + \frac{p}{a_{1y} + a_{2y}p} \tag{7}$$

$$\text{Maximum strength surface } \Delta\sigma_m = a_0 + \frac{p}{a_1 + a_2 p} \tag{8}$$

$$\text{Residual strength surface } \Delta\sigma_r = \frac{p}{a_{1f} + a_{2f}p} \tag{9}$$

where the eight parameters, namely, $a_0, a_1, a_2, a_{1f}, a_{2f}, a_{0y}, a_{1y}$ and a_{2y}, for three surfaces can be obtained from the experimental data (e.g., triaxial compression test, biaxial compression test or uniaxial tension/compression test) [14].

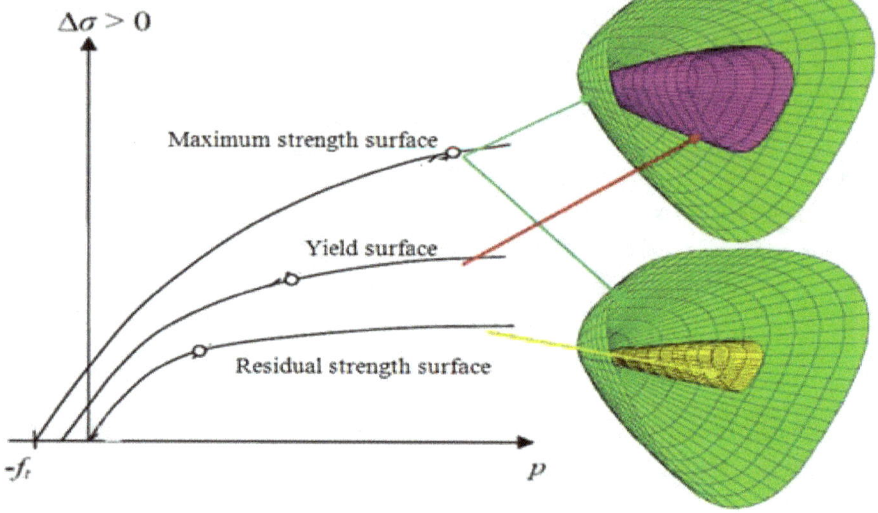

Figure 1. Strength surfaces for the Karagozian and Case concrete model (KCC) material model [14].

2.2. Damage Factor in the KCC Model

After reaching the initial yield surface, but before reaching the maximum strength surface, the current surface can be obtained as a linear interpolation between yield surface $\Delta\sigma_y$ and maximum strength surface $\Delta\sigma_m$:

$$\Delta\sigma = \eta(\Delta\sigma_m - \Delta\sigma_y) + \Delta\sigma_y \tag{10}$$

After reaching the maximum strength surface, the current failure is interpolated between the maximum strength surface $\Delta\sigma_m$ and the residual strength surface $\Delta\sigma_r$, which is similar to the above computation:

$$\Delta\sigma = \eta(\Delta\sigma_m - \Delta\sigma_r) + \Delta\sigma_r \tag{11}$$

where η varies from zero to one depending on the accumulated effective plastic strain parameter λ. The value of η normally starts at zero and increases to unity at $\lambda = \lambda_m$, then decreases back to zero at some larger value of λ. λ_m is the plastic strain at maximum strength surface. The accumulated effective plastic strain λ can be expressed as follows:

$$\lambda = \int_0^{\bar{\varepsilon}^p} \frac{d\bar{\varepsilon}^p}{r_f[1 + p/(r_f f_t)]^{b_1}} \quad \text{for} \quad p \geq 0 \tag{12}$$

$$\lambda = \int_0^{\bar{\varepsilon}^p} \frac{d\bar{\varepsilon}^p}{r_f[1 + p/(r_f f_t)]^{b_2}} \quad \text{for} \quad p < 0 \tag{13}$$

where f_t is the quasi-static tensile strength, $d\bar{\varepsilon}^p$ is effective plastic strain increment and r_f is the dynamic increase factor (DIF) of the material under dynamic loading. The damage factors b_1 and b_2 define the softening behaviour due to compression ($P \geq 0$) and tension ($P < 0$), respectively. Parameter b_1 can be determined by considering compressive energy G_c (area under the compressive stress-strain curve) obtained from the uniaxial compression test in single element simulation. It is obtained iteratively until the area under the stress-stain curve from single element simulation coincides with G_c/h, where h is the element size. The damage factor b_2 in Equation (13) is related to tensile softening of the material and determined from experimental data. The fracture energy G_f can be obtained from the uniaxial tensile test or three-point notched beam test. Then, the single element simulation of uniaxial tensile test is employed to obtain the stress-strain curve from the numerical analysis. Hence, the value of b_2 can be obtained until the area under the tensile stress-stain curve from a single element coincides with G_f/w_c, where w_c is the localization width, and typically, w_c is normally taken as one- to six-times the maximum aggregate size [14].

Based on Equations (10) and (11), the stress softening factors η and λ are governed by the accumulation of effective plastic strain. However, when the stress path is very close to the negative hydrostatic pressure axis, i.e., isotropic tension, wherein the hydrostatic pressure would decrease from zero to $-f_t$, where no deviatoric stress occurred, no damage accumulation would occur based on these equations. However, in such concrete-like materials, damage cannot be avoided even at this state. Therefore, the above condition has to be modified by including pressure-softening effects near or after tensile failures. In this case, a volumetric damage increment is calculated and added to the total damage factor λ whenever the stress path is close to the triaxial tensile path.

2.3. Strain Rate Effect

The material model KCC also includes a radial rate enhancement on the material failure surface. This is because experimental data for concrete-like materials are typically obtained along radial paths from the origin in deviatoric stresses versus hydrostatic pressure via unconfined compressive and tensile tests. It is well known that the strain rate effect is important for concrete-like materials under severe dynamic loading. A typical DIF-strain rate curve for concrete-like materials can be obtained from the servo hydraulic fast loading tests and the split Hopkinson pressure bar (SHPB) test.

2.4. Equation of State

In addition to the strength surface model, an equation of state (EOS) is needed to describe the relationship between hydrostatic pressure and volume change of the material subject to dynamic load. EOS is usually determined using a fly impact (i.e., for steel) test or triaxial compressive test (i.e., for concrete or geomaterials). The isotropic compression portion of the KCC material model consists of pairs of hydrostatic pressure P and corresponding volume strain μ. It is implemented as a piece-wise curve in this model.

3. Determination of Parameters for Asphalt Concrete Material

The KCC model is employed to simulate AC material to capture dynamic response under impact and blast loading. This section describes the determination of the model parameter for the asphalt concrete with the compressive strength of 4.6 MPa, which is used in the following application example. For the asphalt concrete with a different compressive strength, this method can be used as a reference.

3.1. Strength Surface

As mentioned in Section 2.1, the KCC material model has three strength surfaces: strength, residual strength and yield surfaces. These three surfaces can be obtained through curve fitting of suitable experimental data. In this study, due to the few data for triaxial compressive tests of asphalt concrete, available data are extracted from Park et al. [4] with the compressive strength f_c = 0.311 MPa for AC. Figure 2 presents the determination of the three surfaces by curve fitting for AC with f_c = 0.311 MPa. The intersection point of maximum strength surface and residual strength surface is the so-called brittle-to-ductile point. This point can be determined by experimental data under high confining pressure. However, it is difficult to determine this point in the strength surface as no experimental data are available for AC materials. Based on the experimental data for concrete [17], this point was usually taken as p/f_c = 3.878. Considering that the size and strength of aggregates used in AC and concrete material are almost the same, the brittle-to-ductile point for AC is taken to be the same as that for concrete. This value might be conservative for AC due to the higher content of coarse aggregate in AC. However, for simulation purpose, this value is acceptable.

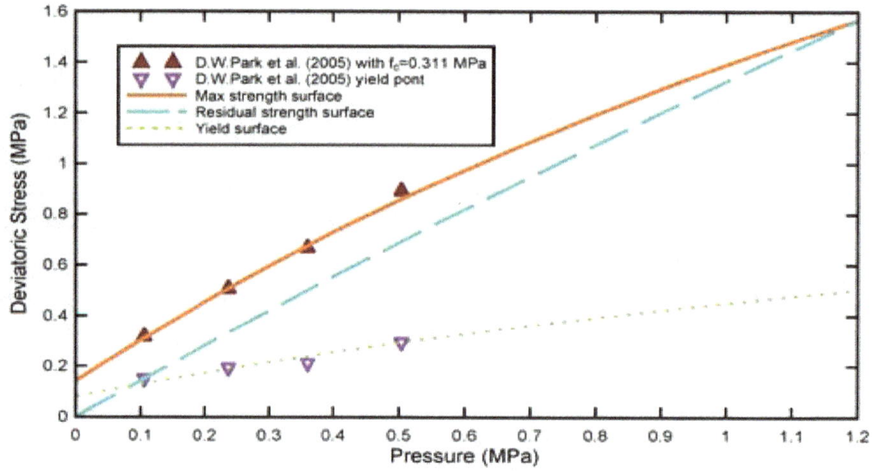

Figure 2. Determination of parameters in KCC from experimental data.

If new asphalt concrete with known unconfined compression strength $f'_{c,new}$ is to be modelled, but its strength surfaces are otherwise unknown, then one way of scaling data from a known material is proposed as follows [14]:

$$r = \frac{f'_{c,new}}{f'_{c,old}} \tag{14}$$

where $f'_{c,old}$ is the unconfined compressive strength for a previously modelled AC. Then, the new material strength surface can be taken as:

$$\Delta\sigma_n = a_{0n} + \frac{p}{a_{1n} + a_{2n}p} \tag{15}$$

in which $a_{0n} = a_0 r, a_{1n} = a_1, a_{2n} = a_2/r$.

The new asphalt concrete with unconfined compressive strength f_c = 0.8 MPa [2] is used to validate the parameters obtained from the scaling method. Figure 3 shows the maximum strength surface determined by the scaling method. It can be seen that the maximum strength surface fit very well with the experimental data, and thus, it can be concluded that the parameters for AC with different compressive strengths can be obtained by the scaling method.

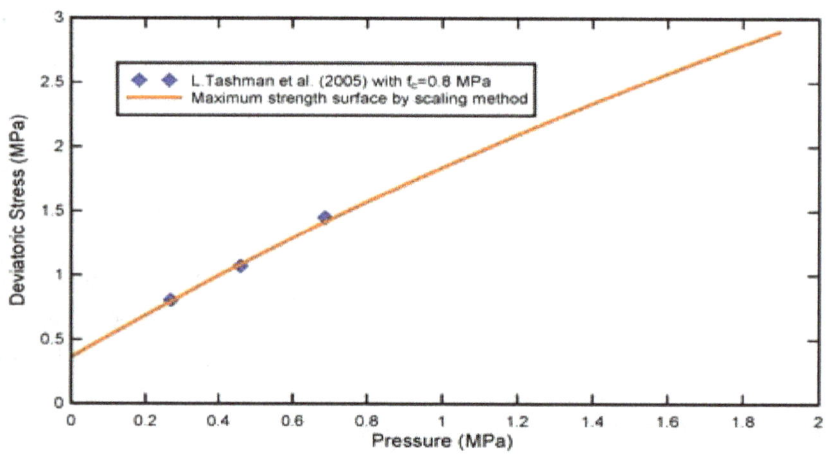

Figure 3. Validation of the failure surface from the scaling law.

In this study, the unconfined compressive strength for AC material is 4.6 MPa, and the tensile strength is 0.7 MPa at 35 °C. By scaling the data from the established curves given in Figure 2 [11], the appropriate strength surface of the current materials can be determined; the strength parameters are given in Table 1.

Table 1. Parameters for asphalt concrete (AC) material with f_c = 4.6 MPa.

Parameter	Value
a_0	2.071
a_1	0.6
a_2	0.0135
a_{0y}	1.183
a_{1y}	2.00
a_{2y}	0.0473
a_{1f}	0.70
a_{2f}	0.0037

3.2. Damage Factor

The strain hardening and softening pairs (η, λ) in Equations (10) and (11) describe the material behaviour transmitted from the yield surface to the maximum strength surface and from the maximum strength surface to the residual strength surface, respectively. During the transmission, parameter η varies from zero to one, depending on the accumulated effective plastic strain parameter λ. However, it is found that the original damage factor pairs (η, λ) in the KCC model are only suitable for concrete and not for the AC material due to AC having higher plastic failure strain. Thus, the input for accumulated effective plastic strain λ should be modified. Based on the uniaxial compressive test for AC, it was found that at peak stress, the corresponding strain was approximately 0.018, and the final failure strain was about 0.1; while for normal concrete, the corresponding strain at peak stress was around 0.0022. Hence, the λ is modified to give the high failure strain for AC in the current study. Additionally, it is found that when λ is adjusted to 10-times the original λ value, the numerical results seemed to show good agreement with the experimental results from the unconfined compressive test for AC. Figure 4 shows the modified and original series of (η, λ) pairs. It can be seen that the modified damage factor provided smoother descending than the original damage factor and had a higher failure strain that matched the behaviour of AC very well.

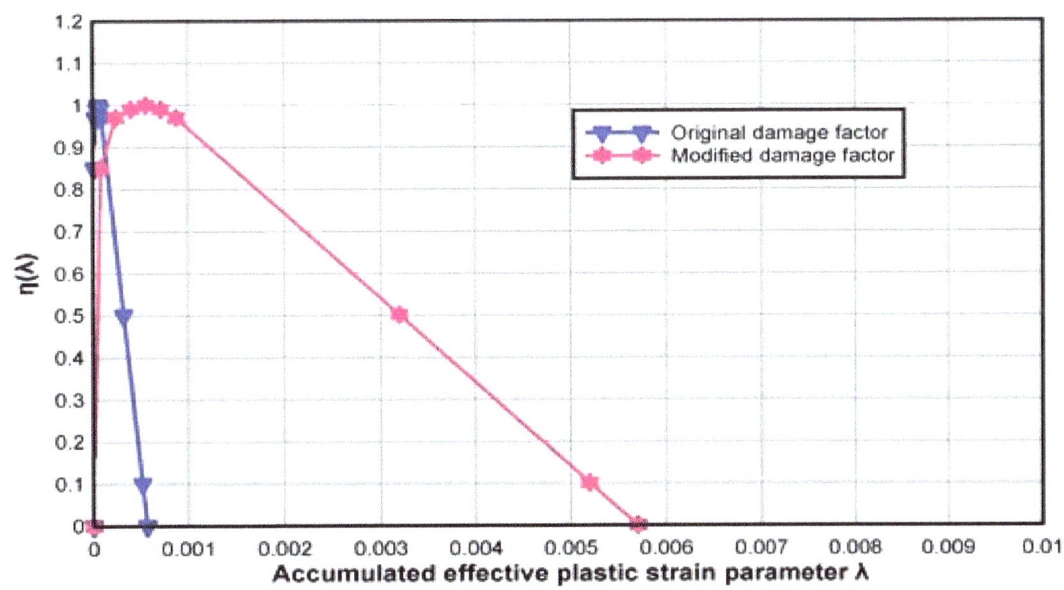

Figure 4. Damage factor used for AC material.

3.3. Equation of State

There are limited EOS data for AC material. The available EOS data are for AC with compressive strength of f_c = 3.8 MPa [5]. In this study, the compressive strength for AC is f_c = 4.6 MPa. Thus, the pressure-volume pairs can be calculated using the volumetric scaling method [18]. In this method, assuming that new data are obtained at the same volumetric strains, thus, the new corresponding pressure (pc_{new}) can be:

$$pc_{new} = pc_{old}\sqrt{r} \tag{16}$$

and the new corresponding unloading bulk modulus ($k\mu_{new}$) is:

$$ku_{new} = ku_{old}\sqrt{r} \tag{17}$$

where r is the scaling factor, which is the ratio of compression strength for new material to the compression strength of the previous material modelled. Hence, the EOS data for f_c = 4.6 MPa are calculated based on Equations (16) and (17), and the EOS inputted in the numerical model is shown in Figure 5.

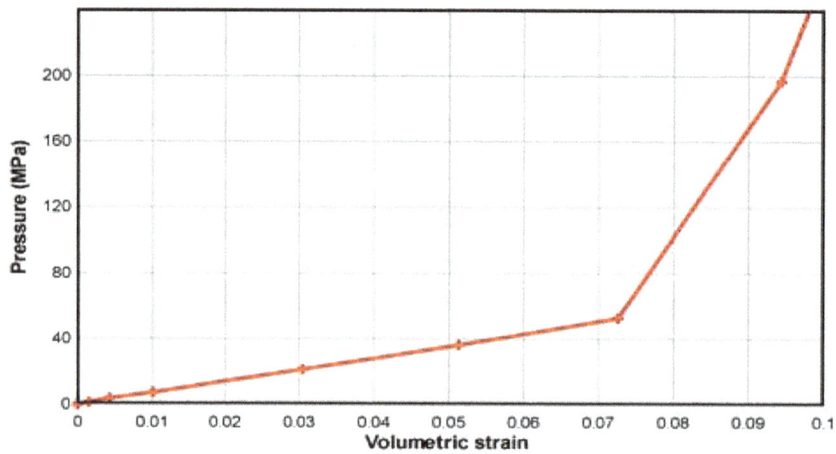

Figure 5. Equation of state (EOS) for AC with f_c = 4.6 MPa.

3.4. Softening Parameters b_1, b_2

The softening parameters (b_1, b_2) shown in Equations (12) and (13) control the material softening behaviour after peak stress. These parameters are obtained from experiments, as detailed below.

- Value of b_1 from the uniaxial compressive test:

  The uniaxial compressive test was conducted for AC according to ASTM 1074. The detailed test results and setup can be further referred to [11]. Based on the test results, it is found that the corresponding strain at peak stress (f_c = 4.6 MPa) was about 0.018, and the final failure strain was about 0.1, which was higher than that of concrete. The Young's modulus obtained from stain gauges attached at the middle height of the sample was 598 MPa. Based on experimental results, the compressive energy G_c was calculated at 15.1 MPa·mm. Hence, for example, the b_1 value for a 10-mm mesh size was calculated as 3.45 using the method stated in Section 2.1.

- Value of b_2 from the fractural test:

  The value of b_2 is determined by fracture energy G_f, which can be obtained from the uniaxial tensile test or the three-point single-edge notched beam test (SNB). In the current study, the SNB test was carried out to evaluate fracture energy G_f for the AC material. The detailed theory about the SNB test can be found in the established literature [19]. Therefore, only the test result is presented here. In the SNB test, the compacted AC beam was fabricated with a dimension of

$400 \times 100 \times 100$ mm^3 depth. A mechanical notch was sawn with a depth of 20 mm, which gave a ratio of notch to beam depth of 0.2. The simply supported sample with a span length of 340 mm was tested under a 35 °C temperature. From the test, fracture toughness K_{IC} can be obtained according to the formula suggested by Karihaloo and Nallathambi [19]. Then, the fracture energy G_f is calculated using:

$$G_f = \frac{(1 - v^2) K_{IC}^2}{E} \tag{18}$$

in which E is the elastic modulus and ν is Poisson's ratio.

The parameter b_2 is further determined by assigning fracture energy G_f in the use of single element simulation of the uniaxial tensile test. The b_2 is then obtained via an iterative procedure until the area under the stress-stain curve from the single element simulation coincides with the value of G_f/w_c. The parameters obtained from SNB and single element simulation for AC (f_c = 4.6 MPa) are summarized in Table 2.

Table 2. Parameters from the single-edge notched beam test (SNB) and single element simulation.

Parameters	Unit	Value
K_{IC}	MPa·mm $^{1/2}$	12.2
ν	-	0.35
E	MPa	598
G_f	MPa·mm	0.221
w_c	mm	40
G_f/w_c	-	0.00554
f_t	MPa	0.7
b_2	-	0.2

3.5. Strain Rate Effect

The DIF curve for AC under different strain rates was obtained using servo hydraulic fast loading tests and the split Hopkinson pressure bar (SHPB) test in the current study. The strain rate produced by the servo hydraulic machine was approximately 10^{-5} to 1 s^{-1}, and the higher strain rate loading was obtained through SHPB testing. The detailed setup and procedure for SHPB and the hydraulic test for AC can be referred to Wu [20]. The DIF value for compressive (DIF_c) and tensile (DIF_t) strength obtained from the test are given as:

$$DIF_c = \frac{f_{dc}}{f_{sc}} = 3.18 + 1.098\log_{10}(\dot{\varepsilon}) + 0.1397\log_{10}{}^2(\dot{\varepsilon}) \quad \text{for} \quad \dot{\varepsilon} \le 100\text{s}^{-1}$$

$$DIF_c = \frac{f_{dc}}{f_{sc}} = 21.39\log_{10}(\dot{\varepsilon}) - 36.76 \quad \text{for} \quad 100\text{s}^{-1} < \dot{\varepsilon} \le 200\text{s}^{-1} \tag{19}$$

$$DIF_t = \frac{f_{dt}}{f_{st}} = 1.86 + 0.1432\log_{10}(\dot{\varepsilon}) \quad \text{for} \quad \dot{\varepsilon} \le 15\text{s}^{-1}$$

$$DIF_t = \frac{f_{dt}}{f_{st}} = 6.06\log_{10}(\dot{\varepsilon}) - 5.024 \quad \text{for} \quad 15\text{s}^{-1} \le \dot{\varepsilon} \le 100\text{s}^{-1} \tag{20}$$

However, for the compressive DIF curve, a numerical modelling of the SHPB test adopting this DIF curve found that the initial segment of this curve matched the experimental results very well, while the numerical model results for strain rate larger than 100 s^{-1} seemed to overestimate the stress. This can be due to the "double counting" of the inertia effect in the numerical modelling when the strain rate exceeded 100 s^{-1}. Hence, in the current model, the second segment in compressive DIF (Equation (19)) is ignored when the strain rate exceeds 100 s^{-1}. Beyond this, the DIF is assumed to remain a constant value. For the tensile DIF curves, in the macro-level numerical model, the

KCC material model cannot capture the aggregate interlocking that propagates the micro-cracking and energy dissipation beyond the localization zone [20–22]. Therefore, the above tensile DIF curve (Equation (20)) with two branches is used in the numerical model. The tensile and compressive DIF curves of asphalt concrete used in numerical model are summarized in Figure 6.

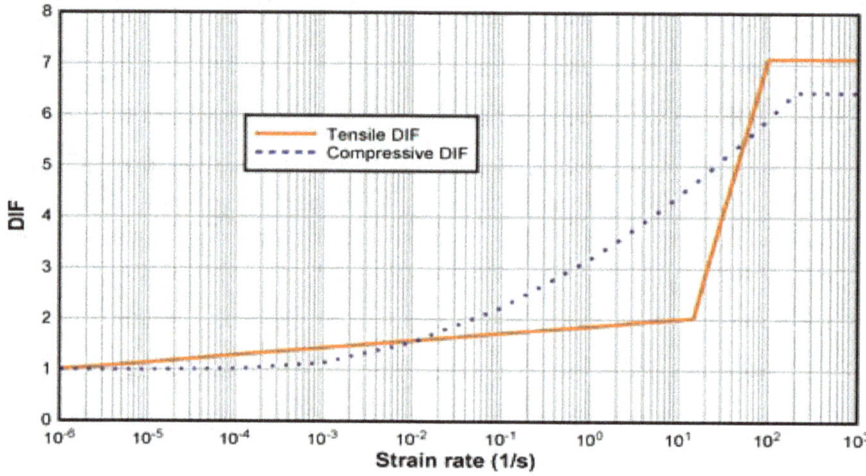

Figure 6. Tensile and compressive dynamic increase factor (DIF) curve used in the numerical model for asphalt concrete.

4. Application Example

4.1. Case for Blast Loading

4.1.1. Field Blast Test

In this section, the AC material is used in the multi-layer pavement. Figure 7 shows the cross-sectional view of the multi-layer pavement slab including 100 mm-thick engineering cementitious composite (ECC) at the bottom, 100 mm-thick high strength concrete (HSC) in the middle and 75 mm-thick AC at the top. The geogrid (GST) was placed at the middle of the AC layer to reinforce the AC material. The numerical model for the multi-layer pavement system under blast load is developed based on the configuration of the full-scale field blast test [23]. Selected key features of the field blast test and numerical model are presented below.

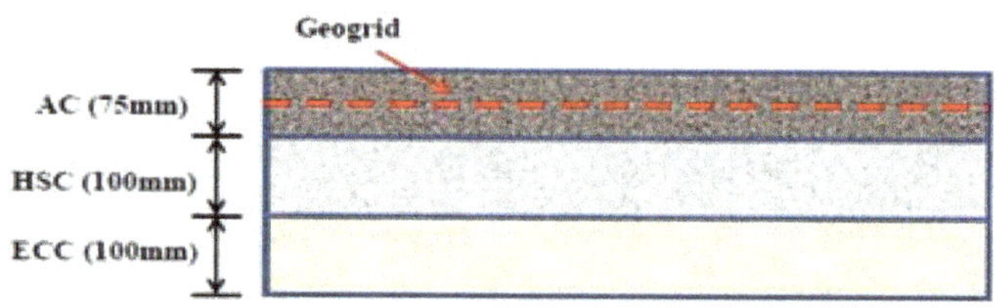

Figure 7. Cross-section of the multi-layer pavement slab. HSC, high strength concrete; ECC, engineering cementitious composite.

In the field blast test, a bomb with the equivalent of a 7.3 kg Trinitrotoluene (TNT) charge weight was selected for testing and placed at about 170 mm above the surface of the pavement. The charge weight was evaluated based on the typical terrorist weapon attack. This multi-layer pavement slab was cast at site with a dimension of 2800 mm × 2800 mm × 275 mm (width × depth × thickness). The pavement slab was anchored to the ground to simulate the practical boundary condition, and a vertical anchor was installed at each corner. Figure 8 presents the multi-layer pavement slab before the blast load.

Figure 8. Plan view of the multi-layer pavement slab before the blast event.

Various instruments were installed onto the slab to measure its responses during blast loading. Figure 9 shows the instrumentation installed on the pavement slab. Four accelerometers were installed at the middle of the side of the slab to measure both vertical (V1 and V2 in Figure 9) and horizontal accelerations (H1 and H2 in Figure 9). The accelerometers were mounted onto steel frames that were cast together with the slab. Three total pressure cells (TPC) (TPC1, TPC 2 and TPC3 in Figure 9) were buried in the soil just below the slab to measure the pressure transferred from the pavement slab.

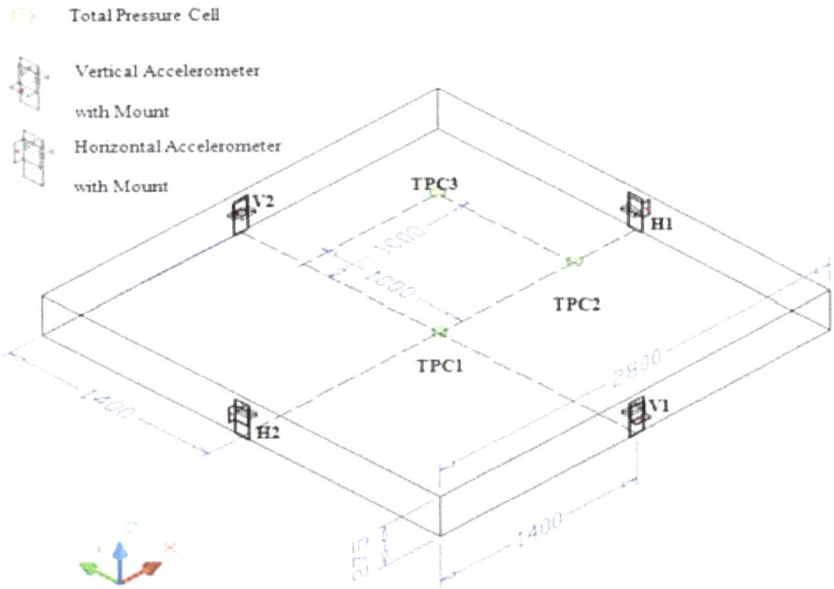

Figure 9. Layout of instrumentation for field blast test. TPC, total pressure cell.

A numerical model is established using LSDYNA package [24]. In this model, the slab and foundation soil are discretised in space with one point Gauss integration eight-node hexahedron Lagrange element. Only a quarter of the slab is modelled with symmetric boundary conditions (as shown in Figure 10). The geogrid is spatially discretised with the shell element, and it is assumed that the geogrid is fully bonded within the AC layer. The anchors on the pavement slab are simulated as the fixed points (fixed in vertical direction) in the corresponding position in the numerical model. The non-reflection boundary is applied on the side and bottom of the foundation soil to model semi-infinite space. Based on the mesh study on the numerical model, the 10-mm element size is adopted for the pavement slab, geogrid and soil mass. The blast pressure is extracted from AUTODYN and used for numerical model analysis. The detailed process of applying pressure to the pavement surface can be referred to elsewhere [20,25].

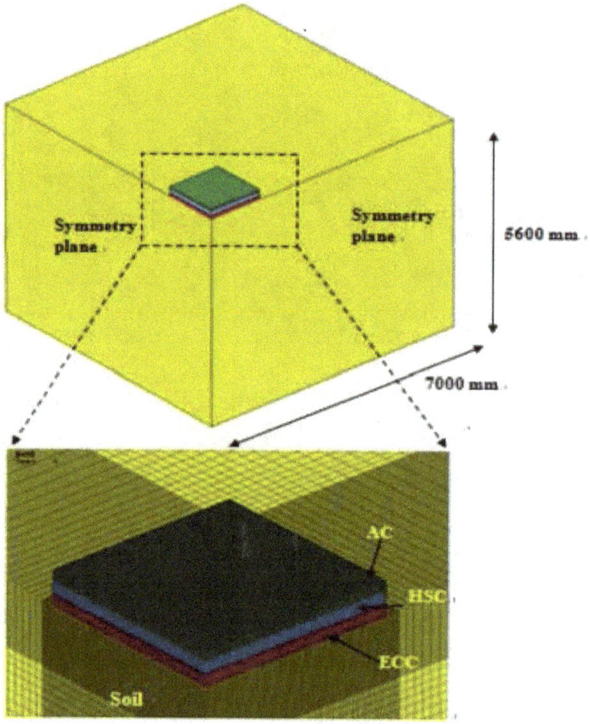

Figure 10. Finite element model of the multi-layer pavement slab.

In the numerical model, AC, ECC and HSC are all grouped as concrete-like materials and modelled by the KCC model [20,23]. The basic parameters for these three materials are listed in Table 3. The parameters for AC material in the KCC model can be found in Section 3. While the process of the determination of HSC and ECC parameters for KCC model is the same as that mentioned in Section 3, the detailed parameters can be referred to Wu [20]. The Drucker–Prager model and plastic-kinematic model [17] are employed to model the foundation soil and geogrid, respectively. The AUTOMATIC_SURFACE_TO_SURFACE contact algorithm is employed to model the interaction between pavement slab and soil. The TIEBREAK contact algorithm in LSDYNA is used to simulate the interface behaviour between HSC and AC layers. The parameters for the foundation soil, geogrid and interface property are in Tables 4–6, respectively.

Table 3. Basic properties of materials in the multi-layer pavement for the blast load.

Parameters	AC	HSC	ECC
Young's modulus E (MPa)	598	33,000	18,000
Compressive strength f_c (MPa)	4.6	55	64
Tensile strength f_t (MPa)	0.7	4.35	5
Poisson ratio ν	0.35	0.2	0.22

Table 4. Parameters for Geogrid MG-100 using the plastic-kinematic model.

Parameters	Symbol	Units	Value
Density	ρ	kg/m^3	1030
Young's modulus	E	MPa	500
Poisson's ratio	ν	-	0.3
Yield stress	σ_y	MPa	7.5
Tangent modulus	E_t	MPa	333
Thickness	t	mm	2.4
Erosion strain	ε_s	-	0.038

Table 5. Material properties of the soil mass.

Parameters	Symbol	Units	Value
Density	ρ	kg/m^3	2100
Shear modulus	G	MPa	13.8
Poisson's ratio	ν	-	0.3
Cohesion	c	kPa	62
Friction angle	ϕ	°	26

Table 6. Parameters for the interface simulation.

Parameters	Value
Contact type	TIEBREAK
Friction for static	0.71
Friction for dynamic	0.56

4.1.2. Numerical Result

The results of the numerical modelling of the multi-layer pavement under blast loading, with the incorporation of the above-mentioned material models, are summarized and compared with the blast test results. In the numerical results, the fringe level in the damage contour is the value for the scaled damage indicator δ, which is defined to describe the damage level of the material [14,20,23]. A scaled damage indicator δ is related to the effective plastic strain λ in the material: (i) at the yield surface, $\lambda = 0$, leading to $\delta = 0$; (ii) at the maximum strength surface, $\lambda = \lambda_m$, leading to $\delta = 1$; and (iii) at the residual strength surface, $\lambda = \lambda_r >> \lambda_m$, leading to $\delta = 1.99 \approx 2$. Thus, the δ value moving from 0 to 1 to 2 indicates that the failure surface migrates from the yield surface to the maximum strength surface and to the residual strength surface, respectively, as the material being stressed. In this study, when the residual strength of material reduces to 20% of its peak strength, the material seems to suffer severe failure. The plastic strain corresponded to that residual strength used to calculate the delta value. Both the laboratory and field test for AC material indicated that when δ was greater than 1.8, the material would be severely damaged.

The damage situation for the multi-layer pavement slab in field blast test is shown in Figure 11. Figure 11a shows that the blast pressure destroyed the upper half of the AC layer above the GST reinforcement. It is also noted that only the centre of the GST piece was burned off during the blast event. Figure 11b shows the resulting damage on the HSC layer after removing the top layer of asphalt. From this figure, it could be seen that the crater was very shallow and did not punch through the whole layer, and a crater of around 700 mm in diameter and a depth of 10 mm was formed on the HSC layer.

(a) (b)

Figure 11. Damage of the multi-layer pavement after blast loading: (**a**) damage pattern of the AC layer in field blast test; (**b**) damage pattern of the HSC layer in field blast test.

The damage contour for the AC layer in numerical model is shown in Figure 12a. Comparing Figures 11a and 12a, it is observed that the damage pattern in the numerical model is symmetrical, while that in the field measurement was skewed. This is because the bomb in the field was not placed at the centre of the slab, and one side of the AC layer was more severely damaged than the other. Shear cracking near the anchor point was observed in the numerical model, which was similar to the experimental observations in the field test. It could be concluded that the basic failure pattern given by the numerical model agrees well with the results obtained from the field-testing. Figure 12b shows the damage pattern for the HSC layer. Comparing Figure 11b with Figure 12b, the damage pattern for HSC is very consistent between field measurement and numerical results. The diameter of the crater was about 750 mm in the numerical model, which is very close to that of the blast test result. As shown in Figure 12b, shear cracks are also observed near the anchor points. Based on the damage pattern in the field blast test, the crater on the top face of the HSC is shown to be shallow and with a thickness of less than 10 mm. However, after cracking occurred at the bottom of the HSC layer, the numerical model shows that the bottom of the HSC layer has experienced severe cracking. This might be due to the combination of the bending of the HSC layer under the blast load and the reflection of the stress wave at the bottom interface. In the numerical model, the interface between HSC and ECC is assumed to be fully bonded. However, ECC is more flexible than HSC, and thus, it would cause tensile stress at the bottom of the HSC layer when deformed together. The compression stress wave from the top face would also travel within the HSC layer and reflect as a tension stress at the interface, which could cause spalling. Based on the damage pattern in the numerical model, the HSC layer might be considered having failed, while the field observation suggests that HSC may have partially failed.

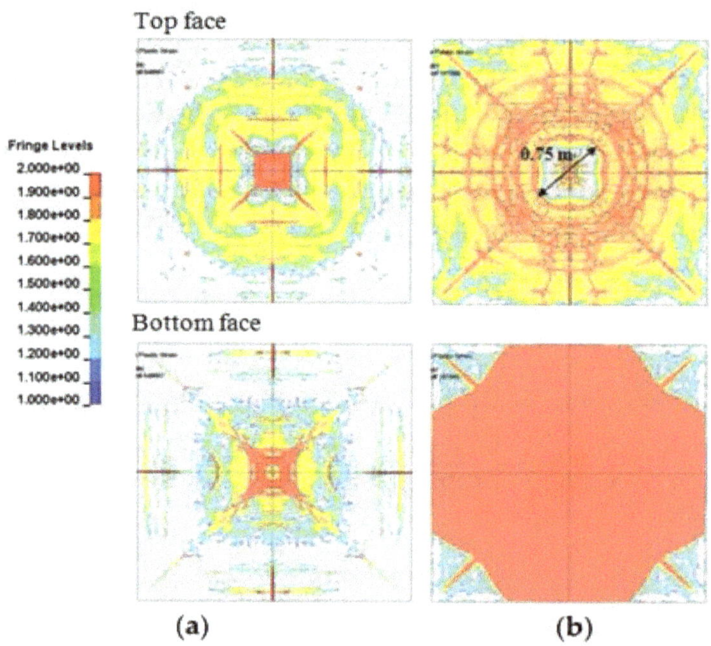

Figure 12. Damage contour for the AC and HSC layer in the multi-layers pavement: (**a**) damage contour of the AC layer in the numerical simulation; (**b**) damage contour of the HSC layer in the numerical simulation.

In the field blast test, the four accelerometers were installed at the mid-side of pavement slab (as shown in Figure 9). These accelerometers were used to measure the vertical and horizontal acceleration of the pavement slab subjected to blast loading. For the horizontal acceleration, due to the centre of the charge being closer to one side of the pavement slab; there were two different horizontal acceleration readings; while in the numerical model, it was assumed that the explosive occurred in the centre of the pavement slab. Thus, in this section, only the vertical acceleration from the field blast test was compared with that of the numerical model. In the numerical model, the raw nodal acceleration contained considerable numerical noise. The ELEMENT_SEATBELT_ACCELEROMETER could be used to eliminate numerical noise and obtain more accurate node acceleration. The vertical acceleration from the blast testing is compared with that of the numerical model as shown in Table 7. The results from both the blast testing and the numerical simulation are comparable. The maximum difference of vertical acceleration between the blast testing and the numerical model is about 10%, and the numerical model predicted slightly higher in the vertical acceleration than that of the blast test.

Table 7. Vertical acceleration of the multi-layer pavement slab.

Item	Field Trial Test	Numerical Result	Deviation from Field Trial Test
Max. vertical acceleration (m/s^2)	35,400	38,870	10%

The pressure values in the corresponding points in the numerical model are compared with pressures obtained from the blast test, as summarized in Table 8. The layout of the total pressure cell in the blast could be referred to Figure 9. The pressure values from the numerical simulation are shown to be close to that from the blast test for TPC2; while for TPC3, it has a 20% discrepancy with the numerical simulation considering the inherent variation in the blast test. TPC1 was damaged during the blast test, and hence, no pressure reading was recorded from it. The numerical model predicts that the pressure might be as high as 13 MPa at that point, which is far beyond the maximum measurement capacity of the pressure cell installed. That can explain why TPC1 was destroyed due to the overwhelming blast loading.

Table 8. Peak reading for the total pressure cell.

Item	Field Blast Test (kPa)	Numerical Result (kPa)	Deviation from Field Trial Test
TPC1	Destroyed	13,393	Sensor destroyed as pressure >> range
TPC2	273	267	2%
TPC3	200	241	20%

4.2. Case for Impact Loading

4.2.1. Laboratory Drop Weight Impact

The multi-layer pavement slab was subjected to an 1181-kg drop weight impact. The cross-section of the multi-layer pavement is the same as that given in Figure 7. The basic mechanical properties of the, HSC, ECC and AC are determined according to ASTM standards, and the results are summarized in Table 9. The drop weight was a cylindrical projectile with a hemispheric head (100 mm in diameter and made with high strength steel), and the pavement slab was subjected to two times of impact from the same drop height of 1.5 m. During the test, the multi-layer pavement slab was placed on the top of compacted soil/sand in a steel strong box. Directly below the slab was the geocell (MiraCell MC-100) which was filled with compacted soil/sand. This was to enhance the strength of the soil/sand layer and to provide a solid sub-base as to simulate the practical condition. The setup for the multi-layer pavement slab is given in Figure 13. Various instruments were also installed to monitor the response of the pavement during the drop weight test. Figure 14 shows the positioning of the potentiometers. A photodiode system was used to trigger the data acquisition system during the test. It consists of two photodiodes and two laser sources placed 100 mm vertically apart. The data acquisition system would be triggered when the falling projectile crosses the top laser emitter. Impact velocity could be determined using the time interval that the projectile took to cross the second laser emitter.

Table 9. Basic properties of the materials in the multi-layer pavement in the drop weight test.

Parameters	AC	HSC	ECC
Young's modulus E (MPa)	598	40,000	18,000
Compressive strength f_c (MPa)	4.6	90	80
Tensile strength f_t (MPa)	0.7	4.35	5
Poisson ratio ν	0.35	0.2	0.22

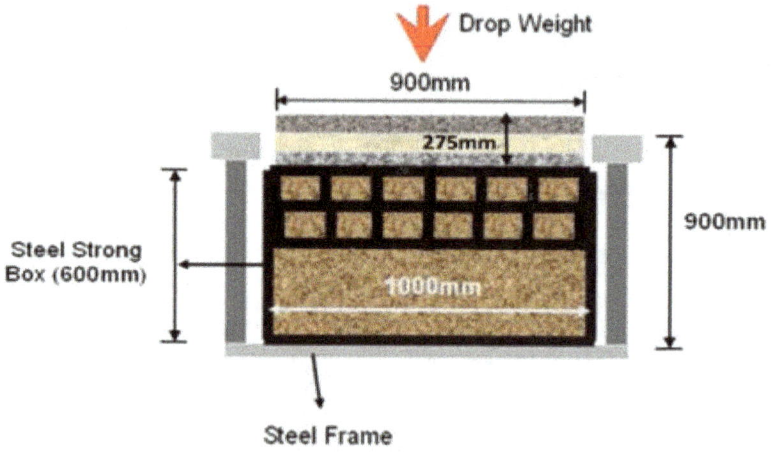

Figure 13. Setup for the impact test.

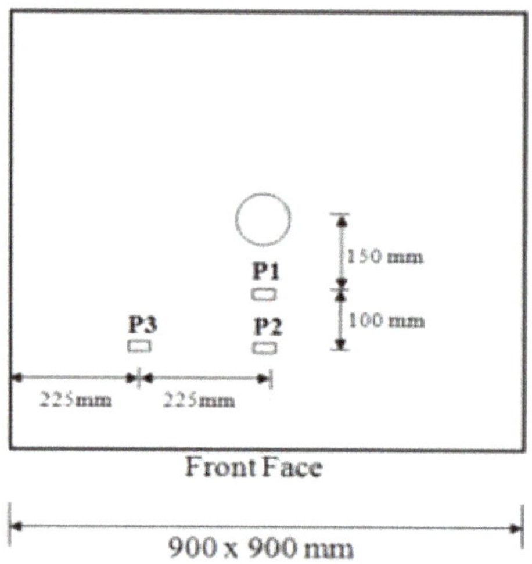

Figure 14. Positioning of the potentiometers.

In the numerical model, the AUTOMATIC_SURFACE_TO_SURFACE contact algorithm is employed to model the interaction between pavement slab and soil. The contact algorithm TIEBREAK is also used to simulate the interface behaviour between the AC and HSC layer, which is the same as that in the numerical model for blast loading as mentioned in Section 4.1.2. As the impactor might penetrate the AC layer, the erosion method is employed in the simulation, in which the maximum plastic failure strain of 0.2 is used to delete the distorted element once the actual strain exceeds this predefined failure strain. The multi-layer pavement slab, the drop weight head and soil mass are discretised in space with one point Gauss integration eight-node hexahedron elements. The geogrid is simulated with the four-node Belytschko–Tsay shell element that allows no bending resistance. In addition, the PLASTIC_KINEMATIC material model is employed to describe the bi-linear behaviour of the geogrid under tensile loading [23]. The parameters used for the geogrid are the same as given in Table 4.

The Drucker–Prager model is used to simulate soil mass. In the laboratory drop weight test, the upper soil layers were compacted and reinforced with geocell material, which would enhance the strength of the soil, and the lower layer has no reinforcement. Hence, in the numerical model, it is necessary to consider the function of the geocell material. From the laboratory test on the geocell-encased sand [26], it is observed that the geocell confinement did not change the friction angle of soil while significant cohesion occurred in the granular soil, which indicated that for the geocell-reinforced sand layer, the strength and stiffness behaviour of the soil would be enhanced. However, in the numerical model, it is difficult to model and mesh the geocell material due to its complex geometry. Hence, it would be preferable to use the composite model to consider the enhancement of the shear strength and stiffness of the geocell-reinforced sand layer. Madhavi et al. [27] purposed an empirical equation to calculate Young's modulus of the geocell-reinforced sand using the secant tensile modulus of the geocell material and Young's modulus parameter of the unreinforced sand, which could be expressed as:

$$E_r = 4(\sigma_3)^{0.7}\left(K_u + 200M^{0.16}\right) \tag{21}$$

in which E_r is the Young's modulus of the geocell-reinforced sand, M is the secant modulus of the geocell material at axial strain 2.5% in kN/m and σ_3 is the confining pressure from the geocell in kPa. k_u is the dimensionless modulus parameter of the unreinforced sand, which is a modulus number in the hyperbolic model developed by Duncan and Chang [28]. The confining pressure σ_3 could be calculated as:

$$\sigma_3 = \frac{2M}{D_0}\left(\frac{1-\sqrt{1-\varepsilon_a}}{1-\varepsilon_a}\right) \tag{22}$$

where D_0 is the initial diameter of the geocell and ε_a is the axial strain of the geocell at failure; the induced cohesion in the geocell-reinforced sand is then related to the increase in the confining pressure σ_3:

$$c_r = \frac{\sigma_3}{2}\sqrt{K_P} \tag{23}$$

in which c_r is the enhanced cohesion and k_p is the coefficient of passive earth pressure.

In the current study, the geocell MC-100 from Polyfelt is used. The geocell dimension is a rhombus with two diagonal lengths of 203 mm and 244 mm. Thus, the equivalent diameter is calculated as about 177.5 mm. The secant modulus M for the geocell is obtained as 278 kN/m from the tensile test [20]. Additionally, the value of ε_a is taken as 4.8%. The modulus parameter k_u for unreinforced sand in the current study is taken as 727 MPa according to the curve fitting from the triaxial test, and hence, the confining pressure, enhanced cohesion and Young's modulus for geocell-reinforced sand could be calculated based on Equations (21)–(23). The parameters for the unreinforced sand and geocell reinforced sand are summarized in Table 10.

Table 10. Material properties of the foundation soil.

Parameters	Reinforced Sand Layer	Unreinforced Sand Layer
Density, ρ (kg/m^3)	1600	1600
Young's modulus E (MPa)	103.5	40
Shear modulus, G (MPa)	39.8	15.4
Poisson's ratio, ν	0.3	0.3
Cohesion, c (MPa)	0.089	0.001
Friction angle, ϕ (°)	40	40

During the impact test, the deformation of the drop weight head was negligible compared to the deformation of the pavement slab. Hence, the drop weight head is modelled with a rigid body in the current study. For the configuration of the drop weight head, the simple cylindrical shape is modelled instead of modelling the head with weight mass (as shown in Figure 15). The simple cylindrical head has a diameter of 100 mm with a length of 1292 mm. The total mass for the simple cylindrical head is about 1181 kg, from which the density of the drop head would be obtained. The properties for the drop head are listed in Table 11. The convergence study is conducted, and it was found that a 5-mm element size gave a stable response, which is therefore applied for the simulation. The numerical model of the multi-layer pavement slab under drop weight impact load is given in Figure 16.

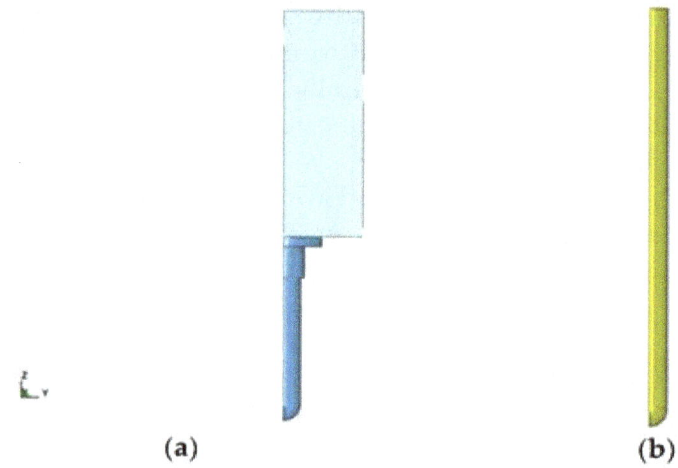

Figure 15. Configuration of the drop weight head (quarter model): (**a**) standard configuration; (**b**) simple configuration.

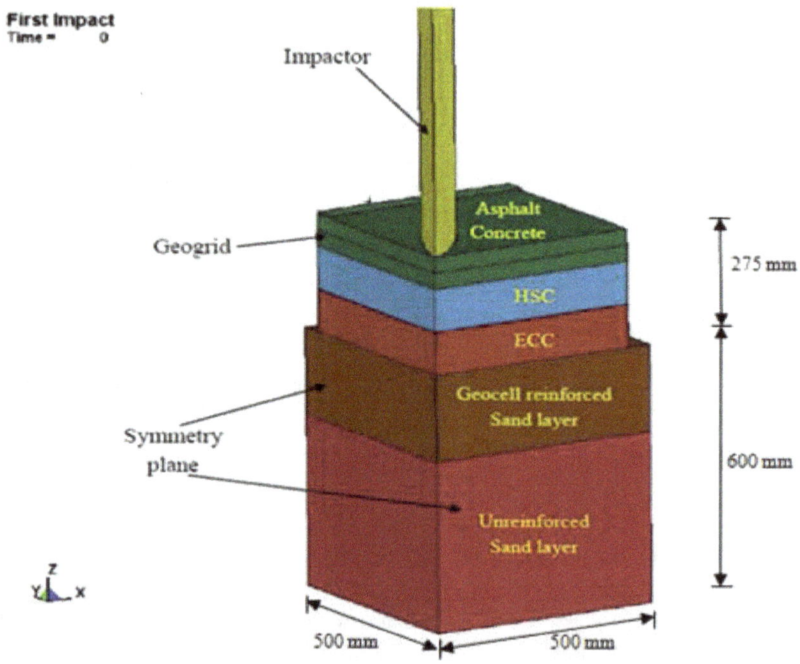

Figure 16. Numerical model for a multi-layer pavement slab under drop weight impact (quarter model).

Table 11. Properties of the drop weight head.

Parameters	Drop-Weight Head
Young's modulus, E (GPa)	207
Yield stress, f_y (MPa)	500
Poisson ratio, ν	0.3
Density, ρ (kg/m^3)	118,000

For the 1.5-m drop weight impact, the drop weight head is assigned with 5.02 m/s for the first impact. For the second impact, due to the penetration of the AC layer in the first impact, the distance between the laser diode system and the face of the pavement slab is increased, while in the numerical model, the impact head is just placed at the position right before reaching the surface. Hence, the initial velocity in numerical model for the second impact is determined as the sum of the experimental recorded velocity and velocity caused by gravity acceleration. Thus, the velocity is calculated as 5.06 m/s for the second impact with the gravity acceleration of 9.8 m/s^2.

After the first impact, the fully restarted method in LSDYNA is used to conduct the second impact simulation. At the beginning of the second impact, the stress and residual velocity within the pavement slab and the velocity from the first impact are set to be zero. The damage factor and plastic strain is retained in order to check the accumulated damaged behaviour after the second impact. The simulation of the second impact is carried out when the downward velocity of impactor reached zero, as it is very time consuming to continue to simulate the vibration of pavement slab after impact. It should be noticed that the pavement slab stopped rebounding, which might bring numerical errors (i.e., energy unbalance).

4.2.2. Numerical Result

The damaged situations for the multi-layer slab under the first drop weight impact for the experiment and numerical model are presented in Figure 17. It is found from the numerical simulation that the AC layer is penetrated through, and the drop weight head is impeded by the HSC layer due to the high compressive strength. Such findings from the numerical model are consistent with the observation in the physical test. It is observed that no severe and moderate damage happened in the ECC layer. Additionally, the integrity of both HSC and ECC layers is kept after the first impact.

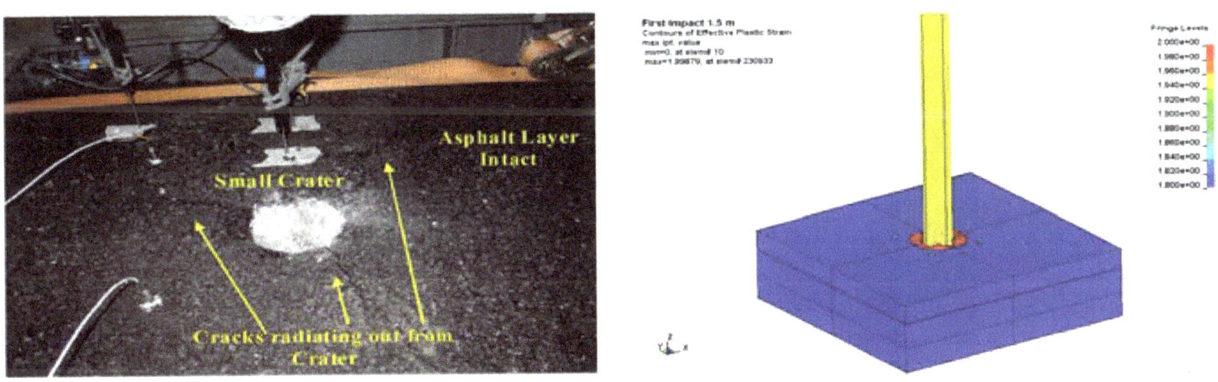

(a) (b)

Figure 17. Damage of the multi-layer slab after the first impact: (**a**) damage pattern of slab in the laboratory test; (**b**) damage pattern of slab in the numerical simulation.

The damage situations for the multi-layer pavement slab under the second drop weight impact and the numerical model are given in Figure 18. The simulation results indicate that the HSC layer stopped the impactor; however, severe damage happens at the top surface and in the middle at the side of the HSC layer. Severe damage occurs at the rear surface of the HSC layer, as well. For the ECC layer, it is shown that the cracking occurs and propagates from the centre at the top surface. However, the damage area is smaller than that of the HSC layer. The cracks are also found in the middle at the side of the ECC layer, and similar to the experiment, the crack does not propagate through the thickness of the layer. Figure 19 compares the numerical results with the experimental results about the rear surface of ECC layer after the second impact for the multi-layer slab. It is found that the damaged contour and cracking pattern were similar. Such results further indicate the reasonable close agreement between the numerical analysis and the experimental results using the KCC model for AC material.

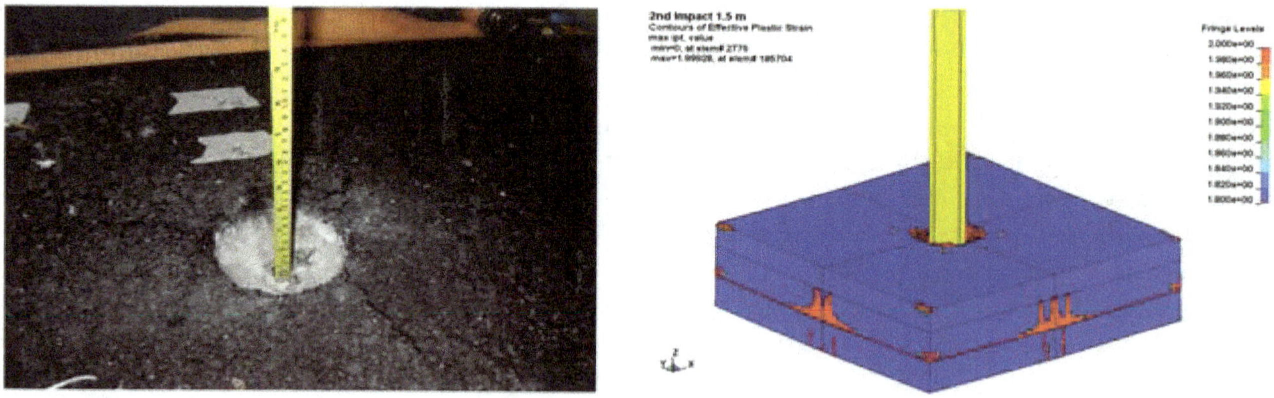

(a) (b)

Figure 18. Damage of multi-layer slab after the second impact: (**a**) damage pattern of slab in the laboratory test; (**b**) damage pattern of slab in the numerical simulation.

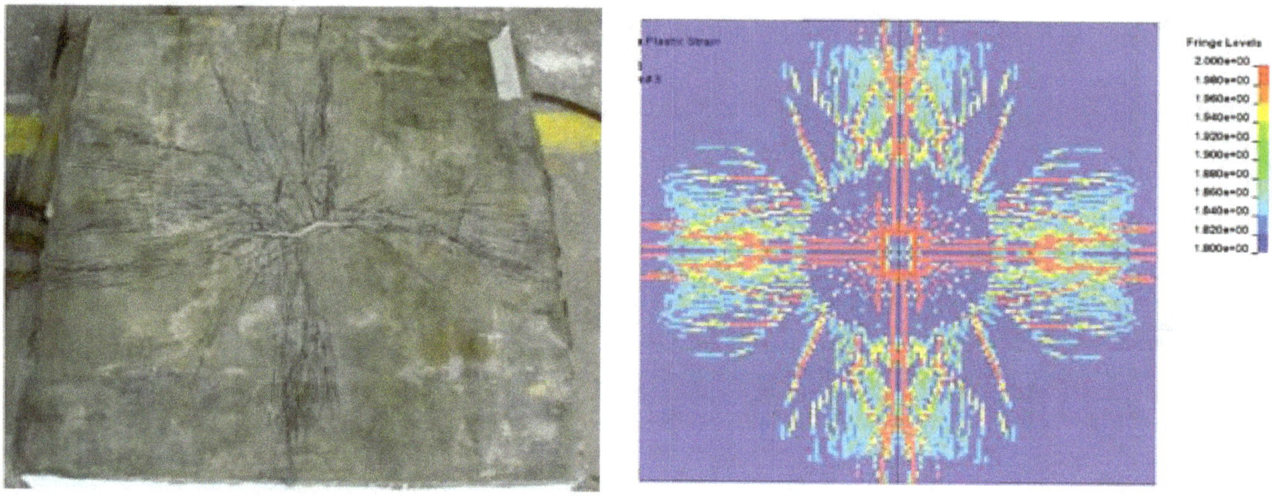

(a) (b)

Figure 19. ECC bottom after the second impact: (**a**) damage pattern of the bottom of ECC layer in the laboratory test; (**b**) damage pattern of the bottom of ECC layer in the numerical simulation.

Table 12 compares the numerical results with the experimental data for the multi-layer slab under two times of the drop weight impact. After the first impact, the settlement from the numerical results is close to that obtained from the experiment at P2. A big deviation is found in between the numerical results and experiment results on the settlement at P3. This might be attributed to the dislocation of the potentiometer at P3 upon impact, which could lead to a misreporting of the settlement at P3. In addition, the settlement at P1 is lower in the numerical results compared with the experimental results. This could be attributed to the erosion process in the numerical model or, more specifically, the energy released phenomena. At the second impact, since no erosion technology is employed for the HSC layer, the numerical results are more reasonable. In the impact simulation, the HSC layer suffered deformation instead of being penetrated, and hence, no erosion technique was adopted to trigger the energy released phenomena. Thus, the settlement profile of the pavement slab decreased with the increase of the radial distance from the centre, which indicated that the whole pavement slab bent during the impact. In general, the numerical models provide reasonable estimations on the settlements of the multi-layer slab under impacts, in particular under the second impact load, though discrepancies exist at certain regions.

Table 12. Summary of the settlements of the multi-layer slab from both numerical and experiment results.

Potentiometer	1st Impact			2nd Impact		
	Test Result	FE Results	FE/Test	Test Result	FE Results	FE/Test
P1 (mm)	—	−1.54	—	−3.62	−7.62	2.10
P2 (mm)	−1.81	−1.70	0.94	−7.60	−6.63	0.87
P3 (mm)	−5.27	−1.86	0.35	−5.58	−4.83	0.87

−ve → downwards

5. Conclusions

In this paper, the formulations of strength surfaces and damage factor in the KCC model are discussed and evaluated. It is found that the KCC model would reproduce the real behaviour of concrete-like material under severe loadings, such as blast and impact loadings. Then, the key parameters controlling the behaviour of AC material under dynamic loadings are discussed, and the parameters of AC materials using the KCC model are determined by laboratory static and dynamic tests. The proposed damage factor in this paper results in smoother descending than the original damage factor and also a higher failure strain that could improve the simulation of the behaviour of AC material. Furthermore, the numerical modelling of multi-layer pavement, in which the AC material served as the surface layer, under blast and impact loading is conducted using the KCC model. These results are compared with that of the field blast test and the laboratory drop weight impact tests, and it is found that the KCC model for the AC material can reproduce the real material behaviour under blast and impact loading.

Acknowledgments: Part of this research was sponsored by the National Basic Research Program of China (No. 2015CB058003), the Beijing Municipal Natural Science Foundation (No. 8172010) and the Youth Teacher Training Scheme from the Shanghai Education Committee (No. ZZGCD15053). Thanks are given here.

Author Contributions: Jun Wu conceived of and designed the experiments. Jun Wu performed the experiments and analysed the data. Liang Li developed the numerical model and prepared the paper. Xiuli Du discussed the numerical results and provided a suggestion on the overall organization of the paper. Xuemei Liu conducted the numerical model of the SHPB test for concrete-like materials and revised the manuscript.

References

1. Huang, Y.H. *Pavement Analysis and Design*; Pearson Education, Inc.: Upper Saddle River, NJ, USA, 2004.
2. Tashman, L.; Masad, E.; Little, D.; Zbib, H. A Microstructure-Based Viscoplastic Model for Asphalt Concrete. *Int. J. Plast.* **2005**, *21*, 1659–1685. [CrossRef]
3. Seibi, A.C.; Sharma, M.; Ali, G.; Kenis, W.J. Constitutive Relations for Asphalt Concrete under High Rates of Loading. *Transp. Res. Rec.* **2001**, *1767*, 111–119. [CrossRef]
4. Park, D.W.; Martin, A.E.; Lee, H.S.; Masad, E. Characterization of Permanent Deformation of an Asphalt Mixture Using a Mechanistic Approach. *KSCE J. Civ. Eng.* **2005**, *9*, 213–218. [CrossRef]
5. Tang, W.; Ding, Y.; Yuan, X. The HJC Model Parameters of an Asphalt Mixture. In Proceedings of the 9th International Conference on the Mechanical and Physical Behaviour of Materials under Dynamic Loading (DYMAT 2009), Brussels, Belgium, 7–11 September 2009; pp. 1419–1423.
6. Tekalur, S.A.; Shukla, A.; Sadd, M.; Lee, K.W. Mechanical Characterization of a Bituminous Mix under Quasi-Static and High-Strain Rate Loading. *Constr. Build. Mater.* **2009**, *23*, 1795–1802. [CrossRef]
7. Chang, G.K.; Meegoda, J.N. Micromechanical Simulation of Hot Mixture Asphalt. *J. Eng. Mech.* **1997**, *123*, 495–503. [CrossRef]
8. Collop, A.C.; McDowell, G.R.; Lee, Y. Use of the Distinct Element Method to Model the Deformation Behavior of an Idealized Asphalt Mixture. *Int. J. Pavement Eng.* **2004**, *5*, 1–7. [CrossRef]
9. Liu, Y.; Dai, Q.; You, Z. Viscoelastic Model for Discrete Element Simulation of Asphalt Mixtures. *J. Eng. Mech.* **2009**, *135*, 324–333. [CrossRef]

10. You, Z.; Liu, Y.; Dai, Q. Three-Dimensional Microstructural-Based Discrete Element Visoelastic Modeling of Creep Compliance Tests for Asphalt Mixtures. *J. Mater. Civ. Eng.* **2010**, *23*, 79–87. [CrossRef]
11. Riedel, W.; Hiermaier, K.T.S. Penetration of reinforced concrete by BETA-B-500-numerical analysis using a new macroscopic concrete model for hydrocodes. In Proceedings of the 9th International Symposium on Interaction of the Effect of Munitions with Structures, Berlin, Germany, 3–7 May 1999; pp. 315–322.
12. Adley, M.D.; Frank, A.Q.; Danielson, K.T.; Akers, S.A.; O'Daniel, J.L.; United States Army Corps of Engineers; Engineer Research and Development Center (U.S.); Geotechnical and Structures Laboratory (U.S.). *The Advanced Fundamental Concrete (AFC) Model*; Technical Report ERDC/GSL TR-10-51; Army Engineer Research and Development Center: Vicksburg, MS, USA, 2010.
13. Sherburn, J.A.; Hammons, M.I.; Roth, M.J. Modeling Finite Thickness Slab Perforation Using a Coupled Eulerian–Lagrangian Approach. *Int. J. Solids Struct.* **2014**, *51*, 4406–4413. [CrossRef]
14. Malvar, L.J.; Crawford, J.E.; Wesevich, J.W.; Simons, D. A plasticity concrete material model for DYNA3D. *Int. J. Impact Eng.* **1997**, *19*, 847–873. [CrossRef]
15. Polanco-Loria, M.; Hopperstad, O.S.; Borvik, T.; Berstad, T. Numercial Predictions of Ballistic Limit for Concrete Slabs Using a Modified Version of the HJC Concrete Model. *Int. J. Impact Eng.* **2008**, *35*, 290–303. [CrossRef]
16. Holmquist, T.J.; Johnson, G.R.; Cook, W.H. A computational constitutive model for concrete subjected to large strains, high strain rates, and high pressures. In Proceedings of the 14th International Symposium on Ballistics, Quebec, QC, Canada, 26–29 Sepetember 1993; pp. 591–600.
17. Chen, W.F. *Constitutive Equations for Engineering Materials*; John Wiley & Sons: Hoboken, NJ, USA, 1982.
18. Malvar, L.J.; Crawford, J.E.; Wesevich, J.W. *A New Concrete Material Model for DYNA3D Release II: Shear Dilation and Directional Rate Enhancements*; Defense Nuclear Agency: Alexandria, VA, USA, 1996.
19. Karihaloo, B.L.; Nallathambi, P. Effective Crack Model for the Determination of Fracture Toughness (Kice) of Concrete. *Eng. Fract. Mech.* **1990**, *35*, 637–645. [CrossRef]
20. Wu, J. Development of Advanced Pavement Materials System for Blast Load. Ph.D. Thesis, National University of Singapore, Singapore, 2012.
21. Li, Q.M.; Meng, H. About the dynamic strength enhancement of concrete-like materials in a split Hopkinson pressure bar test. *Int. J. Solids Struct.* **2003**, *40*, 343–360. [CrossRef]
22. Lu, Y.B.; Li, Q.M. About the dynamic uniaxial tensile strength of concrete-like materials. *Int. J. Impact Eng.* **2011**, *38*, 171–180. [CrossRef]
23. Wu, J.; Chew, S.H. Field Performance and Numerical Modelling of Multi-Layer Pavement System Subject to Blast Load. *Constr. Build. Mater.* **2014**, *52*, 177–188. [CrossRef]
24. Livermore Software Technology Corporation (LSTC). *LS-DYNA Keyword User's Manual*; Livermore Software Technology Corporation (LSTC): Livermore, CA, USA, 2007.
25. ANSYS Inc. *AUTODYN Theory Manual*; Century Dynamics; ANSYS Inc.: Canonsburg, PA, USA, 2005.
26. Rajagopal, K.; Krishnaswamy, N.R. Behavior of Sand Confined with Single and Multiple Geocells. *Geotext. Geomembr.* **1999**, *17*, 171–181. [CrossRef]
27. Madhavi, L.G.; Rajagopal, K. Parametric Finite Element Analyses of Geocell-Support Embankments. *Can. Geotech. J.* **2007**, *44*, 917–927. [CrossRef]
28. Duncan, J.M.; Chang, C. Nonlinear Analysis of Stress and Strain in Soils. *J. Soil Mech. Found. Div.* **1970**, *96*, 1629–1653.

12

Laboratory and On-Site Tests for Rapid Runway Repair

Federico Leonelli [1], Paola Di Mascio [1] , Antonello Germinario [2], Francesco Picarella [3], Laura Moretti [1,*], Mauro Cassata [3] and Alberto De Rubeis [4]

1 Department of Civil, Constructional and Environmental Engineering, Sapienza University of Rome, Via Eudossiana 18, 00184 Rome, Italy; fede.leonelli@gmail.com (F.L.); paola.dimascio@uniroma1.it (P.D.M.)
2 Laboratorio Principale Prove e Sperimentazioni (ITAF Infrastructure Department), Viale di Marino snc, 00043 Ciampino, Italy; laboratorioprove@tiscalinet.it
3 2 Reparto Genio A.M. (ITAF Infrastructure Department), Viale di Marino snc, 00043 Ciampino, Italy; francesco.picarella@aeronautica.difesa.it (F.P.); mauro.cassata@aeronautica.difesa.it (M.C.)
4 Servizio Infrastrutture A.M. (Chief of ITAF Infrastructure Department), Viale dell'Università 4, 00185 Rome, Italy; alberto.derubeis@aeronautica.difesa.it
* Correspondence: laura.moretti@uniroma1.it

Featured Application: The results of this study offer a broad vision for the rapid runway repair, giving a useful quantitative and objective tool for airport management body; moreover, they could be applied also for road pavements.

Abstract: The attention to rapid pavement repair has grown fast in recent decades: this topic is strategic for the airport management process for civil purposes and peacekeeping missions. This work presents the results of laboratory and on-site tests for rapid runway repair, in order to analyse and compare technical and mechanical performances of 12 different materials currently used in airport. The study focuses on site repairs, a technique adopted most frequently than repairs with modular elements. After describing mechanical and physical properties of the examined materials (2 bituminous emulsions, 5 cement mortars, 4 cold bituminous mixtures and 1 expanding resin), the study presents the results of carried out mechanical tests. The results demonstrate that the best performing material is a one-component fast setting and hardening cement mortar with graded aggregates. This material allows the runway reopening 6 h after the work. A cold bituminous mixture (bicomponent premixed cold asphalt with water as catalyst) and the ordinary cement concrete allow the reopening to traffic after 18 h, but both ensure a lower service life (1000 coverages) than the cement mortar (10,000 coverages). The obtained results include important information both laboratory level and field, and they could be used by airport management bodies and road agencies when scheduling and evaluating pavement repairs.

Keywords: rapid runway repair; cement mortar; cold bituminous mixture; expanding resin; aircraft classification number (ACN); pavement classification number (PCN)

1. Introduction

Transport infrastructures ensure transport mobility and accommodate infrastructures which provide the essential needs of the population, as food, energy, telecommunications, waterworks, health and safety networks, sewage systems [1]. Therefore, transport network is a lifeline [2], and its vulnerability exposes people to additional risks. Air transport needs for high priority because it is a strategic infrastructure. Often it is the only alternative to link remote territories, especially when it comes to emergency [1,3]. For example, airports are strategic when other transport infrastructures are not usable as consequence of a natural disaster (e.g., earthquake, flooding, storm) [4] or when tactical transport should support a peacekeeping mission [5]. For a rescue operation to succeed, a fully functioning system is of the essence [6–8]; therefore, repair and maintenance works should be fast and effective to ensure the opportune evenness during the operations [9,10].

For a long time, the Air Force studied in many countries the rapid runway repair (RRR) because its strategic importance; in 2016 the North Atlantic Treaty Organization (NATO) published the Standardization Agreement (STANAG) 2929 about this issue [11]. It provides for data and elements useful also for civil sector when rapid repair needs, and considers a standard pavement damage as a crater with a real diameter of 12 m and a maximum depth of 3 m. The term "real diameter" refers to not only the real crater caused by a warp, but also to the surrounding affected pavement. Under such conditions, two main categories of repairs could be applied: site repairs and repairs with prefabricated elements. Repairs with prefabricated elements refer to application of modular prefabricated elements on roller compacted granular materials. This solution ensures greater strength than the on-site ones, and prevent Foreign Objects Debris (FOD).

Three types of modular repairs are currently used:

- fiberglass mats: the pavement discontinuity is filled with controlled granulometry stones, rolled and levelled with the unpaired part of the pavement (Figure 1). A fiberglass mat, composed of two or more layers of fiberglass impregnated with polyurethane or polyester resins, is laid upon to prevent FOD risk [12]. Finally, the mat is anchored to the pavement with bolts and plugs. This method is simple and rapid: its longest procedure is fixing to the ground [13].
- precast concrete slabs: precast concrete slabs 1.5 × 1.5 m wide and 15 cm thick are laid down on a foundation levelled, which is 15 cm under the final pavement level (Figure 2) [14]. Slabs have a steel containment profile around their perimeter, double internal reinforcement, and two slots for lifting [15]. The system guarantees bearing capacity and durability, but has several operational difficulties: existing pavement should be cut to contain exactly a proper number of slabs; hot mastics, resins or hot bitumen should be applied to finish the joints [16].
- metal mats: prefabricated metal elements are suitable for both recess and covering execution (Figure 3). Usually, aluminium mats 4 cm thick are used because their high strength and low weight. Joints are simple and exact, outside elements are tied to avoid removal of elements and risk of FOD.

Figure 1. Anchoring the fiberglass mat.

Figure 2. Precast concrete slabs with a steel containment profile.

Figure 3. Metal mats.

Site repairs restore pavement continuity creating a structural package filling the crater or the discontinuity. Quickly hardening cement/resin-bounded mortars or bituminous emulsions are used. Three types of interventions are possible:

- recovery by percolation: the crater is filled with part of the material (10–70 mm diameter) spilled from the crater itself, then different materials of suitable granulometry complete the filling up to reach the ground level. Percolation of cementitious or bituminous binders finishes the upper thickness of the pavement. The granulometry of filling material varies with the used binder: it is 10–25 mm when it is bitumen, 25–70 mm when it is cement;
- recovery by surface filling: the crater is filled as in the previous case, but no more than 5–8 cm from the final level; a surface layer of cold bituminous asphalt composes the new upper layer and completes the repair;
- recovery by deep filling: the crater is filled as before, up to 20–30 cm from the final level; the restoration will be realized using ordinary concrete.

Site repairs are more frequent than modular repairs because they are more versatile: they allow repair under various conditions (e.g., extension of pavement to be repaired, volume to be filled ...). Their technique could be used also in road sector, where site repairs are just about the only used and rapid ones are desirable [17–19]. Under such conditions, the study focused on evaluation and identification of the best technical solution for easy and rapid pavement repair. As consequence of this condition, the study examined 12 materials and mixtures having with reference the NATO standard STANAG 2929 [11], which defines the maximum time for airport reopening after a repair work. Laboratory and in situ [20,21] tests have been performed to find the best choice from a technical point of view; finally, the examined materials have been compared respect to their installation costs. The results from this work provide interesting information useful to design a RRR, both for airport and road pavements.

2. Materials and Methods

Four different categories of materials used for repairing airport pavements have been examined in the study: bituminous emulsions, quick-hardening cement mortars, ordinary cement concretes, cold bituminous conglomerates, and an expanding resin (Table 1).

Table 1. Materials examined in the study.

Category	Notation
Bituminous emulsions	E1 E2
Cement mortars	M1 M2 M3
Cement concretes	M4 M5
Cold bituminous mixes	B1 B2 B3 B4
Expanding resin	R

Two bituminous emulsions have been tested:

- E1 is an over-stabilized cationic emulsion composed of 60% styrene-butadiene-styrene (SBS)-modified bitumen. Table 2 lists technical characteristics of bitumen extracted from the emulsion.

Table 2. Technical characteristics of bitumen extracted from the emulsion E1.

Characteristic	Value	Unit of Measure
Penetration at 25 °C	55	dmm
Softening point	62	°C
Fraas breaking point	−16	°C

Its correct temperature use ranges from 5 and 80 °C; moreover, it contains structural regenerative additives, therefore it is suitable for cold state repairs.

- E2 is a bicomponent modified, workable cold bitumen: it is useful for pavement maintenance when temperature ranges between 10 and 30 °C. Its maturity time is not more than 45 min after mixing the two components.

Five quick-hardening cement mortars (see Table 1) have been tested:

- the first and second cement mortars are composed of a Portland cement respectively 32.5 (M1) and 42.5 (M2) compliant with the standard EN 197-1 [22]. Mortars have been mixed with a water/powder (w/p) ratio equal to 0.45;
- the third cement mortar (M3) is a one-component fast setting and hardening cement with silica fume mortar [23–25]. It is fibre reinforced, suitable for smoothing, filling, and repairing concrete surfaces. Its correct w/p ratio is 0.13. Its elastic modulus evaluated according to the standard EN 13412 [26] is 32.6 GPa.
- the fourth cement mortar (M4) is a one-component fast setting and hardening cement mortar with graded aggregates. It is suitable for smoothing, filling, and repairing concrete surfaces; for thickness over 5 cm it is suitable for casting with 6/10 aggregates without segregation. Its correct w/p ratio is 0.125. Its elastic modulus evaluated according to the standard EN 13412 [26] is 32.6 GPa.

- the fifth cement mortar (M5) is a thixotropic, non-shrink, fibre reinforced, fast setting and hardening hydraulic mortar with graded aggregates. It is suitable for filling and repairing concrete surfaces. Its correct w/p ratio is 0.22. Its elastic modulus evaluated according to the standard EN 13412 [26] is 22 GPa.

Figure 4 compares the granulometry curves of M3 and M5.

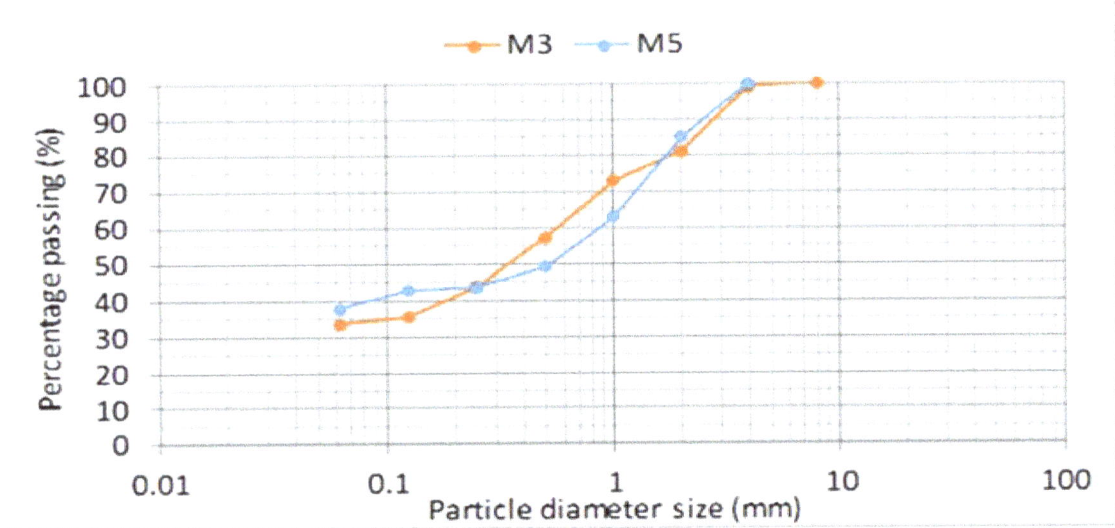

Figure 4. Granulometry curve of M3 and M5.

The cement mortars M3 and M5 have similar granulometric curve: both have over 33% percentage passing at 63 μm, and over 98% percentage passing at 4 mm. Nevertheless, M5 has more fine content than M3, while it has less fine sand than M3: these differences explain the different w/p ratios above listed.

Figure 5 compares the compressive strength of cement mortars M3, M4 and M5 whose time, temperature, and procedures for laying are compliant with those adopted for ordinary cement concrete. Compressive strength has been evaluated after maturity at 20 °C according to the standards: EN 12190 [27] for M3 and M4, and UNI EN 1015-11 for M5 [28].

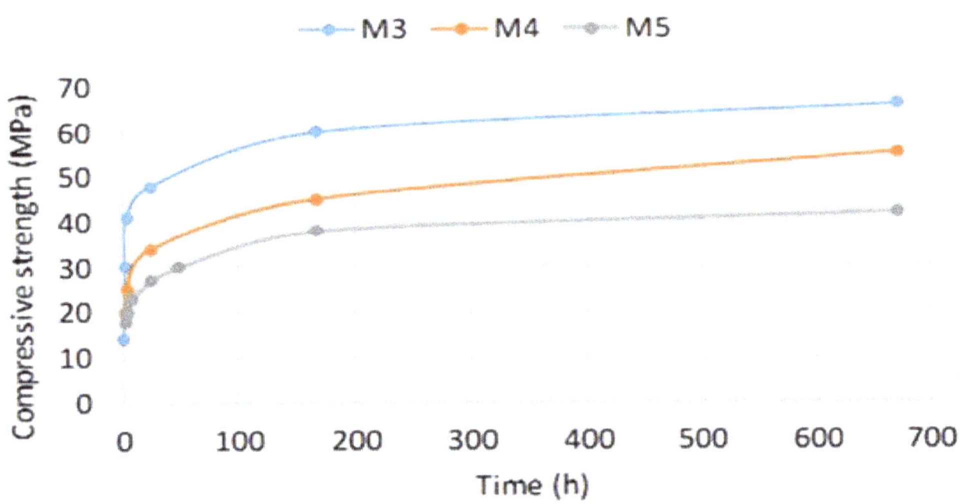

Figure 5. Compressive strength evolution of M3, M4 and M5.

Figure 6 compares the flexural strength of cement mortars M3, M4 and M5. Flexural strength has been evaluated after maturity at 20 °C according to the standards: EN 196-1 [29] and EN 13813 [30] for M3 and M4, and EN 1015-11 [28] for M5.

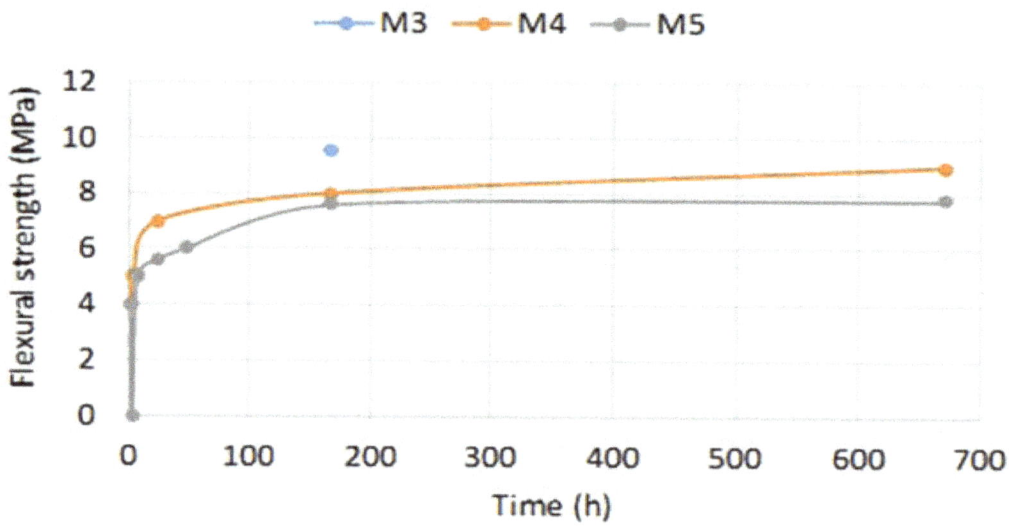

Figure 6. Flexural strength evolution of M3, M4 and M5.

Four cold bituminous mixtures (see Table 1) have been considered. Time, temperature, and procedures for their laying are compliant with those adopted for ordinary asphalt.

- B1 is a premixed cold bituminous mixture composed of fine aggregates and bitumen (6% by volume). After its application, the pavement can be immediately re-opened to traffic. The traffic itself settles the material, ensuring adhesion to the existing paving. This feature reduces time and costs for compaction, which cannot be overlooked using ordinary asphalt mixes [31]. Table 3 lists mechanical characteristics of B1 evaluated according to the standard ASTM D6927 (75 blows on each side) [32].
- B2 is a one-component premixed asphalt: it permits to repair 20–70 mm thick potholes with a single layer application. Table 4 lists technical characteristics of B2.
- B3 is a bicomponent premixed cold asphalt: it permits to repair bituminous and concrete pavements with up to 7 cm thick layers. Water is its catalyst for the hardening process.
- B4 is a premixed cold asphalt composed of bitumen (5.5% by weight of aggregates), vegetal oils, plasticizer additives and aggregates. It permits to repair bituminous and concrete pavements with layers not less than 2 cm and up to 6 cm thick.

Table 5 lists technical characteristics of B4.

Table 3. Mechanical characteristics of bituminous mixture B1.

Characteristic	Value	Unit of Measure
Marshall stability at 25 °C after 24 h	>3	kN
Marshall stiffness at 25 °C after 24 h	>1.5	kN/mm
Residual voids	<10	%
Indirect tensile strength of Marshall specimen at 25 °C after 24 h	>55	kPa

Table 4. Technical characteristics of bituminous mixture B2.

Characteristic	Value	Unit of Measure
Volumetric mass density	2.3	g/cm^3
Aggregate size	0–8	mm
Bitumen content	7.4–8.4	%
Voids content (after 75 blows Marshall)	7–9	%
Marshall stability after 24 h under water at 60 °C	≥4	kN
Marshall flow after 24 h under water at 60 °C	2–5	mm

Table 5. Technical characteristics of bituminous mixture B4.

Characteristic	Value	Unit of Measure
Volumetric mass density	>2.10	g/cm^3
Voids content (after 75 blows Marshall)	<10	%

Figure 7 compares the granulometry curves of B1, B2, B3, and B4.

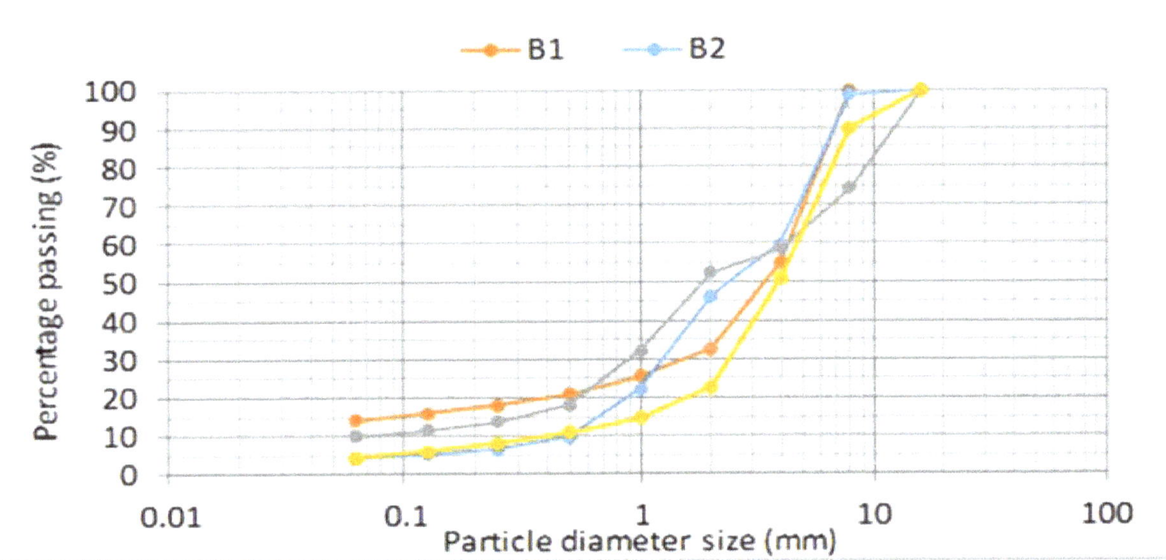

Figure 7. Granulometry curve of B1, B2, B3, and B4.

Expanding resins are used for transport infrastructures when the bearing capacity of soil does not satisfy the design requirements. Their expansion allows consolidation treatment by mean removal of the interstitial water and/or the filling of voids. The consolidation treatment involves executing injections through small metal cannulas placed on a regular mesh on the area to be treated.

In the study, a high-density bi-component polyurethane resin R (see Table 1) has been tested: its starting time is 40 ± 2 s and its expansion time is 85 ± 10 s. Table 6 lists its technical characteristics.

Table 6. Technical characteristics of the examined expanding resin.

Characteristic	Value	Unit of Measure
Compressive strength at 28 days	5.0	MPa
Shear strength at 28 days	5.0	MPa
Percentage closed cells	91.5 ± 1.5	%

In the first phase of the study, the presented 12 materials were tested in the Laboratory of the Italian Air Force (2nd Department of Genio located in Ciampino-Rome) to verify:

- ease of mixing and laying;
- percolability through a D40-70 grain size class (for emulsions and mortars);
- self-levelling properties;
- Marshall stability of cold conglomerates;
- cubic compressive resistance of cement mortars and expanding resin.

According to the need for fast reopening of the airport, mechanical resistance tests were conducted with the timing of 3 h, 6 h, 12 h, 24 h, and 48 h, 3days, 7days, and 28 days to evaluate the maturity of the products.

The texts performed were:

2.1. Percolation Tests with Bituminous Emulsions and Cement Mortars

The percolation test is a laboratory test designed by the 2nd Department of Genio of the Italian Air Force to verify the performance of a granular material bounded by a percolated binder. The procedure involved 7–12 mm granular materials with bituminous emulsions and 40–70 mm aggregates with cementitious mortars. In all cases, the tests aimed to find the binder composition (i.e., content of water for E1, content of cement for E2 and the mortars) which allows a percolation of about 20 cm and therefore it is defined "optimum"; otherwise the binder consistency is "fluid" (if the percolation thickness is more than 20 cm) or "plastic" (if the percolation thickness is lower than 20 cm).

2.2. Structural Strength of Concrete Obtained from Cement Mortars and Standard 32.5/42.5 Cement Concrete

Concrete made from cement mortars was obtained by percolating mortar inside 15 cm cubic moulds where the aggregates were previously located. Specimens were tested according to the standard EN 12390-2 [33].

In this phase specimens with ordinary concrete mixed with cement 32.5 and 42.5 and water/cement ratio equal to 0.500 were also tested. During the mixing process, the authors observed a rather aggressive gripping phenomenon in cement concrete 42.5, which could lead to difficulties during works.

2.3. Marshall Stability of Cold Bituminous Mixtures

Different series of Marshall specimens [34] have been made to evaluate the increments of resistance to time: for each time and cold bituminous mixture, 4 specimens have been tested.

2.4. On-Site Tests

On-site repairs were designed using the software FAARFIELD 1.41 (Federal Aviation Administration Rigid and Flexible Iterative Elastic Layered Design) (Federal Aviation Administration, Washington, DC, USA) [35]. It provides possible configurations of flexible and rigid pavement layers by simulating the number of coverages of the traffic mix. In this study, the layers thickness was calculated considering a reference aircraft, the C-130J (Lockheed Martin, Bethesda, MD, USA), typical for civil and military operations of the Italian Air Force. For this aircraft, the load distribution at the time of landing is 5% on the front gear and 95% on the rear one. 500 annual coverages during 20-year service life were considered: this volume traffic permit to design a permanent repair, as defined by the NATO criteria [36]. The flexible pavement model was used for bituminous materials, while the rigid pavement model was used for cementitious materials and the expanding resin [37,38].

The on-site tests involved the best performing materials found during the laboratory experimentation.

3. Results

3.1. Percolation Tests with Bituminous Emulsions and Cement Mortars

The results of percolation tests are listed in Table 7.

Figure 8 shows some percolation specimens obtained during the study.

Table 7. Results of percolation tests.

Product	Condition and Results			
E1	percentage of cement -	0% fluid	25% fluid	50% optimum
E2	as it is	optimum	-	-
M1	water percentage	40% plastic	45% optimum	50% fluid
M2	water percentage	40% plastic	45% optimum	50% fluid
M3	water percentage	12% plastic	13% optimum	14% fluid
M4	water percentage	12% plastic	12.5% optimum	13% fluid
M5	water percentage	16% plastic	20% plastic	24% plastic

E1 specimens M1 specimens

M5 specimens M4 specimens

Figure 8. Example of percolation specimens.

3.2. Structural Strength of Concrete Obtained from Cement Mortars and Standard 32.5/42.5 Cement Concrete

The results of compression tests on cementitious mixes are listed in Table 8.

Figure 9 highlights M3 and M4 have mechanical features that can be used as solutions for RRR, especially when percolation needs. M1 and M2 exhibited a high percentage of voids and, consequently, lower mechanical characteristics than the ordinary concretes mixed with 32.5 and 42.5 cements (both CEM I type). Ordinary concretes have mechanical characteristics that can be used as solutions for RRR, especially for deep filling.

Table 8. Compressive cubic strength of cementitious mixes.

Product	Compressive Cubic Strength (MPa)							
	Water/Cement Ratio	Time (h)						
		3	6	12	24	72	168	672
M1	0.450	0	0.5	1.8	3.5	13.4	18.6	27.2
M2	0.450	0	0.5	2.5	4.3	13.1	17.6	22.2
M3	0.130	32.9	37.5	41.0	44.8	52.6	57.4	65.8
M4	0.125	21.7	34.7	36.0	38.4	48.1	54.3	64.4
M5	-	-	-	-	-	-	-	-
R	-	4.59	4.60	4.72	4.74	4.77	5.45	5.42
Concrete with cement 32.5	0.500	0.13	1.8	11.5	19.2	30.2	33.5	48.5
Concrete with cement 42.5	0.500	0.31	2.3	12.6	20.4	32.9	38.4	54.0

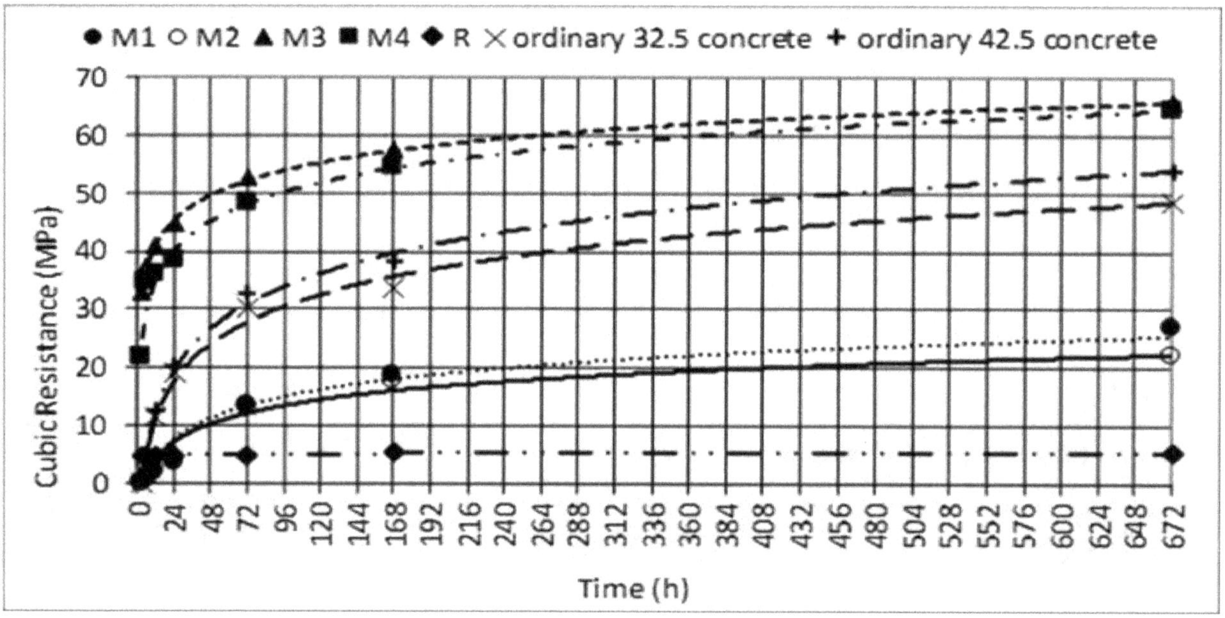

Figure 9. Structural strength depending on time.

3.3. Marshall Stability of Cold Bituminous Mixtures

Table 9 shows the Marshall stability of the examined cold bituminous mixtures at different times.

Table 9. Marshall stability.

Product	Marshall Stability (daN)					
	Time (h)					
	3	6	12	24	48	72
B1	222	238	241	240	266	270
B2	36	-	-	-	-	-
B3	599	850	1041	1232	1327	1213
B4	-	-	-	-	-	-

Figure 10 shows that B1 had insufficient mechanical characteristics to allow the rapid repair of the runways; B2 had only one specimen able to be tested; B4 did not have consistency: its specimen broke even before being inserted into the press. Only B3 showed mechanical characteristics appropriate for rapid repair.

At the end of this laboratory experimentation, the more reliable products which could be used for the runway rapid repair were: M4, B3, and M1. These materials and the expanding resin were used for on-site tests.

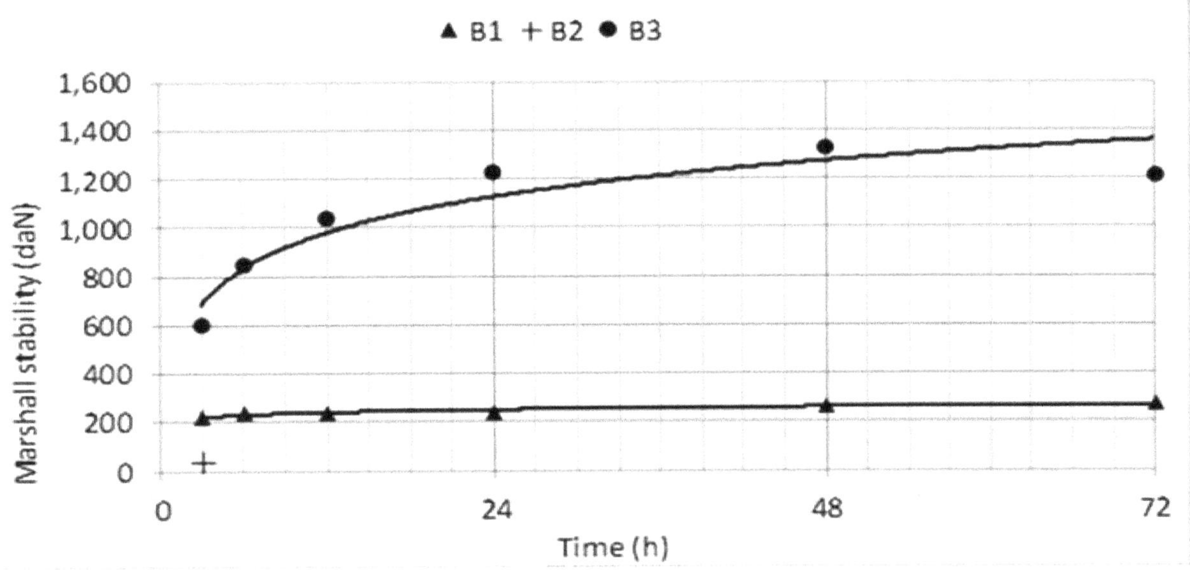

Figure 10. Marshall stability depending on time.

3.4. On-Site Tests

The on-site experimentation involved four test fields to be repaired, arranged in a 12-m diameter circle (Figure 11).

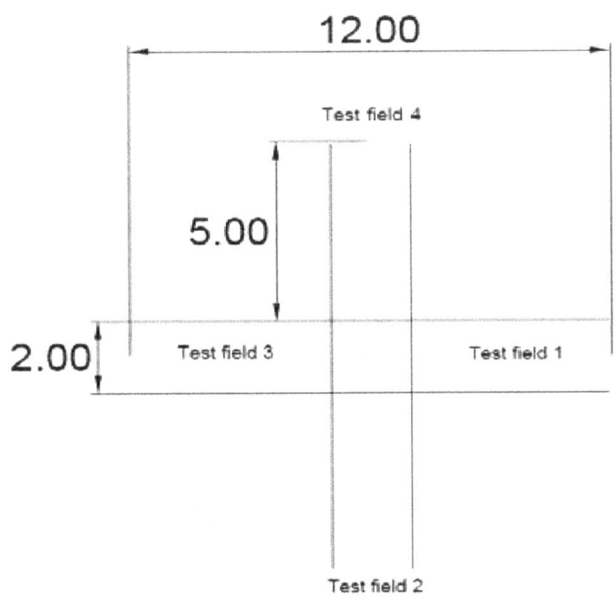

Figure 11. Diagrams of test fields (unit of measure: m).

The test fields were sized to avoid the disturbance effects due to the loads from the adjacent test fields. Their characteristics were:

- Test field 1: consisting of a D40-70 stone foundation and B3 wearing (flexible pavement);

- Test field 2: consisting of a D40-70 stone foundation and M1 wearing (rigid pavement);
- Test field 3: consisting of a D40-70 stone foundation up to 20 cm from the ground level, overlaid by a geotextile and a further layer of crushed stone with M4 used by percolation, (rigid pavement);
- Test field 4: consisting of an expanding resin injected into a stone foundation (40–70 mm) to increase its bearing characteristics and to create a support for concrete slabs.

During the execution of the expanding resin injection into the layer of stone, there was a strong expansion of the mixture, even up to 20 cm (Figure 12).

Figure 12. Expansion of expanded resin.

The on-site experimentation allowed the verification of the load bearing during time through Light Weight Deflectometer (LWD) and Heavy Weight Deflectometer (HWD) tests. LWD was used to evaluate the pavement Deformation Modulus M_d under the resin-treated area [39]. HWD technology allowed to back calculate the elastic modules of the repair layers by mean the software Elmod® 6.1.75 (Dynatest: Søborg, Denmark) [40]. According to the measured site deflection data, the software gave back an approximated deflection basin with decay curves according to the material under study.

Table 10 lists the results of LWD tests on three resin-treated points represented in Figure 13.

Figure 13. Resin-treated points of light weight deflectometer (LWD) tests.

Table 10. Deformation modulus of the resin-treated area.

Time (h)	M_d (MPa)							
	0	3	6	24	48	72	168	672
Point 1	7.37	2.87	4.19	4.5	2.9	3.41	4.1	2.71
Point 2	13.38	6.9	9.58	10.3	10.9	8.66	7.4	7.83
Point 3	14.99	10.13	11.42	7.6	9.8	10.73	15.3	10.11

The results listed in Table 10 highlight high variable values of M_d, as confirmed by the statistical results listed in Table 11. The set of values for conducting this analysis corresponds to the M_d values of the different tested points at different times. For every testing time, the mean, standard deviation, and coefficient of variation (CV, defined as the ratio of the standard deviation to the mean) were calculated.

Table 11. Statistical analysis of the M_d results of the resin-treated area.

Time (h)	0	3	6	24	48	72	168	672
Average (MPa)	11.91	6.63	8.40	7.47	7.87	7.60	8.93	6.88
Standard deviation (MPa)	4.02	3.64	3.76	2.90	4.34	3.77	5.76	3.79
CV	33.7%	54.8%	44.7%	38.9%	55.1%	49.6%	64.4%	55.1%

The high values of CV reveal that this technology is not reliable, as confirmed by the decreasing trend of M_d in Figure 14.

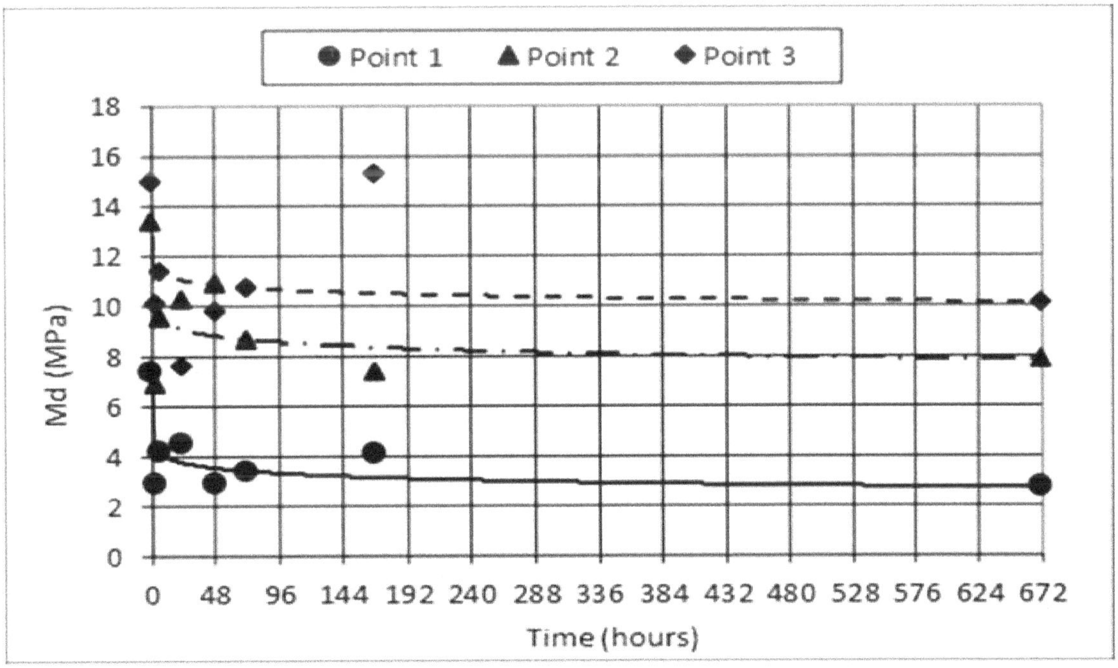

Figure 14. Deformation moduli after the resin injection.

The observed decreasing of M_d is caused by the viscous properties of the resin and its pronounced expansion which mobilizes the granular material. Indeed, this process alters the lithic skeleton because the resin replaces the stones instead of occupying the gaps between them. The mechanical performances shown in Table 10 and Figure 14 prevent the use of the expanding resin as material useful for RRR.

As done for LWD tests, the results of HWD tests and back calculation were performed for different times. During the analysis of the data some inconsistencies were found in the evaluation of the elastic moduli. For a more accurate evaluation of the moduli, filtering of data measured by the tenth geophone, which was always on the extreme edge of the test fields, was carried out (Figure 15).

Figure 15. The HWD machine over the M4-treated area.

The filtering process gave more precise and realistic results in terms of elastic modulus E of the pavement layers. In the first measurements (3 h), only the contribution of five geophones was filtered, according to the Dynatest guidelines [41].

A layered elastic model has been modelled to compute stresses, strains and deflections caused by surface load at any point in the pavement structure. According to the layered elastic theory [42], the model assumed that each pavement structural layer was homogeneous, isotropic, and linearly elastic. Table 12 lists the geometrical properties of the pavements modelled with the software Elmod® 6.1.75 [41].

Table 12. Geometrical properties of pavement models.

Layer	Test Field 1		Test Field 2		Test Field 3	
	Name	Thickness (cm)	Name	Thickness (cm)	Name	Thickness (cm)
Top layer	$E_{1,B3}$	10	$E_{1,M1}$	20	$E_{1,M4}$	20
Bottom layer	$E_{2,B3}$	90	$E_{2,M1}$	80	$E_{2,M4}$	80
Subgrade	E_{sub}	infinite	E_{sub}	infinite	E_{sub}	infinite

Top layers are composed of tested binders percolated into D40-70 granular bottom layers. The subgrade is the natural material underneath the test fields.

Tables 13–15 list the elastic moduli obtained from back analysis respectively for test field with B3, M1 and M4 materials.

Table 13. Elastic moduli of test field with B3.

Time (h)	3	6	24	72	168	672
$E_{1,B3}$ (MPa)	1193	1398	2797	2070	2607	3336
$E_{2,B3}$ (MPa)	450	450	450	450	450	450
E_{sub} (MPa)	350	340	323	322	301	350

Table 14. HWD test on test field with M1.

Time (h)	3	6	24	72	168	672
$E_{1,M1}$ (MPa)	-	-	14,783	21,724	28,193	38,609
$E_{2,M1}$ (MPa)	-	-	450	450	450	450
E_{sub} (MPa)	-	-	350	324	336	350

Table 15. Elastic moduli of test field with M4.

Time (h)	3	6	24	72	168	672
$E_{1,M4}$ (MPa)	15,146	25,614	27,000	27,000	27,150	28,461
$E_{2,M4}$ (MPa)	350	384	405	450	407	427
E_{sub} (MPa)	300	330	398	347	398	360

In the test fields 1 and 2, the values of the bottom layers elastic modulus ($E_{2,B3}$ and $E_{2,M1}$) were the same, equal to 450 MPa. This value of E_2 has been calculated before laying the top materials B3 and M1, and it has been assumed as constant during the subsequent analysis. This choice avoided anomalous back calculation results due to the small tests areas (10 m^2 each). Indeed, under ordinary conditions, the HWD tests are performed on areas larger than those prepared in this study, therefore the wave transmission in the study differs from usual. The most important differences involve the lower layers, whose elastic moduli are deduced from the responses of the geophones most distant from the loading plate. In this case, the response of the most distant geophones is also the most affected by side effects. On the other hand, this assumption has not been possible due to the technical characteristics of the third test field. M4 is a percolated material, thus the thicknesses and mechanical properties of layers are not defined and constant as those of test fields with materials B3 and M1.

Figure 16 shows the evolution of the elastic modulus of top layers paved with M1, M4 and B3.

Figure 16 highlights the mechanical performance of B3 are not comparable with those obtained using the ordinary cement concrete (M1) and the cementitious mortar M4. M4 has the most rapid rate of increase of elastic modulus: its E value is about constant after 1 day. On the contrary, for M1 the increase of the elastic modulus is slower, but its elastic modulus is comparable to that of M4 at 7 days and is growing until at least 28 days, when the difference is appreciable (38.6 GPa for M1 versus 28.4 GPa for M4).

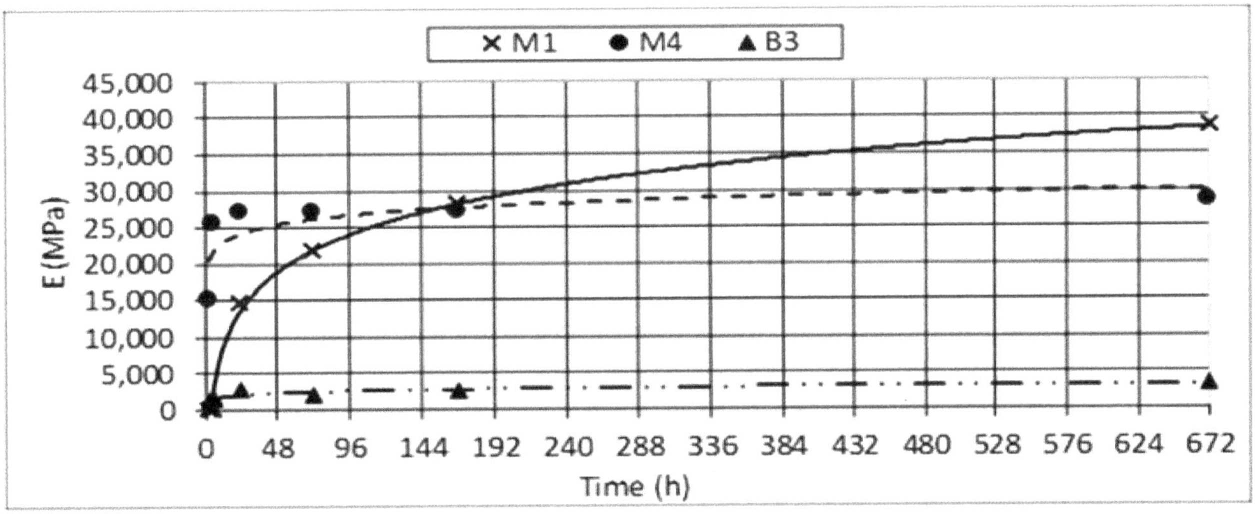

Figure 16. Elastic moduli of top layers.

4. Discussion

At the end of HWD tests, the ACN/PCN method was used to evaluate the load bearing capacity of the tested materials [43,44]. It is a system of rating airport pavements designated by the International Civil Aviation Organization [45] to compare the airport pavement strength (Pavement Classification Number, PCN) to the operation conditions of the traffic mix considering its gross weight and the subgrade bearing capacity (Aircraft Classification Number, ACN). The ACN value is twice the derived single-wheel load expressed in thousands of kilograms, with single-wheel tire pressure standardized at 1.25 MPa, which requires the same pavement thickness required by the examined aircraft. The calculation methods of ACN consider for flexible pavements 10,000 coverages and for rigid pavements 2.75 MPa concrete working stress [44]. Both values are representative of typical airport pavements.

Two criteria allow the evaluation of PCN: the "Using" aircraft or the "Technical" evaluation methods [44]. The first one is based on the experience, the second one on analytical procedures. In this study, the technical method proposed by the software Elmod® 6.1.75 has been used to calculate the PCN value. It considers the elastic modulus of layers (evaluated by mean back calculation), the configuration of the main gear of the design aircraft, and the number of coverages during the service life.

An airport pavement is verified when Equation (1) is satisfied:

$$PCN \geq ACN. \tag{1}$$

Data obtained in the HWD tests allowed the calculation of the PCN, using the software Dynatest Elmod® 6.1.75 [40]: different values were calculated for B3, M1, and M4 test fields at different maturity ages and for three traffic levels. 100, 1000 and 10,000 coverages during the service life were considered to simulate expedient, temporary, and permanent repairs according to the standard STANAG 7208 [36].

In the study, the subgrade under the test fields (Figure 11) was classified as "C" (low strength), therefore the ACN of the reference aircraft is 35.

Tables 16–18 list PCN values obtained for test field respectively with B3, M1 and M4 materials: the red cells indicate PCN<ACN, while green ones PCN ≥ ACN.

Table 16. Pavement Classification Number (PCN) values test field with B3.

Time (h)	3	6	24	72	168	672
Allowable Coverages	-	-	-	-	-	-
100	14	18	32	42	57	63
1000	10	11	26	35	42	53
10,000	3	7	14	22	33	44

Table 17. Pavement Classification Number (PCN) values test field with M1.

Time (h)	3	6	24	72	168	672
Allowable Coverages	-	-	-	-	-	-
100	-	-	39	43	50	50
1000	-	-	32	35	40	41
10,000	-	-	26	28	35	36

Table 18. Pavement Classification Number (PCN) values test field with M4.

Time (h)	3	6	24	72	168	672
Allowable Coverages	-	-	-	-	-	-
100	35	53	57	61	57	62
1000	29	44	47	50	47	51
10,000	24	37	39	42	39	43

The results reveal M4 has the most rapid PCN evolution: after 6 h maturity time, the verification ACN/PCN is satisfied. B3 and M1 ensure comparable results, but their reopening time (when Equation (1) is satisfied) is longer (7.5 days) than that of M4.

At the end of the study, an economic analysis has been carried out to estimate the unit costs for repairing a runway discontinuity [46,47]. Unit costs derive from the official price list used by the Italian Air Force (internal document) (Rome, Italy). Table 19 lists the results of the economic analysis.

Table 19. Unit cost of the examined rapid runway repair (RRR).

Material	Cost (€/m^2)
B3	180
M1	50
M4	75

According to the results listed in Table 19, M4 is the material that best suits needs for RRR: its technical and economic characteristics balance the conflicting objectives of budget and performance. Using M4, the airport could be reopened 3 h after the conclusion of the emergency repair (100 coverages during the service life), while 6 h after the RRR works it is possible to achieve a permanent repair, which allow 10,000 coverages during the service life.

5. Conclusions

The need for rapid repairing runway pavements has grown fast over the years, both for civil purposes and for peacekeeping missions. Therefore, there is the need for a technology that could be applied balancing conflicting objectives of resistance and rapidity. At this purpose, on-site repairs are more versatile than modular repairs currently used in the military sector because they can be adapted to different conditions. The specific sector of runway construction currently needs the comparison of technical performances offered by the innovative materials used for on-site repairs. In Italy, the Laboratory of the Italian Air Force (2nd Department of Genio) analysed technical and economic performances of several types of rapid runway repairs having the C-130J aircraft was as design aircraft. 12 materials currently used for pavement repair have been tested to evaluate their laboratory and on-site performances at different times.

This paper presents the first experimental results obtained on the following materials:

- 2 bituminous emulsions,
- 5 cement mortars,
- 4 cold asphalt mixes,
- 1 expanding resin.

The experimental study involved two phases:

1. laboratory tests on all materials except resin to focus on:
    - ease of mixing and application;
    - percolability (for emulsions, mortars and concretes);
    - Marshall stability (for cold asphalt mixes).
2. on-site experiments on the materials which performed the best for repairing. A cold asphalt mix, a cement mortar, an ordinary cement concrete and the expanding resin were used to repair four 10 m^2 widespread areas. All three tested mixes required time, temperature, and procedures compliant with those used to lay ordinary cement and bituminous runway pavements. On the contrary, the expanding resin required dedicated instruments to be injected. The bearing performances have been evaluated using the ACN/PCN system (for the three mixes) or LWD (for the expanding resin).

The on-site results highlighted that:

- it is not possible to repair runway pavement with expanding resin for two reasons: its strong expansion, even up to 20 cm over the desired level, and its bearing performances. Indeed, the LWD tests showed a decreasing value of deformation moduli over maturity time;
- the tested cement mortar M4 (one-component fast setting and hardening cement mortar with graded aggregates) ensures the fastest runway reopening in presence of a permanent repair of the damaged concrete pavement: 6 h after the work, the runway can be opened to traffic for 10,000 coverages of the design aircraft;
- surface filling (up to 6 cm) with the cold bituminous mixture B3 (bicomponent premixed cold asphalt with water as catalyst) and deep filling repair with the concrete M1 (ordinary concrete mixed with CEM I 32.5) have similar performances: both allow the reopening to traffic after 18 h maturity time, but for only 1000 coverages of the C-130J aircraft. Indeed, M1 and B3 require 7.5 days-age to support 10,000 coverages. B3 and M1 are both suitable for repairing flexible and rigid pavements.

Finally, it was also observed that the unit costs of the examined materials widely vary: B3 has a higher economic impact than that of M1 and M4, mainly due to the material since the costs of machines and works to have the rehabilitated pavement are comparable.

The results also provide a framework and a reference for any further study into specific cases using additional materials. All test methodologies can be applied to simulate with the software FAARFIELD (Version 1.41, Federal Aviation Administration, Washington, DC, USA) a fleet of almost any civil aircraft, therefore the procedure could be replicated for further analysis. As observed in this study, the contribution of each considered material to the overall repair work can be specialized to maximize the performances:

- percolated resin can be used in bottom layers as foundation and base,
- cold bituminous mixtures and cement mortars can be used for top layers.

Author Contributions: Alberto De Rubeis, Mauro Cassata, Paola Di Mascio and Laura Moretti conceived and designed the experiments; Federico Leonelli, Antonello Germinario and Francesco Picarellaperformed the experiments; all the authors analyzed the data and wrote the paper.

References

1. Organisation for Economic Co-Operation and Development. *Infrastructure to 2030. Volume 2 Mapping Policy for Electricity, Water and Transport*; Organisation for Economic Co-Operation and Development: Paris, France, 2007.
2. Cirianni, F.; Fonte, F.; Leonardi, G.; Scopelliti, F. Analysis of lifelines transportation vulnerability. *Procedia Soc. Behav. Sci.* **2012**, *53*, 29–38. [CrossRef]
3. Tong, S.Q.; Wang, N.; Song, N.Q. Emergency evacuation capability evaluation and optimization for an offshore airport: The case of Dalian Offshore Airport, Dalian, China. *Saf. Sci.* **2017**, *92*, 128–137. [CrossRef]
4. Shibayama, T. Japan's transport planning at national level, natural disasters, and their interplays. *Eur. Transp. Res. Rev.* **2017**, *9*, 44. [CrossRef]
5. Ripley, T. *Runway Repair Delays threaten Dutch F-16 deployment*; Jane's Defence Weekly: London, UK, 2006; pp. 353–354.
6. Loprencipe, G.; Zoccali, P. Comparison of methods for evaluating airport pavement roughness. *Int. J. Pavement Eng.* **2017**, 1–10. [CrossRef]
7. Loprencipe, G.; Cantisani, G. Evaluation methods for improving surface geometry of concrete floors: A case study. *Case Stud. Struct. Eng.* **2015**, 14–25. [CrossRef]
8. Bonin, G.; Cantisani, G.; Loprencipe, G.; Ranzo, A. Dynamic effects in concrete airport pavement joints. *Ind. Ital. Cem.* **2007**, *77*, 590–607.
9. Miccoli, S.; Finucci, F.; Murro, R. Assessing project quality: A multidimensional approach. *Adv. Mater. Res.* **2014**, *1030–1032*, 2519–2522. [CrossRef]

10. Miccoli, S.; Finucci, F.; Murro, R. Criteria and procedures for regional environmental regeneration: A European strategic project. *Appl. Mech. Mater.* **2014**, *675–677*, 401–405. [CrossRef]
11. NATO. *STANAG 2929, AATMP-03 (Allied Air Traffic Management Publication) "Airfield Damage Repair (ADR) Capability"*; North Atlantic Treaty Organization: Brussels, Belgium, 2016.
12. Dukes, P.E. Rapid runway repair: Fiberglass mats pass the test. *Mil. Eng.* **1988**, *80*, 453–455.
13. Dover, D.; Anderson, M.; Brown, R.W. Recent Advances in Matting Technology for Military Runways. In Proceedings of the 27th Annual International Air Transport Conference, Orlando, FL, USA, 30 June–3 July 2002.
14. Ashtiani, R.S.; Jackson, C.J.; Saeed, A.; Hammons, M.I. *Pre-Cast Concrete Panels for Rapid Repair of Airfield Rigid Pavements*; Air Force Research Laboratory Materials and Manufacturing Directorate Airbase Technologies Division: Washington, DC, USA, 2012.
15. Bull, J.W.; Woodfors, C.H. Design of precast concrete pavement units for rapid maintenance of runways. *Comput. Struct.* **1997**, *64*, 857–864. [CrossRef]
16. Sander, T.C.; Roesler, J.R. Case study: Runway 12L-30R keel suction rehabilitation, Lambert—St. Louis International Airport. In Proceedings of the 2006 Airfield and Highway Pavement Specialty Conference, Atlanta, GA, USA, 30 April–3 May 2006; pp. 872–884.
17. Wang, Y.; Kong, L.; Chen, Q.; Lau, B.; Wang, Y. Research and application of a black rapid repair concrete for municipal pavement rehabilitation around manholes. *Constr. Build. Mater.* **2017**, *150*, 204–213. [CrossRef]
18. Guan, Y.; Gao, Y.; Sun, R.; Won, M.C.; Ge, Z. Experimental study and field application of calcium sulfoaluminate cement for rapid repair of concrete pavements. *Front. Struct. Civ. Eng.* **2017**, *11*, 338. [CrossRef]
19. Shanahan, N.; Bien-Aime, A.; Buidens, D.; Meagher, T.; Sedaghat, A.; Riding, K.; Zayed, A. Combined effect of water reducer-retarder and variable chloride-based accelerator dosage on rapid repair concrete mixtures for jointed plain concrete pavement. *J. Mater. Civ. Eng.* **2016**, *28*, 04016036. [CrossRef]
20. Kavussi, A.; Abbasghorbani, M.; Moghadas Nejad, F.; Bamdad Ziksari, A. A new method to determine maintenance and repair activities at network-level pavement management using falling weight deflectometer. *J. Civ. Eng. Manag.* **2017**, *23*, 338–346. [CrossRef]
21. Chou, C.P.; Wang, S.Y.; Tsai, C.Y. Methodology of applying heavy weight deflectometer for calculation of runway pavement classification number. *Transp. Res. Rec.* **1990**, 57–64. [CrossRef]
22. EN (European Committee for Standardization). *EN 197-1: 2000. Cement Part 1: Composition, Specifications and Conformity Criteria for Common Cements*; European Committee for Standardization: Brussels, Belgium, 2000.
23. Sadowski, Ł.; Stefaniuk, D. Microstructural evolution within the interphase between hardening overlay and existing concrete substrates. *Appl. Sci.* **2017**, *7*, 123. [CrossRef]
24. Williams, M.; Ortega, J.M.; Sánchez, I.; Cabeza, M.; Climent, M.A. Non-destructive study of the microstructural effects of sodium and magnesium sulphate attack on mortars containing silica fume using impedance spectroscopy. *Appl. Sci.* **2017**, *7*, 648. [CrossRef]
25. Ortega, J.M.; Esteban, M.D.; Rodríguez, R.R.; Pastor, J.L.; Ibanco, F.J.; Sánchez, I.; Climent, M.A. Influence of silica fume addition in the long-term performance of sustainable cement grouts for micropiles exposed to a sulphate aggressive medium. *Materials* **2017**, *10*, 890. [CrossRef] [PubMed]
26. EN (European Committee for Standardization). *EN 13412:2006. Products and Systems for the Protection and Repair of Concrete Structures. Test Methods. Determination of Modulus of Elasticity in Compression*; European Committee for Standardization: Brussels, Belgium, 2006.
27. EN (European Committee for Standardization). *EN 12190:1998. Products and Systems for the Protection and Repair of Concrete Structures. Test methods. Determination of Compressive Strength of Repair Mortar*; European Committee for Standardization: Brussels, Belgium, 1998.
28. EN (European Committee for Standardization). *EN 1015-11:2007. Methods of Test for Mortar for Masonry. Determination of Flexural and Compressive Strength of Hardened Mortar*; European Committee for Standardization: Brussels, Belgium, 2007.
29. EN (European Committee for Standardization). *EN 196-1:2016. Methods of Testing Cement—Part 1: Determination of Strength*; European Committee for Standardization: Brussels, Belgium, 2016.
30. EN (European Committee for Standardization). *EN 13813:2002. Screed Material and Floor Screeds—Screed Materials—Properties and Requirements*; European Committee for Standardization: Brussels, Belgium, 2002.
31. Cantisani, G.; D'Andrea, A.; Di Mascio, P.; Loprencipe, G. RILEM Book series. In *8th RILEM International Symposium on Testing and Characterization of Sustainable and Innovative Bituminous Materials*; Canestrari, F., Partl, M., Eds.; Springer: Dordrecht, The Netherlands, 2016; Volume 11. [CrossRef]

32. ASTM (American Society for Testing and Materials International). *ASTM D6927. Standard Test Method for Marshall Stability and Flow of Asphalt Mixtures*; ASTM International: West Conshohocken, PA, USA, 2015.
33. EN (European Committee for Standardization). *EN 12390-2:2009. Testing Hardened Concrete. Part 2: Making and Curing Specimens for Strength Tests*; European Committee for Standardization: Brussels, Belgium, 2009.
34. ASTM (American Society for Testing and Materials International). *ASTM D6926. Standard Practice for Preparation of Asphalt Mixture Specimens Using Marshall Apparatus*; ASTM International: West Conshohocken, PA, USA, 2016.
35. Federal Aviation Administration. *AC 150/5320-6F—Airport Pavement Design and Evaluation*; Federal Aviation Administration: Washington, DC, USA, 2016.
36. NATO. *STANAG 7208 Airfield Pavement Design Criteria—Study Draft*; North Atlantic Treaty Organization: Brussels, Belgium, 2016.
37. Tiago Bonucci, P. Airfield rigid pavement structural design—A review of main aspects and methods of analysis. In *AIP Conference Proceedings, Proceedings of the 2nd International Symposium on Computational Mechanics, ISCM II, and the 12th International Conference on the Enhancement and Promotion of Computational Methods in Engineering and Science, EPMESC XII, Hong Kong–Macau, China; 30 November–3 December 2009*; American Institute of Physics: College Park, MD, USA, 2010; Volume 1233, pp. 1339–1344.
38. Chordá Ayela, A. Flexible pavements in the Barcelona airport enlargement (Pavimentos flexibles en la ampliación del aeropuerto de Barcelona). *Carreteras* **2010**, *4*, 29–42.
39. Di Mascio, P.; Loprencipe, G.; Maggioni, F. Visco-elastic modeling for railway track structure layers [Modellazione del comportamento visco-elastico degli strati della sede ferroviaria]. *Ing. Ferrov.* **2014**, *3*, 207–222.
40. *Elmod Software Manual*; Dynatest: Søborg, Denmark, 2015.
41. *Dynatest FWD/HWD Test Systems*; Owners Manual; Dynatest: Søborg, Denmark, 2014.
42. Boussinesq, J. *Application des Potentials a L'Etude de L'Equilibre et du Mouvement des Solides Elastiques*; Gauthier-Villars: Paris, France, 1885.
43. Norman, J.A.; Mumayiz, S.; Wright, P.H. *Airport Engineering: Planning, Design and Development of 21st Century Airports*, 4th ed.; John Wiley & Sons: Hoboken, NJ, USA, 2011; ISBN 978-0-470-39855-5.
44. Federal Aviation Administration. *AC 150/5335-5C, Standardized Method of Reporting Airport Pavement Strength—PCN*; Federal Aviation Administration: Washington, WA, USA, 2014.
45. ICAO. *Annex 14, Volume I Aerodromes*; International Civil Aviation Organization: Montreal, QC, Canada, 2013.
46. Di Mascio, P.; Moretti, L.; Panunzi, F. Economic Sustainability of Concrete Pavements. *Procedia Soc. Behav. Sci.* **2012**, *53*, 125–133.
47. Moretti, L. Technical and economic sustainability of concrete pavements. *Mod. Appl. Sci.* **2014**, *8*, 1–9. [CrossRef]

13

Laboratory Evaluation of Rejuvenating Agent on Reclaimed SBS Modified Asphalt Pavement

Jie Wang * , Wei Zeng *, Yongchun Qin, Songchang Huang and Jian Xu

Research Institute of Highway Ministry of Transport, Beijing 100088, China; yc.qin@rioh.cn (Y.Q.); sc.huang@rioh.cn (S.H.); j.xu@rioh.cn (J.X.)

* Correspondence: j.wang@rioh.cn (J.W.); w.zeng@rioh.cn (W.Z.)

Academic Editors: Zhanping You, Qingli (Barbara) Dai and Feipeng Xiao

Abstract: To evaluate the effect of rejuvenating agent on reclaimed SBS (styrene-butadiene-styrene) modified asphalt pavement (RSMAP) material, different tests of asphalt and mixtures were conducted. Firstly, the improvement effect of rejuvenating agents on the aged modified asphalt was tested at macroscopic and microscope level. Then the properties of hot mix asphalt (HMA) with different RSMAP contents (0%, 30%, 50% and 70%) were evaluated by conducting freeze-thaw split, semi-circular bending (SCB) and dynamic modulus (DM). The results indicate that rejuvenating agent can recycle the properties of aged modified asphalt effectively. The initial phase structure of the aged modified asphalt is not changed by adding rejuvenating agents. Moreover, the SBS particles area ratio of modified asphalt blends is significantly correlated with tenacity as the proportion of rejuvenating agent increases. For mixtures, RSMAP is harmful to moisture susceptibility and low-temperature cracking resistance of recycled mixture, especially with high RSMAP content. Moreover, the two properties can be improved by adding rejuvenating agents, but the recycled mixtures with high RSMAP content are not easy to recover to HMA mixture level. In general, the rejuvenating agent has an effect on the stiffness of the recycled mixture, but this is effect not obvious. When loading frequency reaches a higher value, the DM of recycled mixtures and HMA mixture tend to be consistent including high RSMAP content.

Keywords: road engineering; reclaimed SBS modified asphalt pavement; rejuvenating agent; phase structure; fracture energy; fracture toughness; stiffness

1. Introduction

Polymer modified asphalt has been widely applied in the construction and maintenance of asphalt pavement in China since the 1990s in order to adapt to the heavy load of modern traffic, and effectively resist the adverse effect of environmental conditions [1,2]. (styrene-butadiene-styrene) SBS modified asphalt is most widely used as it has the advantages of adequate stability under high temperature and low temperature, and satisfactory elastic resilience [3–5]. Especially in the 21st century, SBS modified asphalt has been extensively used in China for the surface layer construction of asphalt pavement, and it is also used for the construction of many asphalt middle layers [6,7].

A large number of SBS modified asphalt pavements have entered into or are approaching the overhaul period considering their designed life goal of 15 years, given the high traffic volume, and overloaded vehicles in China. The maintenance of modified asphalt pavements is facing the problem of disposal of reclaimed SBS modified asphalt pavement (RSMAP). If RSMAP cannot be used rationally, it will lead to serious environmental pollution and economic losses. Therefore, the recycling of RSMAP has become a focus of highway maintenance in China [8,9]. Currently, there are many in-depth studies on the asphalt recycling [10–12] with many successful applications in China. However, the studies on SBS modified asphalt recycling need to be further expanded. Some questions that need to be answered include whether the aged modified asphalt can be effectively recycled; the role of rejuvenating agent in aged modified asphalt; the effect of different types and proportion of rejuvenating agents on the properties of aged modified asphalt and HMA with different RSMAP contents, especially for high content [13–15]. In addition, the studies of the low-temperature cracking resistance, which is one of the most properties of recycled mixtures, are relatively few in number [16]. These questions make the recycling of aged modified asphalt uncertain.

SBS modified asphalt is a multiphase blend composed of polymer SBS and asphalt [17,18]. The two kinds of materials exist with differences in molecular weight and chemical structure as a thermodynamically incompatible system. Because of the interface interaction between different components, it is difficult for the blends to reach a homogeneous level and just achieve physical structure compatibility. The differences of compatibility has lead to a great differences in the properties of modified asphalt [19]. Therefore, the phase structure of SBS modified asphalt can generally be divided into three types: (1) single-phase continuous structure with SBS polymer as dispersed phase and asphalt phase as continuous phase. This type is more common in highway construction; (2) single-phase continuous structure with SBS polymer as continuous phase and asphalt phase as dispersed phase; (3) two-phase continuous structure with SBS polymer and asphalt phase both as continuous phase. The different phase structure has a significant impact on the properties of modified asphalt [20,21]. Ignoring the phase structure to study modified asphalt recycling is not appropriate. This study takes into consideration the phase characteristics of modified asphalt. It focuses on the role of rejuvenating agent in aged modified asphalt, and the effect of rejuvenating agent on properties of recycled mixtures, including high RSMAP content. This study is intended to further clarify the recycling mechanism of rejuvenating agent on RSMAP.

2. Materials and Methods

2.1. Materials

The RSMAP material used in this paper was obtained from Beijing, and the aged modified asphalt extraction from the mixture was performed by Abson method [22]. The aged asphalt content of RSMAP is 3.6%, and the main properties of the aged modified asphalt and virgin SBS modified asphalt used in the study are shown in Table 1.

Table 1. Main properties of modified asphalt.

Properties		Aged Modified Asphalt	Virgin Modified Asphalt
Penetration (25 °C)/0.1 mm		36.3	48.1
Softening point/°C		68.8	69.3
Ductility (5 °C)/cm		0.1	30.7
Viscosity (135 °C)/Pa·s		2.24	1.05
Toughness test	Maximum tension/N	599.9	332.7
	Toughness/N·m	17.74	13.52
	Tenacity/N·m	6.65	9.87

Three rejuvenating agents with different viscosity were selected in this study in order to evaluate the recycling effect on the aged modified asphalt, and represented by H, S, G respectively. The main properties of the rejuvenating agents were tested according to ASTM D4552 and the results are shown in Table 2.

Table 2. Main properties of rejuvenating agent.

Type	Viscosity (60 °C)/10^{-3} Pa·s	Flash Point/°C	Saturates Content/%	Aromatic Content/%	Density/g·cm^{-3}	Tests on Residue from TFOT	
						Viscosity Ratio *	Weight Change/%
H	123	256	17.07	62.07	1.002	2.3	−3.34
S	1202	267	23.38	44.35	1.013	1.8	−2.62
G	2607	278	25.98	49.81	1.028	1.4	−2.73

* Viscosity ratio = Viscosity (60 °C) of residue from TFOT/Virgin viscosity (60 °C).

In the process of aggregate gradation design of recycled mixtures with different RSMAP contents, the gradation should be as similar as possible. AC-13 is selected as the type of recycled mixtures. The virgin coarse aggregate is basalt, the crushing value is 14.3%, and the Los Angeles abrasion value is 13.5% [23]. The virgin fine aggregate is limestone. Table 3 shows the aggregate gradation and asphalt content of recycled mixtures with different RSMAP contents (0%, 30%, 50% and 70%). The type of rejuvenating agent for recycled mixtures used in this study is S.

Table 3. Aggregate gradation and asphalt content of recycled mixtures with different reclaimed SBS (styrene-butadiene-styrene) modified asphalt pavement (RSMAP) contents.

Sieve Size/mm	Type of Recycled Mixtures			
	0%	30%	50%	70%
16	100.0	100.0	100.0	100.0
13.2	97.3	97.2	97.0	96.9
9.5	75.1	75.0	73.4	72.6
4.75	41.9	41.9	40.4	40.7
2.36	28.1	29.1	28.2	28.4
1.18	18.2	18.7	18.4	18.6
0.6	13.6	13.8	13.7	14.0
0.3	10.2	10.3	10.3	10.6
0.15	8.0	7.9	8.0	8.4
0.075	6.3	6.2	6.4	6.7
Virgin asphalt content/%	0.0	3.7	3.1	2.4
Total asphalt content/%	4.8	4.8	4.9	5.0

2.2. Methods

The aged modified asphalt was blended with three types of rejuvenating agents with the proportion of 3%, 6% and 9% by mass of aged modified asphalt, respectively. The tests include the macroscopic properties of 25 °C penetration, softening point, 5 °C ductility, 135 °C Brookfield viscosity, toughness and tenacity [22]. The phase characteristics of the modified asphalt were evaluated using fluorescence microscope.

The properties of recycled mixtures in this study mainly focus on moisture susceptibility, low-temperature cracking resistance and stiffness. Conducted tests included freeze-thaw split [22], semi-circular bending (SCB) and dynamic modulus (DM).

2.3. Semi-Circular Bending Test

The SCB test is used as a method for evaluating asphalt mixture cracking resistance [24]. The test configuration and sample size are illustrated in Figure 1. Test samples were fabricated using SGC at 7 ± 0.5% air void. SGC-compacted samples were 10 cm in height and 15 cm in diameter; each sample was trimmed to half-disc-shaped samples 15 cm in diameter and 5 cm thick. The notch is cut in the middle point of the lower surface of the sample to ensure that the crack will initiate at the desired point. In this study, the notch depth was 1.5 cm. The test was conducted at −20 °C and −10 °C at constant loading rate of 1 mm/min.

In order to evaluate the crack resistance of recycled mixtures, fracture energy (G_f) is used, which can be obtained by dividing fracture work with ligament area [25], as provided in Equation (1). The total fracture work can be calculated by the area under the P-u curve. Figure 1d shows a typical SCB load versus displacement curve.

$$G_f = \frac{W_f}{A_{lig}} \tag{1}$$

where G_f = fracture energy (J/m^2); A_{lig} = ligament area (m^2) = $(r - a) \times t$; r = specimen radius (m); a = notch depth (m); t = specimen thickness (m); and W_f can be calculated as follows in Equation(2).

$$W_f = \int Pdu \tag{2}$$

where P = applied load (kN); and u = load line displacement (mm).

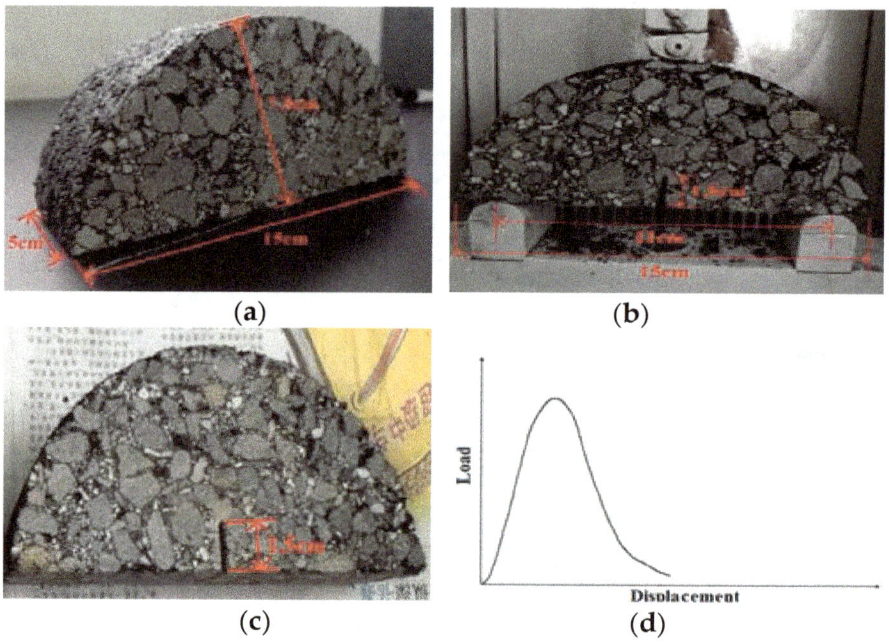

Figure 1. (**a**) Sample dimensions; (**b**) semi-circular bending (SCB) test configuration; (**c**) samples with a notch depth; (**d**) typical SCB load versus displacement (P-u) curve.

Fracture toughness (K_c) is considered as the stress intensity factor at the critical load (P_c), and provides a single parameter characterization that includes the effect of specimen configuration, boundary conditions, and load. In addition, K_c is qualified as an intrinsic material property, describing the ability to resist cracking [26]. In this paper, K_c is calculated using the following equation.

$$K_c = Y_I\, \sigma_c \sqrt{\pi a} \tag{3}$$

where $\sigma_c = P_c/2rt$; $Y_I = 4.782 + 1.219(a/r) + 0.063\exp[7.045(a/r)]$; P_c = peak load; Y_I = normalized stress intensity factor.

2.4. Dynamic Modulus Test

Dynamic modulus (E^*) is a fundamental property of asphalt mixture that defines the viscoelastic stiffness as a function of loading rate and temperature. In this study, test samples were fabricated using SGC at 7 ± 0.5% air void. SGC-compacted samples were 17 cm in height and 15 cm in diameter, and the samples were cored and trimmed to 15 cm in height and 10 cm in diameter. DM tests were conducted at three temperatures (4 °C, 20 °C, 40 °C) and six frequencies (25 Hz, 10 Hz, 5 Hz, 1 Hz,

0.5 Hz, 0.1 Hz) using asphalt mixture performance test (AMPT, IPC Global, Victoria, Australia) and following AASHTO T342-11, as shown in Figure 2.

The stress level and the number of loading cycles in the test are shown in Tables 4 and 5 respectively.

(a) (b)

Figure 2. (**a**) Three parallel sample; (**b**) asphalt mixture performance test (AMPT) equipment.

Table 4. The stress level (kPa) of dynamic modulus (DM) test.

Loading Frequency/Hz	Test Temperature/°C		
	4	**20**	**40**
25	1400	700	250
10	1260	630	228
5	1120	560	206
1	980	490	184
0.5	840	420	162
0.1	700	350	140

Table 5. The number of loading cycles of DM test.

Loading Frequency/Hz	Number of Loading Cycles
25	200
10	200
5	100
1	20
0.5	15
0.1	15

Using the dynamic modulus test results measured at three different temperatures and six different loading frequencies, the master curves of recycled mixtures based on the time-temperature superposition principle were constructed for a reference temperature of 20 °C [27]. The sigmoidal function is used to fit and develop the master curve, as provided in Equation (4).

$$\log|E^*| = \delta + \frac{a}{1+\exp(\beta+\gamma(\log f_r))} \tag{4}$$

where $|E^*|$ = dynamic modulus (MPa); f_r = reduced frequency at reference temperature; δ = minimum value of $|E^*|$; $\delta + \alpha$ = maximum value of $|E^*|$; β, γ = fitting parameters.

Equation (5) shows the temperature shift factor.

$$\log(f_r) = \log(a(T)) + \log(f) \tag{5}$$

where $a(T)$ = temperature shift factor; f = frequency at a particular temperature.

The temperature shift factor may be written in the form of Equation (6) by using the Arrhenius time-temperature superposition model [28].

$$\log(a(T)) = c\left(\frac{1}{T_k} - \frac{1}{T_{ref}}\right) \quad (6)$$

where T_k = the test temperature in Kelvin temperature; T_{ref} = the reference temperature in Kelvin temperature; c = nonlinear fitting curve coefficient.

A nonlinear optimization program in 1stOpt software was used to determine the master curve coefficients.

3. Results and Analyses

3.1. Macroscopic Properties of Rejuvenating Agents on Aged Modified Asphalt

It is well known that the stiffness of modified asphalt will increase after aging, resulting in attenuation of the low-temperature property and improvement for the high-temperature property. Consequently, modified asphalt recycling should focus on the recovery of low-temperature properties. Table 6 and Figures 3 and 4 show the results of different types of rejuvenating agents on aged modified asphalt.

Table 6. The effect of rejuvenating agent on aged modified asphalt.

Indexes	Aged Modified Asphalt	H			S			G		
		3%	6%	9%	3%	6%	9%	3%	6%	9%
Penetration/0.1 mm	36.3	50	85	108	46	50	62	39	49	57
Softening point/°C	68.8	64.9	59.4	59.2	64.5	64.3	62	67.2	64.2	62.3
Ductility/cm	0.1	19.3	25.9	39.3	17.8	21.2	29.6	13.7	20.7	35.7
Viscosity/Pa·s	2.24	1.07	0.83	0.64	1.07	1.01	0.88	1.26	1.01	0.81

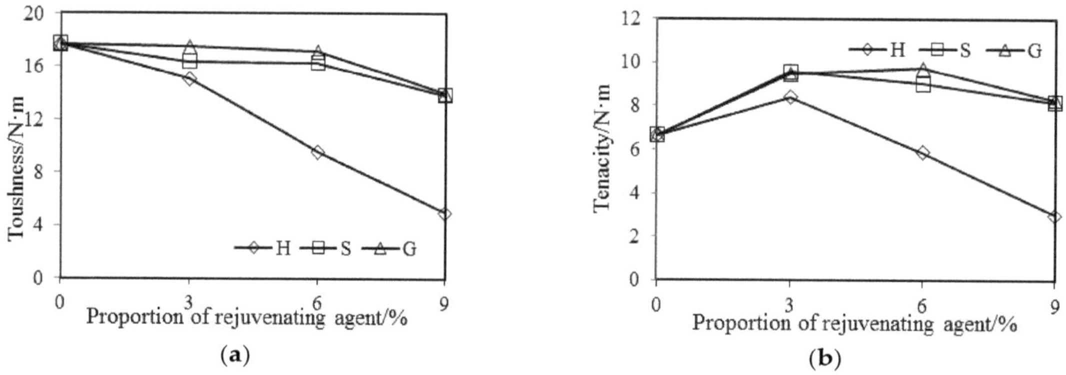

Figure 3. Relationship between (**a**) toughness, (**b**) tenacity and proportion of Rejuvenating Agent.

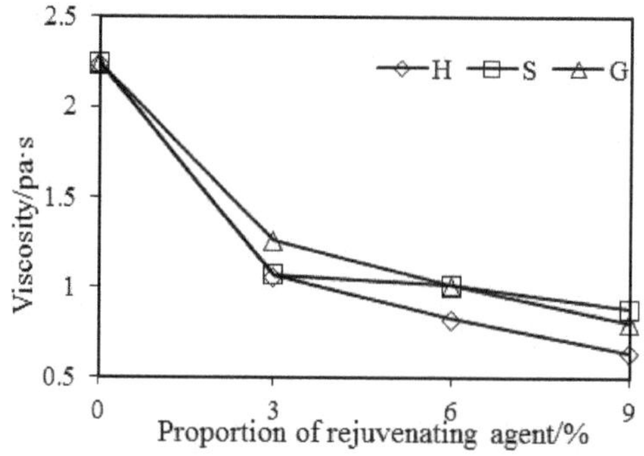

Figure 4. Relationship between penetration and viscosity.

Table 6 shows that: with the increasing of rejuvenating agent proportion, softening point and viscosity decrease, while penetration and ductility increase. When the proportion of H, S, G is 9%, the ductility of modified asphalt blends increases from brittle fracture to 39.3 cm, 29.6 cm and 35.7 cm respectively, which indicates that the rejuvenating agent can recover the low-temperature property of modified asphalt. Satisfactory results can be achieved which indirectly prove the sufficient feasibility of modified asphalt recycling.

From Figure 3, the toughness decreases continuously as the rejuvenating agent proportion increases, and the trend of tenacity is different. When the proportion of rejuvenating agent is small, the tenacity increases at first, and then decreases when the proportion increases continuously, probably because the rejuvenating agent is beneficial to the improvement of swelling effect between polymer SBS and asphalt. Moreover, it can partially improve the mechanical properties of modified asphalt. The tenacity increase can also indirectly prove the feasibility of modified asphalt recycling. When the proportion of rejuvenating agent increases further, the tenacity of modified asphalt blends begins to decrease due to reduced viscosity.

Figure 4 indicates that the penetration of modified asphalt blends increases with the decreasing of viscosity, but the trend is not linear. When the viscosity of modified asphalt blends is more than 1.1 Pa·s, the change of penetration along with the viscosity is not obvious. However, the penetration increases noticeably with the decreasing of viscosity, roughly showing linear relationship when the viscosity of modified asphalt blends is less than 1.1 Pa·s. This phenomenon indicates that the viscosity must be less than a certain value for the change of penetration to be obvious, and this result is similar to the conclusion of reference [29].

In conclusion, the rejuvenating agent can partially or completely recycle the properties of aged modified asphalt. Increasing the proportion of rejuvenating agent can better improve the low-temperature ductility. However, the correlation of penetration, ductility and softening point indicates that increasing the proportion of rejuvenating agent to improve ductility will lead to a sharp increase of penetration and a decrease of softening point. For example, when the proportion of H is 6%, the penetration of modified asphalt blend is 85 (0.1 mm). However, the ductility is still relatively small. The results indicate that the recycling of penetration and ductility is not synchronous and it is not appropriate to use the penetration and ductility to carry on the design of modified asphalt recycling at the same time.

3.2. *Anti-Aging Property of Modified Asphalt Blends*

To examine the anti-aging property of modified asphalt blends, the thin film oven test (TFOT) was conducted. According to the recycling effect and economy of the above-mentioned three rejuvenating agents H, S, G, the proportion of rejuvenating agent for TFOT were set to 3%, 6%, 6% respectively. The test results are shown in Table 7.

Table 7. Anti-aging property of modified asphalt blends.

Properties		Specification Requirement	Modified Asphalt Blends		
			3%H	6%S	6%G
Penetration/0.1 mm		40~60	50.0	50.3	49.3
Softening point/°C		≥60	64.9	64.3	64.2
Ductility/cm		≥20	19.3	21.2	20.7
Tests on residue from TFOT	Penetration ratio/%	≥65	78.0	78.3	79.3
	Ductility/cm	≥15	10.8	8.1	9.7
	Weight change/%	≤±1.0	0.302	0.438	0.369

Table 7 shows that the penetration ratio and weight change of modified asphalt blends after TFOT meet the requirements of China's specifications, except in terms of ductility. The results indicate that, although the properties of aged modified asphalt blended with rejuvenating agent can be recycled, the low-temperature property often cannot meet the requirements after aging, indicating that the modifier has lost part of the original modified effect.

Therefore, modified asphalt recycling should focus on its low-temperature property, requiring rejuvenating agent not only to effectively recover the colloidal structure of aged modified asphalt, but also to repair and supplement the aged modifier. Ultimately, the objective is to recover both the low-temperature cracking resistance and high-temperature rutting resistance of modified asphalt.

3.3. Phase Characteristics of Rejuvenating Agents on Aged Modified Asphalt

Polymer SBS in asphalt absorbs light oil to form the polymer phase. Under the fluorescence excitation, the polymer phase reflects longer wavelength light, appearing white. In contrast, the asphalt phase does not reflect any light, appearing black. So the asphalt phase and polymer phase are clearly distinguished by fluorescence microscope. As the reflected light field is applied, there is no damage to the polymer phase distribution in the asphalt. Hence, the fluorescence microscope gets highly reproducible asphalt micrographs. By analyzing the morphological structure and dispersion state of polymer SBS in asphalt, it is possible to effectively evaluate the role of rejuvenating agent on the aged modified asphalt, and establish the relationship between macroscopic mechanical properties and microscopic characteristics of modified asphalt blends.

To ensure the reliability of fluorescence images and avoid the particularity of individual samples, three slides were made in parallel for each sample. Four representative fluorescence images were taken for each slide.

Table 8 shows the fluorescent micrographs of aged modified asphalt and modified asphalt blends with different types and proportions of rejuvenating agents.

Table 8 shows that the polymer SBS in aged modified asphalt are spherical particles distributed in the asphalt, and this modified asphalt belongs to a single-phase continuous structure, with polymer SBS as dispersed phase and asphalt phase as continuous phase. Compared with modified asphalt blended with different types and proportions of rejuvenating agents, recycling did not change its phase structure, still belonging to single-phase continuous structure. The SBS particle's uniformity is satisfactory. With the naked eye only, one cannot see the difference of SBS particle in appearance and shape. The images have to be further processed to quantify potential differences.

Professional image processing software (Image-Pro Plus) is applied to analyze and process the fluorescence images of modified asphalt blends, which can segment, count and measure the images. The area ratio of polymer SBS particles is the quantitative evaluation indexes to describe the phase characteristics and recycling mechanism of modified asphalt. Here the area ratio refers to the percentage of polymer SBS particles area to total area, indirectly reflecting the SBS swelling degree in the asphalt.

Table 8. Fluorescence micrographs of modified asphalt.

Type		Aged Modified Asphalt	
Sample figure		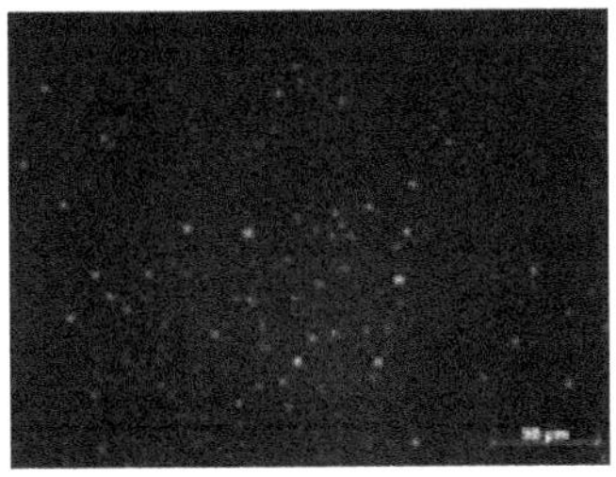	
Type	3%H	6%H	9%H
Sample figure	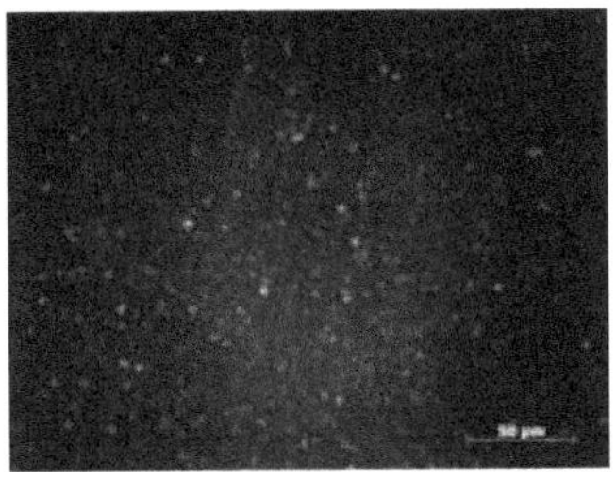	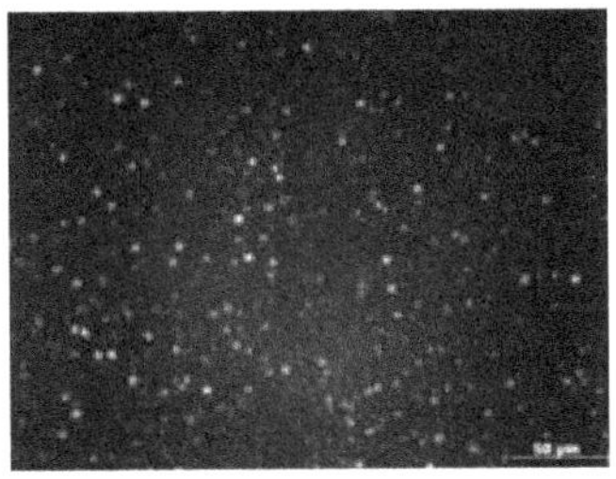	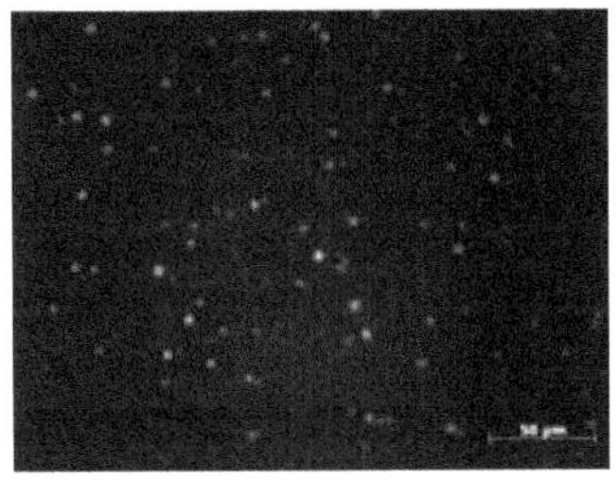
Type	3%S	6%S	9%S
Sample figure			
Type	3%G	6%G	9%G
Sample figure	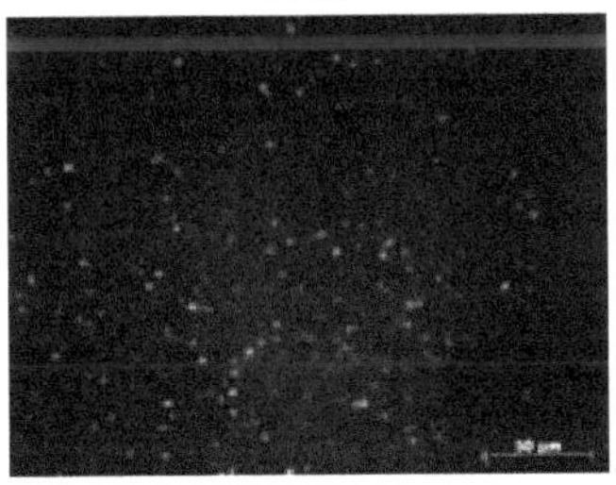		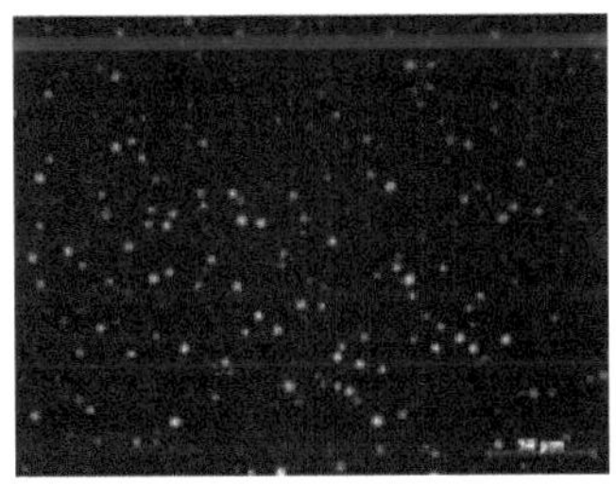

Table 9 shows the phase characteristics of modified asphalt blended with different proportions of rejuvenating agents.

Table 9. Phase characteristics of modified asphalt blends.

Indexes	Aged Modified Asphalt	Types of Modified Asphalt Blends								
		3%H	6%H	9%H	3%S	6%S	9%S	3%G	6%G	9%G
Area ratio of SBS particles/%	2.753	3.679	2.980	2.030	3.832	3.672	3.147	4.207	4.539	4.081
Standard deviation/%	0.38	0.51	0.29	0.31	0.54	0.57	0.48	0.46	0.59	0.32

From Table 9, the area ratio of polymer SBS particles increases at first and then decreases, presumably because the proportion of rejuvenating agent was small, and the swelling effect of SBS particles continued with the action of light oil. With the increasing of the proportion of rejuvenating agent, the swelling effect reaches a certain extent, and excess of rejuvenating agent decreases the relative content of SBS particles in asphalt, resulting in the area ratio of SBS particles beginning to decrease.

A comparison was made of the relationship of macroscopic properties shown in the Section 3.1 and the microscopic characteristics of modified bitumen blends shown in Table 9 based on the connection points of rejuvenating agent proportion. The comparison indicates that the variation of area ratio of SBS particles is consistent with the variation in tenacity.

It is concluded that with an increase of proportion of rejuvenating agent, the trend of area ratio is similar to tenacity, but this may be just an apparent phenomenon. Software statistical analysis was conducted at bivariate significance level of 0.05. The results of modified asphalt blends are shown in Table 10.

Table 10. Correlation of area ratio of SBS particles and tenacity with proportion of rejuvenating agents.

Rejuvenating Agent Type	Correlation Coefficient	Significance
H	0.953	0.047
S	0.983	0.017
G	0.957	0.043

Table 10 shows that the correlation coefficient between area ratio of SBS particles and tenacity are all over than 0.95, indicating that the positive correlation is strong. The significance parameters are all less than 0.05, indicating that the two indexes have significant correlation.

According to the above research into penetration, softening point, 5 °C ductility, viscosity, and tenacity of recycled asphalt and economy, the proportion of rejuvenating agent S for recycled mixtures is set to 6%.

3.4. Freeze-Thaw Split Test Results

The moisture susceptibility of recycled mixtures is evaluated by freeze-thaw split test, and the tensile strength ratio (TSR) is used as the evaluation index. The results are shown in Table 11, and * in the study represents the addition of rejuvenating agent.

Table 11. Freeze-thaw split test results of recycled mixtures.

RSMAP Content	No Freeze-Thaw Split Strength/MPa	Freeze-Thaw Split Strength/MPa	TSR/%
0%	0.95	0.87	91.1
30%	1.40	1.12	80.0
30% *	1.28	1.12	87.2
50%	1.28	1.02	79.7
50% *	1.31	1.19	90.8
70%	1.33	0.89	66.9
70% *	1.09	0.9	82.6

*: the addition of rejuvenating agent.

Table 11 shows that TSR of recycled mixtures decreases significantly with increasing of RSMAP contents. When the RSMAP content is 70%, the TSR is 66.9%, which could not meet China's specification. By adding rejuvenating agent, TSR can increase significantly. This indicates that RSMAP will decrease moisture susceptibility, especially with high RSMAP contents; adding rejuvenating agent can improve the moisture susceptibility of recycled modified mixtures including high RSMAP contents. However, the moisture susceptibility of recycled mixtures is not easy to recover to HMA level.

3.5. Semi-Circular Bending Test Results

Figure 5 graphically illustrates the fracture energy (G_f) and fracture toughness (K_c) of recycled mixtures in SCB test, and RA in the Figure 5 represents rejuvenating agent.

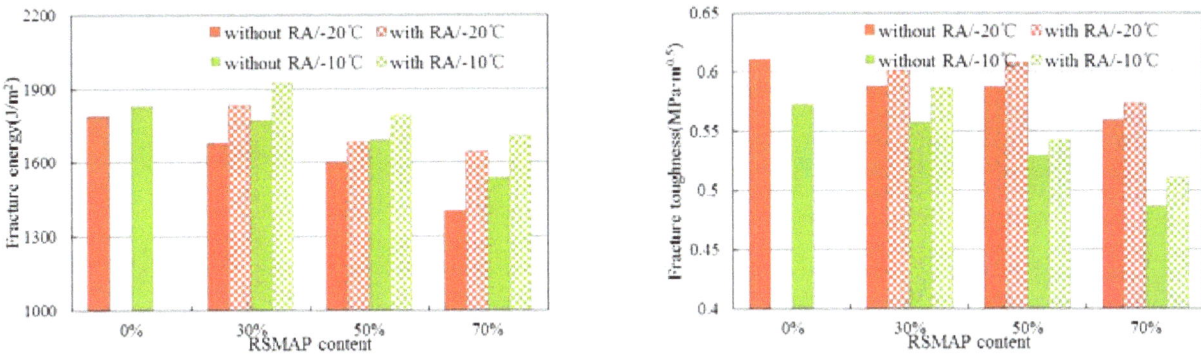

(a) (b)

Figure 5. SCB test results of recycled mixtures. (**a**) Fracture energy; (**b**) Fracture toughness.

Figure 5 shows that the trends for SCB fracture energy and fracture toughness are decreasing with increasing RSMAP content in the mixture at the two temperatures. As the test temperature increases, the fracture energy increases and fracture toughness decreases. When the RSMAP content is 70%, the fracture energy and fracture toughness of recycled mixture without rejuvenating agent decrease by 22% and 9% compared with the control mixtures (RSMAP is 0%). By adding rejuvenating agent, both indexes are improved. However, the two indexes of recycled mixtures with high RSMAP contents are not easy to recover to HMA level.

This indicates that RSMAP has a negative impact on low-temperature properties of recycled mixtures, and the lower the temperature is, the greater the negative impact is. The rejuvenating agent can improve the fracture energy and fracture toughness of recycled mixtures including high RSMAP contents.

3.6. Dynamic Modulus Test Results

The dynamic modulus master curve model parameters (Equation (4)), developed for different recycled mixtures, are presented in Table 12.

Table 12. Dynamic modulus master curve model parameters of recycled mixtures.

RSMAP Content	$\|E^*\|$ Master Curve Parameters					Correlation Coefficient
	δ	α	β	γ	c	R^2
0%	2.074	2.311	−0.986	−0.689	8516.5	0.99
30%	1.573	2.935	−1.109	−0.487	8356.7	0.99
30% *	1.901	2.609	−0.809	−0.515	7249.7	0.99
50%	1.116	3.458	−1.360	−0.389	7943.0	0.99
50% *	1.717	2.809	−1.094	−0.452	8084.1	0.99
70%	0.338	4.223	−1.603	−0.339	8728.1	0.99
70% *	1.358	3.107	−1.229	−0.39	8218.9	0.99

*: the addition of rejuvenating agent.

From Table 12 and based on the nonlinear fitting statistics, it is evident that the dynamic modulus models used for developing the master curves are highly related. The temperature shift factors for a reference temperature 20 °C are shown in Table 13.

Table 13. Temperature shift factors (log$a(T)$) of recycled mixtures.

RSMAP Content	Temperature Shift Factors (log$a(T)$)		
	4 °C	20 °C	40 °C
0%	1.678	0	−1.857
30%	1.647	0	−1.822
30% *	1.429	0	−1.581
50%	1.566	0	−1.732
50% *	1.594	0	−1.763
70%	1.721	0	−1.903
70% *	1.620	0	−1.792

*: the addition of rejuvenating agent.

The dynamic modulus master curves of recycled mixtures with different RSMAP contents are provided in Figure 6. To facilitate comparison, the master curve of control mixture is also provided in Figure 6.

From Figure 6a–c, it's observed that dynamic modulus of all mixtures tested increases with the increase of the loading frequency. In general, the dynamic modulus values of recycled mixtures are significantly higher than control mixture, and the difference is greater with higher RSMAP content. By adding rejuvenating agent, the dynamic modulus decreases slightly. Figure 6d shows that at higher loading frequency, there is no notable difference between the dynamic modulus of the recycled mixtures and control mixture. The results indicate that the rejuvenating agent has an effect on the stiffness of the recycled mixtures generally, but the effect is not obvious. In addition, when the loading frequency reaches a greater value, the difference of recycled mixtures and control mixture tend to be consistent, and include high RSMAP content.

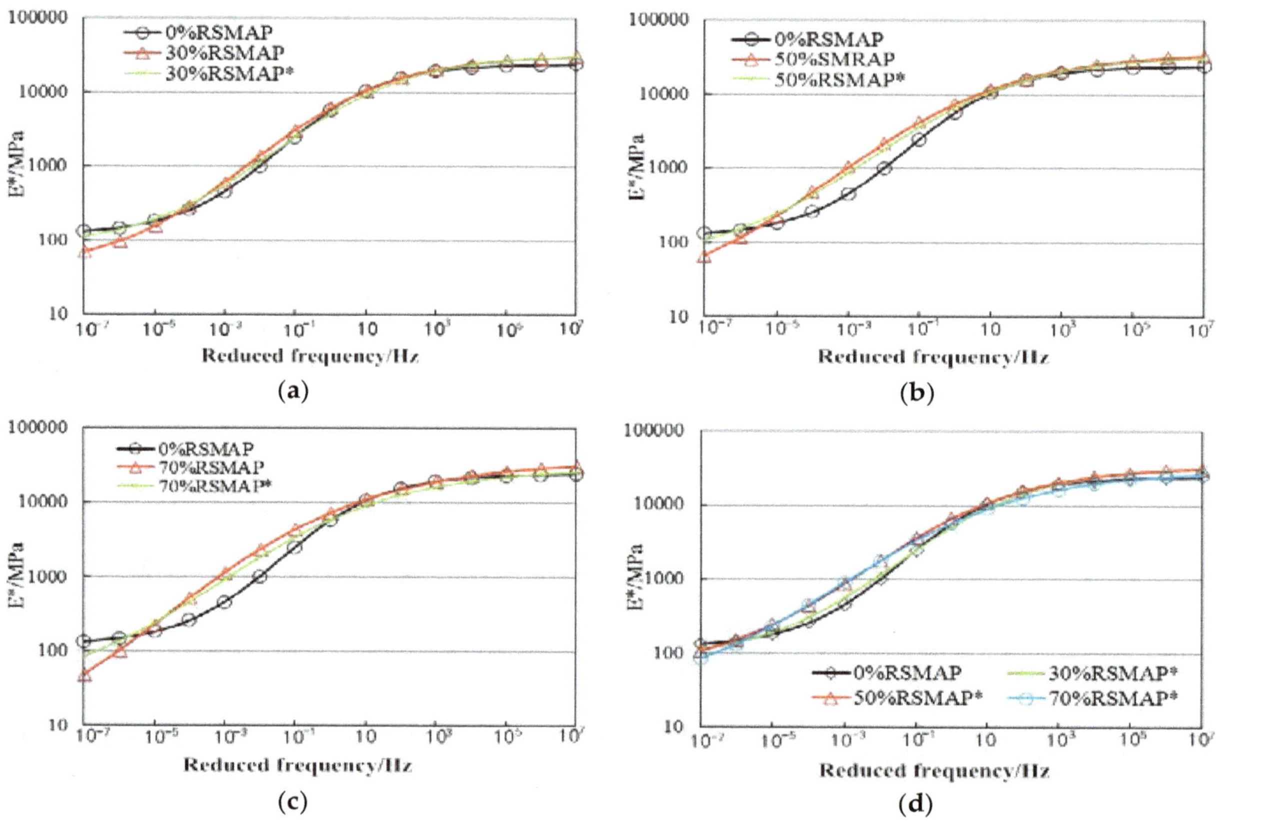

Figure 6. (**a**) Master curves of recycled mixtures with 30% RSMAP; (**b**) Master curves of recycled mixtures with 50% RSMAP; (**c**) Master curves of recycled mixtures with 70% RSMAP; (**d**) Master curves of recycled mixtures with all RSMAP contents.

4. Conclusions

Rejuvenating agent can effectively recycle the properties of aged modified asphalt, and improve the swelling effect between modifier and asphalt. However, the anti-aging property of modified asphalt blends is poor. The relationship between penetration and viscosity of modified asphalt blends shows that the change of penetration will be obvious when the viscosity is less than a certain value.

The recovery degree of penetration and ductility of modified asphalt blends is poorly synchronized when applying rejuvenating agent. It is not appropriate to apply the viscosity and ductility tests to carry out the recycling design at the same time.

Rejuvenating agent cannot change the phase structure of aged modified asphalt, and the SBS particle's uniformity is still satisfactory. On the basis of microscopic analysis, the SBS particles area ratio of modified asphalt blends is significantly correlated with tenacity as the proportion of rejuvenating agent increases.

RSMAP can decrease moisture susceptibility of HMA mixture, especially with high RSMAP content, and adding rejuvenating agent can improve the property. However, the property of recycled mixtures with high RSMAP content is not easy to recover to HMA mixture level.

RSMAP has a negative impact on low-temperature cracking resistance of recycled mixtures, and the lower the temperature is, the greater the negative impact is. Rejuvenating agent can improve the fracture energy and fracture toughness of recycled mixtures including high RSMAP contents. However, the recycled mixtures with high RSMAP content are not easy to recover to HMA mixture level.

In general, the rejuvenating agent has an effect on the stiffness of the recycled mixture, but not an obvious one. When the loading frequency reaches higher value, the difference of recycled mixtures and control mixture tend to be consistent, and include high RSMAP content.

Acknowledgments: This research was supported by the national Science & Technology Support Program "The technology, equipment and demonstration for the application of reclaimed asphalt pavement (No. 2014BAC07B01)" and "Low-carbon construction technology and project demonstration of asphalt pavement (No. 2014BAC07B05)". We state that we have received funds for covering the costs to publish in open access.

Author Contributions: Jie Wang, Yongchun Qin, Songchang Huang and Jian Xu conceived and designed the experiments; Jie Wang and Wei Zeng performed the experiments; Jie Wang and Wei Zeng analyzed the data; Wei Zeng contributed materials; Jie Wang wrote the paper.

References

1. Jin, H.; Gao, G.; Zhang, Y.; Zhang, Y.X.; Sun, K.; Fan, Y.Z. Improved properties of polystyrene-modified asphalt through dynamic vulcanization. *Polym. Test.* **2002**, *21*, 633–640. [CrossRef]
2. Zhang, F.; Yu, J. The research for high-performance SBR compound modified asphalt. *Constr. Build. Mater.* **2010**, *24*, 410–418. [CrossRef]
3. Khodaii, A.; Mehrara, A. Evaluation of permanent deformation of unmodified and SBS modified asphalt mixtures using dynamic creep test. *Constr. Build. Mater.* **2009**, *23*, 2586–2592. [CrossRef]
4. Khodaii, A.; Moghadas Nejad, F.; Forough, S.A.; Ahari, A.S. Investigating the effects of loading frequency and temperature on moisture sensitivity of SBS-modified asphalt mixtures. *J. Mater. Civ. Eng.* **2014**, *26*, 897–903. [CrossRef]
5. Kök, B.V.; Çolak, H. Laboratory comparison of the crumb-rubber and SBS modified bitumen and hot mix asphalt. *Constr. Build. Mater.* **2011**, *25*, 3204–3212. [CrossRef]
6. Huang, W.D.; Zheng, M.; Tang, N.P.; Shan, Z. Comparson of evaluation parameters for high temperature performance of SBS modified asphalt. *J. Build. Mater.* **2017**, *20*, 139–144.
7. Fu, H.Y.; Xie, Z.D.; Yu, M.; Dou, D.Y.; Li, L.F.; Yao, S.D. Dynamic shear rheologic properties of SBS modified asphalt. *J. Highw. Transp. Res. Dev.* **2005**, *22*, 9–12.
8. Hu, Z.W.; Sun, H.W.; Ma, S.J. Hot in-place recycling performance of SBS modified SMA mixture. *J. Highw. Transp. Res. Dev.* **2012**, *29*, 29–47.

9. He, Z.Y.; Ran, L.F.; Cao, Q.X. Mechanism study of recycle of SBS modified asphalt based on spectrum analysis. *J. Build. Mater.* **2015**, *18*, 900–904.
10. Jamshidi, A.; Hamzah, M.O.; Shahadan, Z.; Yahaya, A.S. Evaluation of the rheological properties and activation energy of virgin and recovered asphalt binder blends. *J. Mater. Civ. Eng.* **2015**, *27*. [CrossRef]
11. Huang, S.C.; Turner, T.F. Aging Characteristics of RAP Blend Binders: Rheological Properties. *J. Mater. Civ. Eng.* **2014**, *26*, 966–973. [CrossRef]
12. Zhou, Z.G.; Yang, Y.P.; Zhang, Q.P.; Gao, J.Y. Recycling behavior of recycling agent on aged asphalt. *J. Traffic Transp. Eng.* **2011**, *11*, 10–16.
13. Cong, P.; Luo, W.; Xu, P.; Zhao, H. Investigation on recycling of SBS modified asphalt binders containing fresh asphalt and rejuvenating agents. *Constr. Build. Mater.* **2015**, *91*, 225–231. [CrossRef]
14. Sun, L.; Wang, Y.; Zhang, Y. Aging mechanism and effective recycling ratio of SBS modified asphalt. *Constr. Build. Mater.* **2014**, *70*, 26–35. [CrossRef]
15. Cong, P.; Zhang, Y.; Liu, N. Investigation of the properties of asphalt mixtures incorporating reclaimed SBS modified asphalt pavement. *Constr. Build. Mater.* **2016**, *113*, 334–340. [CrossRef]
16. Winkle, C.V.; Mokhtari, A.; Lee, H.; Williams, R.C.; Schram, S. Laboratory and field evaluation of HMA with high contents of recycled asphalt pavement. *J. Mater. Civ. Eng.* **2017**, *29*. [CrossRef]
17. Goli, A.; Ziari, H.; Amini, A. Influence of Carbon Nanotubes on Performance Properties and Storage Stability of SBS Modified Asphalt Binders. *J. Mater. Civ. Eng.* **2017**, *29*, 654–660. [CrossRef]
18. Kang, A.H.; Kou, C.J.; Liu, X.Y. Evaluation methods of dispersed phase uniformity in SBS modified asphalt based on digital image analysis technology. *J. Sichuan Univ.* **2014**, *46*, 172–176.
19. Huang, W.; Sun, L. Phase structure of modified asphalt. *J. Highw. Transp. Res. Dev.* **2001**, *18*, 1–3.
20. Sengoz, B.; Isikyakar, G. Evaluation of the properties and microstructure of SBS and EVA polymer modified bitumen. *Constr. Build. Mater.* **2008**, *22*, 1897–1905. [CrossRef]
21. Zhang, D.; Zhang, H.; Shi, C. Investigation of aging performance of SBS modified asphalt with various aging methods. *Constr. Build. Mater.* **2017**, *145*, 445–451. [CrossRef]
22. JTG E20-2011. *Standard Test Methods of Bitumen and Bituminous Mixtures for Highway Engineering*; China Communication Press: Beijing, China, 2011.
23. JTG E42-2005. *Test Methods of Aggregate for Highway Engineering*; China Communication Press: Beijing, China, 2005.
24. Zegeye, E.T.; Moon, K.H.; Turos, M.; Clyne, T.R.; Marasteanu, M.O. Low temperature fracture properties of polyphosphoric acid modified asphalt mixtures. *J. Mater. Civ. Eng.* **2012**, *24*, 1089–1096. [CrossRef]
25. Wu, Z.; Mohammad, L.N.; Wang, L.B.; Mull, M.A. Fracture resistance characterization of superpave mixtures using the semi-circular bending test. *J. ASTM Int.* **2006**, *2*, 1–15. [CrossRef]
26. Li, X.J.; Marasteanu, M.O.; Kvasnak, A.; Jason, B. Factors study in low-temperature fracture resistance of asphalt concrete. *J. Mater. Civ. Eng.* **2010**, *22*, 145–152. [CrossRef]
27. Khosravifar, S.; Haider, I.; Afsharikia, Z.; Schwartz, C.W. Application of time–temperature superposition to develop master curves of cumulative plastic strain in repeated load permanent deformation tests. *Int. J. Pavement Eng.* **2015**, *16*, 214–223. [CrossRef]
28. Ghabchi, R.; Singh, D.; Zaman, M.; Hossain, Z. Laboratory characterisation of asphalt mixes containing RAP and RAS. *Int. J. Pavement Eng.* **2016**, *17*, 829–846. [CrossRef]
29. Kupolati, W.K. Characterization of bitumen extracted from used asphalt pavement. *Eur. J. Sci. Res.* **2010**, *7*, 1663–1672. [CrossRef]

14

Fatigue Life Prediction of High Modulus Asphalt Concrete-Based on the Local Stress-Strain Method

Mulian Zheng [1,*], Peng Li [1,*], Jiangang Yang [2], Hongyin Li [3], Yangyang Qiu [4] and Zhengliang Zhang [5]

1 Key Laboratory for Special Area Highway Engineering of Ministry of Education, Chang'an University, South Erhuan Middle Section, Xi'an 710064, Shaanxi, China
2 Civil Engineering Materials Laboratory, East China Jiaotong University, Shuang Gang East Street, Nanchang 330000, Jiangxi, China; mjgchd@163.com
3 Shandong Highway Administration Bureau, Shungeng Road of Jinan City, Jinan 250000, Shandong, China; chdmaqiang@163.com
4 Jinhua Traffic Planning and Design Institute, Songlian Road of Jin Dong District, Jinhua 321015, Zhejiang, China; qiu_yangyang@126.com
5 Anhui Highway Administration Center, Da Bie Shan Road of Hefei City, Hefei 230088, Anhui, China; xuhaileilie@163.com
* Correspondence: zhengml@chd.edu.cn (M.Z.); lp042820321@126.com (P.L.)

Academic Editors: Zhanping You, Qingli (Barbara) Dai and Feipeng Xiao

Abstract: Previously published studies have proposed fatigue life prediction models for dense graded asphalt pavement based on flexural fatigue test. This study focused on the fatigue life prediction of High Modulus Asphalt Concrete (HMAC) pavement using the local strain-stress method and direct tension fatigue test. First, the direct tension fatigue test at various strain levels was conducted on HMAC prism samples cut from plate specimens. Afterwards, their true stress-strain loop curves were obtained and modified to develop the strain-fatigue life equation. Then the nominal strain of HMAC course determined using finite element method was converted into local strain using the Neuber method. Finally, based on the established fatigue equation and converted local strain, a method to predict the pavement fatigue crack initiation life was proposed and the fatigue life of a typical HMAC overlay pavement which runs a risk of bottom-up cracking was predicted and validated. Results show that the proposed method was able to produce satisfactory crack initiation life.

Keywords: pavement engineering; high modulus asphalt concrete; local stress-strain; modified Neuber equation; fatigue life

1. Introduction

Fatigue damage refers to the accumulation of damage in asphalt concrete, causing cracks and fracture extensions, and ultimately resulting in loading failure. Fatigue failure includes the formation of the crack nucleation stage (fatigue), crack extension (stability) of crack growth, and fracture failure (unstable crack propagation). Because fracture failure happens rapidly, the fatigue life of pavement generally includes crack formation and propagation. The term "stage of fatigue crack initiation" usually refers to the fatigue crack nucleation and propagation, which require examination during engineering processes. In road engineering the period from crack nucleation to the crack reaching a length of 5 mm is treated as the life of the initiation of the fatigue crack.

There are different approaches to quantify the fatigue crack process in asphalt concrete mixtures. Among the different approaches, practitioners primarily use the strain approach, which was first introduced by Monismith, and the energy approaches because these approaches are simple and consist of experimentally oriented procedures [1–3]. The most common fatigue model adopted by the Mechanistic Empirical Pavement Design Guide (MEPDG) is a strain approach that considers the material's properties using the value of dynamic modulus forms [4]. Al-Qadi described the dissipated energy changes during loading and proposed four parameters to define fatigue performance for 13 mm dense-graded Hot Mix Asphalt (HMA) using bending beams at various strain amplitudes [5]. Molayem used continuum damage theory to characterize the fatigue properties of asphalt binders and their effect on fatigue resistance. This proposed model can be used as an alternative to time-consuming beam fatigue tests [6].

Metcalf evaluated the Strategic Highway Research Program (SHRP) fatigue crack initiation life prediction model with crack measurements in laboratory slab fatigue tests. The study showed that the SHRP crack initiation life model significantly underestimates the fatigue crack initiation life for slabs [7].

Kim developed a system for predicting the fatigue life of asphalt mixtures. This work indicates that the fatigue life of a mixture subjected to sinusoidal strain loading can be determined through the model's damage principles using IDT (indirect tensile test) test results [8]. Ali Khodaii employed a large scale experimental setup to examine the influence of the most important parameters on delaying reflection cracking in geogrid reinforced overlay in bending mode. Regression equations were developed to estimate fracture parameters which can be used to predict the crack growth rate in order to design asphalt overlay that is protected from reflective cracking [9].

One of the major achievements of the SHRP asphalt mix research projects is the development of a fatigue life prediction model based on the results of intensive bending beam fatigue tests. Several researchers have conducted fatigue testing on asphalt concrete in the laboratory to investigate fatigue life through bending beam fatigue tests [10–16]. Although this model covers a wide range of asphalt mix related fatigue variables, it has to apply a shift factor to deal with the stress concentration of cracks in High Modulus Asphalt Concrete (HMAC), including the state of stress. It would be helpful to determine how real stress state contributes to fatigue life, because local stress and direct tensile tests more realistically simulate the multiaxial state of stress and cracking in pavements than bending beam tests.

A basic principle of the local stress-strain method is that the nominal stress and strain of a structure under various loadings are firstly converted into local stress and strain at the most unfavorable points (i.e., the stress concentration points) based on the stress-strain loop curve obtained from the fatigue test. Combined with the established strain-fatigue life curve, these local stresses and strains are then corrected and used to estimate fatigue life of the structure.

Local stress-strain fatigue analysis focuses on determining local stress and strain. In the local stress concentration area, there is a complex nonlinear relationship between the stress, strain, and load. Methods for determining the local strain and stress include the test method, the elastic-plastic finite element method, and the approximate calculation method.

In this paper, we conduct direct tensile fatigue tests and cyclic stress-strain tests at different strain levels to determine the fatigue properties of HMAC, and propose a method to predict the fatigue crack initiation life of HMAC. This method is based on a modified Neuber equation and the local stress-strain method.

2. Process for Fatigue Life Estimation Based on the Local Stress-Strain Method

The fatigue life prediction method consists of three aspects: fatigue performance of the material, transformation from nominal stress and strain into local stress and strain, and fatigue cumulative damage theory. A modified Neuber equation was used to transform nominal stress-strain to local stress-strain. The Miner theory and Manson-Coffin equation have been widely used to characterize cumulative fatigue damage. Therefore, the cyclic stress-strain curve and strain-life curve can be combined as the local stress-strain method to estimate the fatigue crack life.

The raw data needed for calculation includes the elastic modulus, the steady cyclic stress–strain curve, the strain-life curve, the theoretical stress concentration factor, and load history.

Steps for fatigue life estimation, based on the local stress-strain method, are shown in Figure 1.

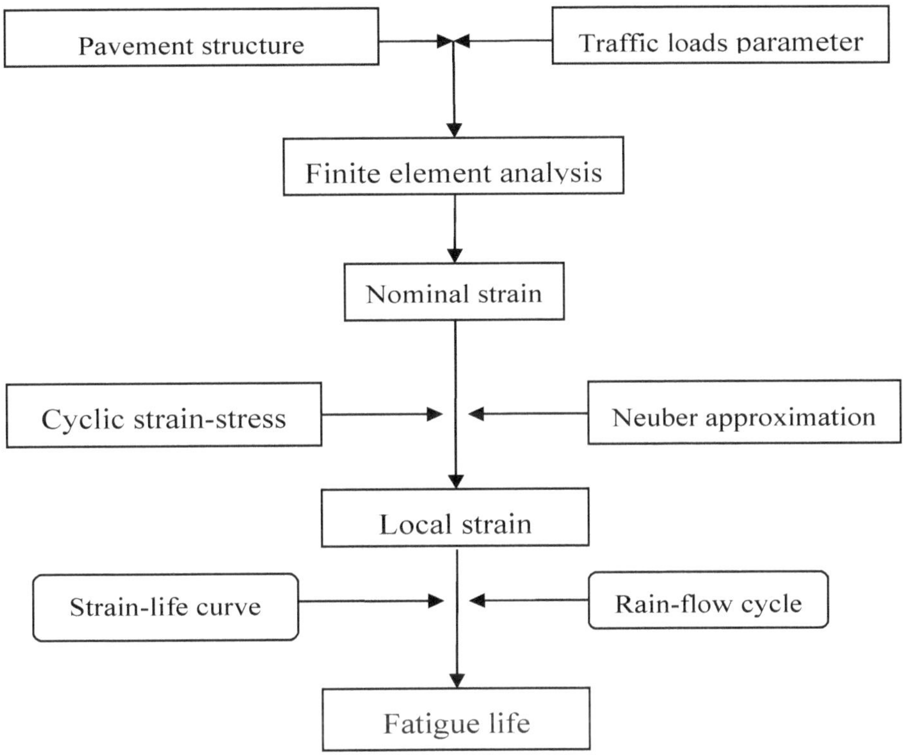

Figure 1. Process for fatigue life estimation based on local stress-strain method.

3. Direct Tensile Test of HMAC

3.1. Materials

90# asphalt was used to compound the high-modulus asphalt in the research. Details of the materials are listed in Table 1.

Table 1. Test results of 90# asphalt.

Items		Unit	Value
25 °C Penetration		0.1 mm	83
Penetration Index		...	−0.84
Ductility (10 °C)		cm	42
Ductility (15 °C)		cm	>150
Softening Point		°C	49.3
Flashing Point		°C	282
Paraffin Content (distillation)		%	0.91
Solubility (C_2HCl_3)		%	99.8
Density (15 °C)		g/cm^3	0.982
TFOT (163 °C, 5 h)	Quality Change	%	−0.005
	Penetration Ratio	%	68.5
	Ductility (15 °C)	cm	38.6
	Ductility (10 °C)	cm	9.6

PR PLASTS was chosen as the HMAC-20 additive with a dosage of 0.6%. The mineral aggregate gradation of the HMAC-20 is shown in Figure 2.

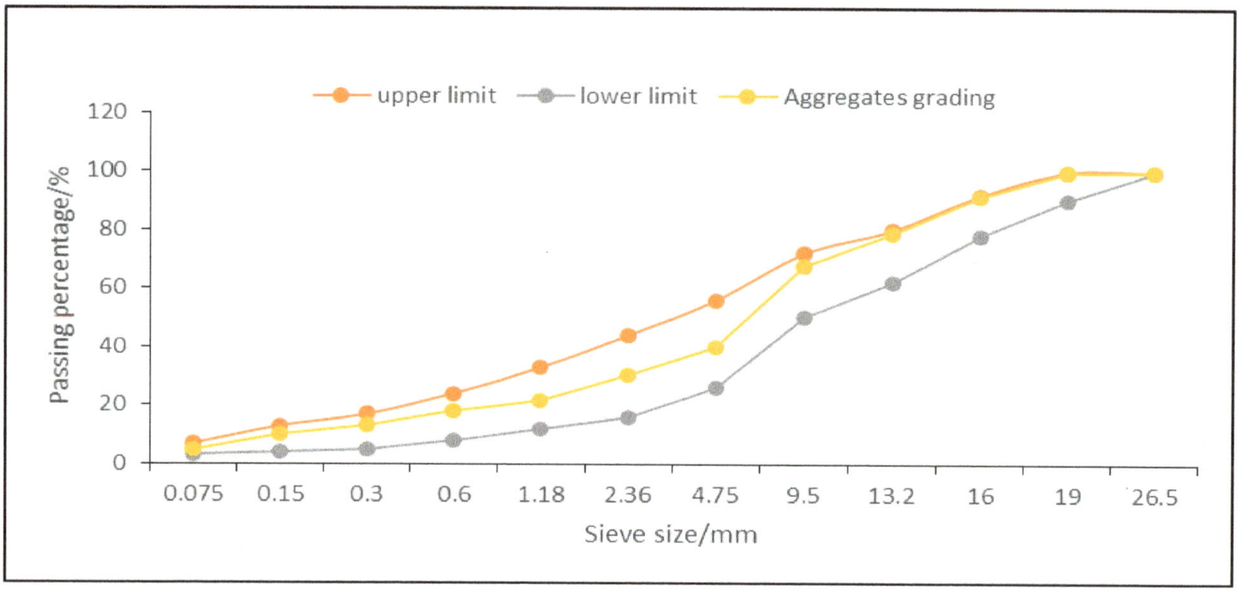

Figure 2. Gradation of the aggregates used in this study.

A plate was made according to the rutting test specifications and then cut into a suitable prism specimen.

3.2. Test Matrix

As shown in Figure 3, fatigue tests were performed by Material Test System (MTS) with a strain-controlled load on asphalt concrete prisms and were conducted in the laboratory.

Figure 3. Material Test System (MTS).

Specimens must be a suitable shape and size for strain control and to ensure the reliability of test results. Several 300 mm × 300 mm × 50 mm plate specimens were firstly prepared and then they were cut into 250 mm × 50 mm × 50 mm prism samples in accordance with test procedures of direct tension testing in the literature [17].

As shown in Figure 4, both ends of the specimen were bonded with a steel plate, and a spherical hinge was set at the end of the active loading position, which ensured the central tension on the tip of the specimen. The strain-controlled cyclic loads were applied on a steel plate, and the spherical hinge could be automatically set back when tension was applied to the specimen. In order to avoid uncontrolled tension, the load was set on the middle of the prism.

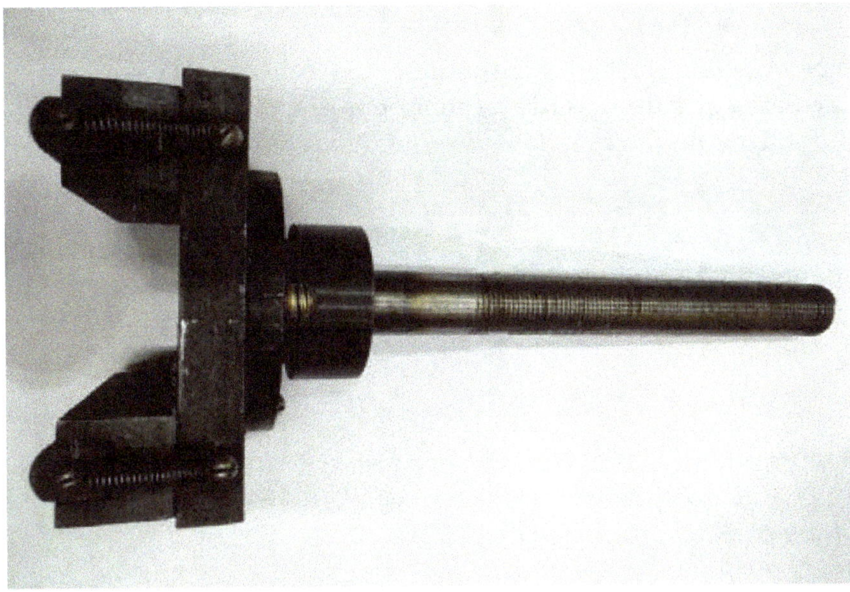

Figure 4. Specimen clamp for tests.

3.3. Loading Frequency

In order to make the experiment feasible and to comply with road surface conditions, the test was conducted at a variety of frequencies (1, 2, 5, 10 Hz). A load frequency of 2 Hz was determined for the test by taking into consideration influential factors including simulation results, the test time, and fixture performance.

3.4. Loading Waveform

In consideration of the fixed method of the specimen clamp, a gentle tug was exerted to avoid displacing the fixture gap. A non-intermittent triangular wave for the loading waveform was adopted (see Figure 5).

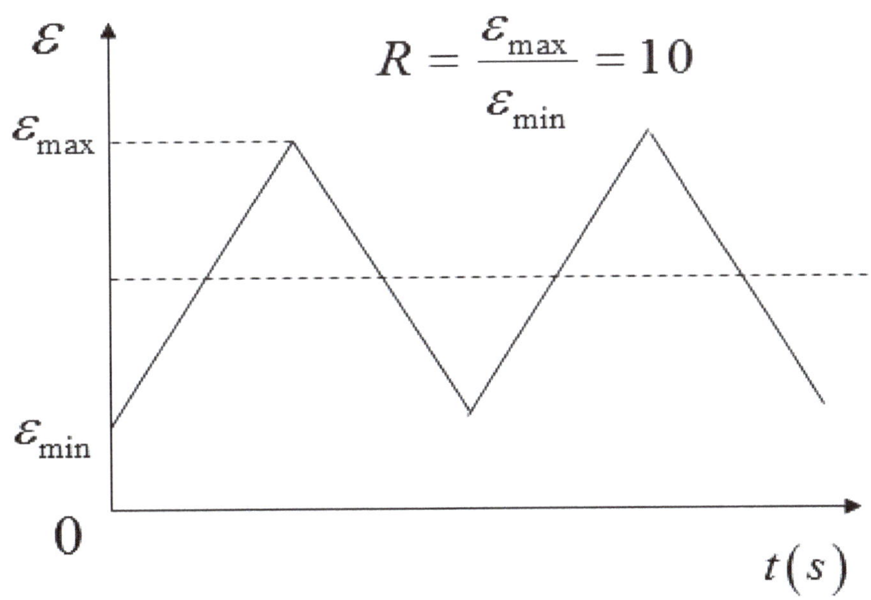

Figure 5. Loading waveform for fatigue test of High Modulus Asphalt Concretes (HMAC).

3.5. Test Temperature

In order to reflect the actual working state of high modulus asphalt pavement, a fatigue equivalent temperature was calculated and used as the test temperature.

SHRP presented the method to calculate the equivalent temperature of the fatigue test [18,19]:

$$T_{eff}(FC) = 0.8(MAPT) - 2.7 \tag{1}$$

where: $T_{eff}(FC)$—equivalent temperature of the fatigue test (°C); $MAPT$—annual average surface temperature in 1/3 Pavement depth (°C); 15 °C was determined to be the test temperature for the HMAC fatigue test.

3.6. Failure Criteria

Load descent failure criterion for the prism specimen was adopted after taking into consideration the asphalt mixture characteristics and test maneuverability. The specimen failed when the load decreased to 75% of the peak loading capacity [20,21].

The axial strain control mode was chosen and the high-low strain ratio of the specimens was set at 10. Four strain levels (0.3, 0.4, 0.5, 0.6) were adopted with three specimens tested in each level. Test conditions were determined, including a 2 Hz load frequency, a 1 mm/min stretching rate, a 15 °C test temperature, and a non-intermittent triangular wave for the loading waveform. These conditions were set in conjunction with the simulation effects, the test time, and clamp performance.

4. True Cyclic Stress-Strain Curve

The cyclic stress-strain curve, which is made by the vertices of the stable hysteresis loop at different strain levels, is a smooth curve. The following methods were used to determine the cyclic stress-strain curves:

(1) Multistage test method. Cyclic loading under several strain amplitude levels was conducted on the specimen, each cycle was loaded a number of times to achieve stability, and a smooth curve was drawn using the overlapping and stable hysteresis loop through the top. This loop was named the cyclic stress-strain curve. The method required few specimens and had a quick measurement speed; however, this method has low precision and it easily produces fatigue damage.

(2) Relegation-enhancement test method. When the specimen was loaded, the strain amplitude was gradually relegated and then enhanced. The other steps of this method are similar to the multistage test method

(3) Stretching after cycle stability method. The specimen underwent a series of strain decreases and increases, reducing the level of strain amplitude after cycle stability was reached. A specimen was then loaded to tensile failure to determine the stress-strain curve. Results of this method are in good agreement with those in the above two methods, however, it has low precision.

(4) Multistage multi-sample method. Multiple experiments controlled by a constant strain were conducted on specimens in multistage strain stages in order to obtain a stable hysteresis loop, then the top of the hysteresis loop was connected and a smooth cyclic stress-strain curve was obtained. This method can accurately reflect the cycle stress-strain properties of materials with high precision, but it requires more specimens and a longer test period than the other methods. The strength coefficient and cyclic strain hardening index are calculated using the cyclic stress-strain curve [22,23]:

$$\frac{\Delta\varepsilon}{2} = \frac{\Delta\sigma}{2E} + \left(\frac{\Delta\sigma}{2K'}\right)^{\frac{1}{n'}} \tag{2}$$

where: $\Delta\varepsilon$ = cyclic strain amplitude, $\Delta\sigma$ = cyclic stress amplitude.

The following equation can be obtained using the plastic component calculation:

$$\varepsilon = \frac{\sigma}{E} + \left(\frac{\sigma}{K'}\right)^{\frac{1}{n'}} \quad (3)$$

where: σ = stress, ε = strain, n' = Cyclic strain hardening exponent, K' = Cyclic stress hardening exponent.

The strain fatigue curve shows that the values of the hysteresis loop apex were calculated when fatigue life reached 50% of the specimen at strain levels ranging from 0.3 to 0.6 in order to determine the true strain and true stress amplitude at different strain levels and the plastic strain component (see Table 2).

Table 2. True stress amplitude and plastic strain component at different strain levels.

Strain Levels	$\Delta\varepsilon$	$\Delta\sigma$/MPa	$\Delta\varepsilon_p$
0.6	0.0014	0.213	0.00041
0.5	0.0013	0.184	0.00038
0.4	0.0011	0.176	0.00036
0.3	0.00081	0.143	0.00027

Table 2 shows the different fitting algorithms that were adopted to fit Equation (3). The fitting parameters K' = 26.847 and n' = 0.725 were then obtained and used in the following formula that predicts the strain of HMAC in the strain-controlled mode:

$$\varepsilon = \frac{\sigma}{E} + \left(\frac{\sigma}{26.847}\right)^{1.3793} \quad (4)$$

Stress can be obtained using Equation (4) with strains ranging from 0.0001~0.001; the results are presented in Figure 6.

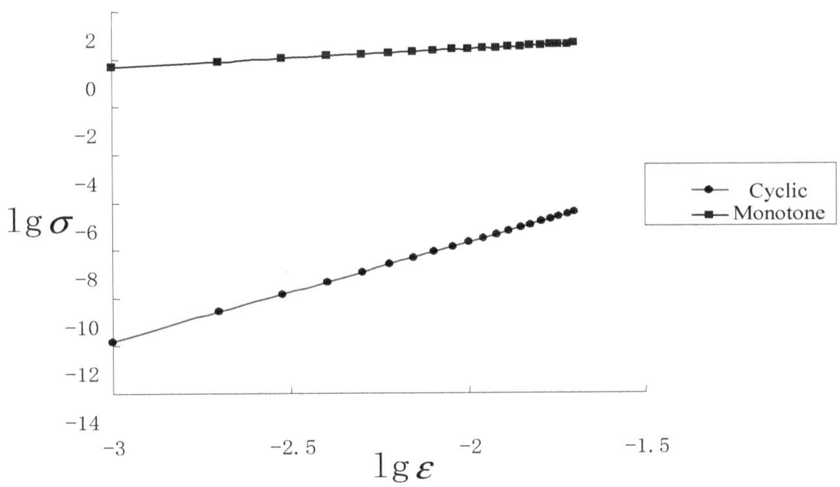

Figure 6. Comparison between cyclic curves and monotone curve.

The parameters K, n, and the HMAC monotone curve are compared, respectively, with the parameters K', n', and the cyclic curves. The cyclic curves are far higher than the monotone curve. Based on the position of the curve, HMAC was determined to be a cyclic hardening material [24], indicating that the stress required for producing the same strain during cyclic loading is higher than when under a static load. However, the hardening phenomenon gradually disappears when the cycling times increase.

Figure 7 shows the relationship between stress and time in the cyclic stress strain test with constant strain controlled. The results show that stress fluctuates as the length of time increases, whereas the general trend is that strain decreases along with the increasing length of time. It can be seen that the HMAC has cyclic relaxation properties [25,26], meaning strain gradually decreases when constant strain is controlled.

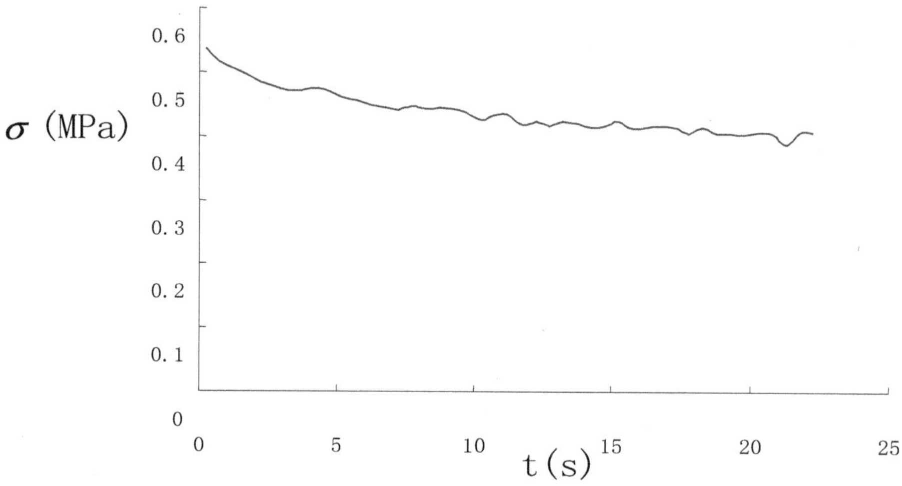

Figure 7. Stress-time history under control of strain.

5. Determination of Local Stress and Strain

5.1. Neuber Equation

The Neuber equation, a concise method for calculating local stress-strain of a notch [27,28], is as follows:

$$K_t = (K_\varepsilon K_\sigma)^{\frac{1}{2}} \tag{5}$$

$$K_\varepsilon = \frac{\varepsilon}{e} \tag{6}$$

$$K_\sigma = \frac{\sigma}{s} \tag{7}$$

where: K_t—theoretical stress concentration factor of notch, K_ε—strain concentration factors, K_σ—factor of stress concentration, e—nominal strain, S—nominal stress.

K_ε and K_σ were placed in Equation (5) and converted into an incremental form, and the Neuber hyperbolic equation was then obtained:

$$\Delta\sigma\Delta\varepsilon = K_t^2 \Delta S \Delta e \tag{8}$$

If the nominal stress is in the elastic range, then $\Delta e = \frac{\Delta S}{E}$ can be obtained, and the equation is as follows:

$$\Delta\sigma\Delta\varepsilon = (K_t \Delta S)^2 / E \tag{9}$$

The Neuber equation is derived from shear stress analysis of a prism. In order to be widely applied in the design and development of HMAC [29,30], the Neuber equation is modified by replacing the theoretical stress concentration factor K_t with the fatigue notch factor K_f. The general method for determining K_f is to fit the regression equation through fatigue test data of the specimen notch, and K_f is typically treated as a constant for most influencing factors. K_t is usually used when there is a lack of K_f data, which is a conservative method for estimating fatigue life. Values of K_t at different strain levels are shown in Table 3.

Table 3. Theoretical stress concentration factor of notch.

Strain Levels	**0.6**	**0.5**	**0.4**	**0.3**
K_t	1.973	1.947	2.046	1.938

5.2. Neuber Equation Correction

To conform to the plane strain problem, the cycle stress-strain curve of a small smooth specimen in the uniaxial stress state test must be modified. Moreover, parameters K_ε and K_σ in the biaxial stress state are different from those in the uniaxial stress state, meaning that the Neuber equations should be modified.

5.2.1. Cyclic $\sigma - \varepsilon$ Curve Correction

Hooke's law was applied to the elastic strain component, whereas the plastic deformation theory was used for the plastic component:

$$\sigma_a{}' = \sigma_a / \left(1 - \mu_1 + \mu_1^2\right)^{\frac{1}{2}} \tag{10}$$

$$\varepsilon_a{}' = \varepsilon_a\left(1 - \mu_1^2\right) / \left(1 - \mu_1 + \mu_1^2\right)^{\frac{1}{2}} \tag{11}$$

$$\mu_1 = \frac{\mu + E\varepsilon_{pa}/(2\sigma_a)}{1 + E\varepsilon_{pa}/\sigma_a} \tag{12}$$

where: σ'_a—the maximum main stress under plane strain condition, ε'_a—the maximum main strain under plane strain condition, μ_1—generalized Poisson's ratio, $\mu_1 = \frac{\mu + E\varepsilon_{pa}/(2\sigma_a)}{1 + E\varepsilon_{pa}/\sigma_a}$, μ—Poisson's ratio, ε_{pa}—plastic strain amplitude of cyclic $\sigma - \varepsilon$ curve under uniaxial stress condition, E_a—elastic modulus of cyclic stress-strain curve.

The equation is achieved by σ'_a—ε'_a fitting as follows:

$$\varepsilon'_a = \sigma'_a\left(1 - \mu^2\right)/E + \left(\sigma'_a/K'\right)^{\frac{1}{n'}} \tag{13}$$

where K'' is corrections of K', n'' is corrections of n'.

The parameter data found in Table 4 were put into Equation (13), and then the revised cyclic $\sigma - \varepsilon$ curve equation was obtained via Equation (14):

$$\varepsilon'_a = \frac{\sigma'_a(1 - \mu^2)}{E_a} + \left(\frac{\sigma}{590.716}\right)^{1.014} \tag{14}$$

Table 4. Correction parameters of cyclic $\sigma - \varepsilon$ curve.

Strain Levels	ε'_a	σ'_a**/MPa**	μ_1	K''	n''
0.6	0.682	0.14	0.427	590.716	0.986
0.5	0.594	0.097	0.419		
0.4	0.473	0.083	0.417		
0.3	0.348	0.076	0.423		

5.2.2. Neuber Equation Correction

To show the plastic penetration level and equivalent stress and strain levels, the right end of the Neuber equation was multiplied by two coefficients, and the following corrections were made to the Neuber equation:

$$\Delta\sigma \cdot \Delta\varepsilon = m \cdot g \cdot \left(K_f \cdot \Delta S\right)^2 / E \tag{15}$$

$$m = \left(1 - \mu' + \mu'^2\right)^{\frac{1}{2}} \tag{16}$$

$$g = m / \left(1 - \mu'^2\right) \tag{17}$$

$$\mu' = 0.5 - (0.5 - \mu) E_s / E \tag{18}$$

where: m—plastic penetrable degree, g—coefficient of equivalent stress and strain, μ'—elastic-plastic Poisson's ratio, considering the compressibility of material, μ—Poisson's ratio, E—elasticity modulus, E_s—secant modulus, determined by the monotone tensile curve.

Parameters μ' = 0.4221, m = 0.87, g = 1.059 can be obtained by putting μ = 0.25, E_s = 374 MPa, and E = 1200 MPa into Equation (15), which becomes the following final form of the Neuber equation that predicts the fatigue life of HMAC in the strain-controlled mode:

$$\Delta\varepsilon \cdot \Delta\sigma = \frac{0.92\left(K_f \cdot \Delta S\right)^2}{E} \tag{19}$$

As nominal stress is in the elastic range, $\Delta e = \frac{\Delta S}{E}$ and K_t = 1.976 were substituted into Equations (9) and (19) to develop Equation (20):

$$\varepsilon\sigma = 0.92 \times 1.976^2 se \tag{20}$$

6. Strain-Life Fatigue Equation

The strain fatigue life equation is characterized by the fatigue curve taking the following form:

(1) $\varepsilon_a - N_f$ curve.

Strain fatigue tests were conducted at different strain levels and under equal strain control to obtain the fatigue data, of the control variables ε_a is the true strain amplitude.

(2) $\varepsilon_{eq} - N_f$ or $\sigma_{eq} - N_f$ curve.

Fatigue data are obtained from fatigue tests that combine stress and strain. The control variables are not equivalent strain amplitude ε_{eq} or equivalent stress amplitude σ_{eq}, which contain the mean stress.

(3) $S - N_f$ curve.

The fatigue data comes from existing data from the S-N curve. The influence of notch and strain fatigue should also be considered.

The second method needs a large amount of experimental data, and the third method was only applied to make a rough estimation, because of its low accuracy. The main $\varepsilon_a - N_f$ curves are shown in Table 5.

Table 5. Strain fatigue curve.

Forms	Author	Expression
$\Delta\varepsilon \sim N_f(R_e=-1)$	M. A. Manson L. F. Coffin	$\varepsilon_a = \frac{\sigma'_f}{E}\left(2N_f\right)^{2b} + \varepsilon'_f\left(2N_f\right)^{2c}$
	H. M. Fu	$\frac{C}{(\Delta\varepsilon_t - \Delta\varepsilon_o)^m} = N_f$
	R. W. Landgraf	$\frac{1}{N_f} = 2\left(\frac{\sigma'_f}{\varepsilon'_f E}\right)\left(\frac{\varepsilon_{pa}}{\varepsilon_{pa}}\right)\frac{\sigma'_f}{\sigma'_f - \sigma_m}$
	Ostergren	$\frac{1}{N_f} = 2\left(\frac{\sigma_{max}\varepsilon_{pa}}{\sigma'_f\varepsilon'_f E}\right)^{1/(a+b)}$
	K. N. Smith	$\sigma_{max}\varepsilon_a = \frac{\sigma'^2}{E}\left(2N_f\right)^{2b} + \sigma'_f\varepsilon'_f\left(2N_f\right)^{b+c}$
	Bergmann	$(\sigma_a + k\sigma_m)\varepsilon_a = \frac{\sigma'^2}{E}\left(2N_f\right)^{2b} + \sigma'_f\varepsilon'_f\left(2N_f\right)^{b+c}$
	F. Erdogan	$\sigma'_a\sigma_{max}^{(1-r)} - \varepsilon_a = \frac{\sigma'^2}{E}\left(2N_f\right)^{2b} + \sigma'_f\varepsilon'_f\left(2N_f\right)^{b+c}$
	V. Adrov	$\frac{1}{N_f} = 2\left(4\sigma'_f\varepsilon'_f/A\right)^{b+c}$ $A = \frac{\beta}{a}I, a = f(\sigma_{max}, \sigma_{min})$ $I = \sum_{k=1}^{k_1}(j+1-2k)\left(\Delta\sigma_k - \Delta\sigma_{j+1} - k\right)\Delta\varepsilon$

The Manson-Coffin equation is widely used for all of the $\varepsilon_a - N_f$ curves by adopting a power function to characterize fatigue performance. The main parameter strain is calculated using the following equation:

$$\varepsilon_a = \frac{\sigma'_f}{E}\left(2N_f\right)^b + \varepsilon'_f\left(2N_f\right)^c \quad (21)$$

Based on the proposed Manson-Coffin equation, direct tensile fatigue tests for HMAC were conducted to obtain the fatigue parameters. Estimation methods for fatigue parameters included the general slope method, modified four-points correlation, and four-points correlation [31,32] (see Table 6).

Table 6. Estimation method for fatigue constant.

Methods	Fatigue Constant	
General slope method	b	-0.12
	C	-0.16
	σ'_f	$1.75\sigma_b/E$
	ε'_f	$0.5\varepsilon_f^{0.006}$
Four-points correlation	b	$-[0.0792 + 0.179\lg(\sigma_f/\sigma_b)]$
	c	$-\lg(3.37\varepsilon_f^{0.25}\left[1 - 81.8\frac{\sigma_b}{E}\left(\frac{\sigma_f}{\sigma_0}\right)^{1.179}\right]^{-1/3})$
	σ'_f	$1.12\sigma_b(\sigma_f/\sigma_b)^{0.893}$
	ε'_f	$0.413\varepsilon_f\left[1 - 81.8\frac{\sigma_b}{E}\left(\frac{\sigma_f}{\sigma_0}\right)^{1.179}\right]^{-1/3}$
Modified four-points correlation	b	Same as four-points correlation
	c	Same as four-points correlation
	σ'_f	$1.18\sigma_b(\sigma_f/\sigma_b)^{0.0946}$
	ε'_f	$0.538\left[1 - 81.8\frac{\sigma_b}{E}\left(\frac{\sigma_f}{\sigma_0}\right)^{1.179}\right]^{-1/3}$

Fatigue test regulation for Manson-Coffin equation requires that $R_e = -1$, otherwise, the equation should be modified, (see Table 7).

Table 7. Modified Manson-Coffin equation when $R_e \neq -1$.

Methods	Modified Manson-Coffin Equation
Morrow elastic stress Linear correction	$\varepsilon_a = \frac{\sigma'_f - \sigma_m}{E}\left(2N_f\right)^b + \varepsilon'_f\left(2N_f\right)^c$
Gerber elastic stress curvature correction	$\varepsilon_a = \frac{\sigma'^2_f - \sigma^2_m}{E\sigma'_f}\left(2N_f\right)^b + \varepsilon'_f\left(2N_f\right)^c$
Marrow total strain correction	$\varepsilon_a = \frac{\sigma'_f - \sigma_m}{\sigma'_f}\frac{\sigma'_f}{E}\left(2N_f\right)^b + \varepsilon'_f\left(2N_f\right)^c$
Sachs plastic correction	$\Delta\varepsilon_a = \frac{\sigma'_f}{E}\left(1 - \frac{\sigma_m}{\sigma_b}\right)^b\left(2N_f\right)^c$

In accordance with the first method, an equal strain fatigue test with strain levels ranging from 0.3 to 0.6 and a monotone direct tensile test were conducted on HMAC using the dynamic tester MTS810 to calculate the $\varepsilon_a - N_f$ curve of asphalt concretes [33]. The $\varepsilon_a - N_f$ curve was characterized by the Mason-Coffin equation, and the power function was adopted to describe the strain life curve:

$$\varepsilon_a = \frac{\Delta\varepsilon}{2} = \frac{\Delta\varepsilon_e}{2} + \frac{\Delta\varepsilon_p}{2} = \left(\frac{\sigma'_f}{E}\right)\left(2N_f\right)^b + \varepsilon'_f\left(2N_f\right)^c \tag{22}$$

In the above equations, σ'_f and ε'_f are approximately equal to the fracture stress σ_f and fracture strain ε_f [34]. Fracture stress and fracture strain for HMAC were listed in the monotone direct tensile test (see Table 8).

Table 8. HMAC parameters of monotone direct tensile test.

Specimen	Fracture Strain	Fracture Stress/MPa	Secant Modulus/MPa
SJ-1	0.00308	0.1588	351.55
SJ-2	0.00384	0.13672	385.60
SJ-3	0.0023	0.13465	413.54
SJ-4	0.0023	0.10162	345.18
Average	0.00288	0.13295	373.97

As seen in Table 8, σ'_f of HMAC is 0.13295 MPa and ε'_f is 0.00288. Constant strain fatigue tests at different stress-strain levels were conducted on HMAC, and the cyclic $\sigma - \varepsilon$ curve at the strain level of 0.6 is shown in Figure 8.

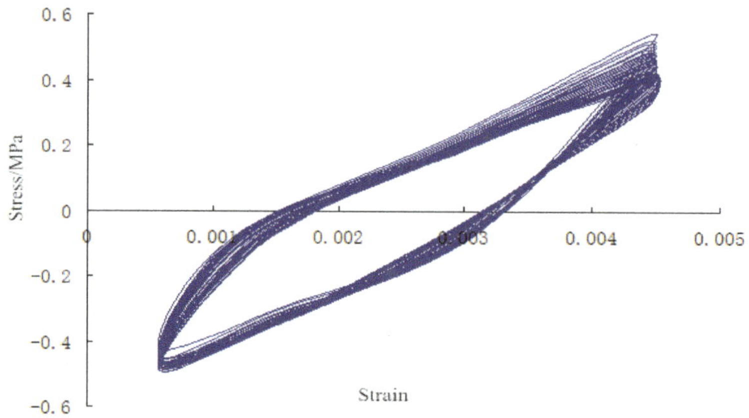

Figure 8. Cyclic $\sigma - \varepsilon$ curve at the strain level of 0.6.

The fatigue parameters strain and fatigue life of the specimen were then obtained, as shown in Table 9.

Table 9. Strain amplitude and fatigue life of HMAC in equal strain fatigue test.

Specimen	**S1**	**S2**	**S3**	**S4**	**S5**	**S6**	**S7**	**S8**
HMAC								
ε_a	0.32	0.53	0.74	0.91	1.15	1.392	1.56	1.66
N_f	18242	15124	7692	4543	3692	2420	1502	1031
Matrix asphalt concretes								
ε_a	1.28	2.17	3.04	3.75	4.18	5.37	6.17	6.89
N_f	4862	3941	1920	1137	1972	1143	108	84

As seen in Table 9, four common fitting algorithms were used to fit the data from Equation (1). The relationship between the fitting parameters and the equation was studied, as seen in Table 10.

Table 10. Parameters b and c using different fitting algorithms.

Fitting Algorithms	***b***	***c***	**R^2**
Marquardt method	−0.10234	−0.10263	0.981687
Quasi-newton method	−0.10233	−0.10264	0.981688
Evolution algorithm	−0.10231	−0.10267	0.981681
Max inherit optimization	−0.10237	−0.10261	0.981687

As shown in Table 10, the fitting parameters (b, c) of the four fitting methods are nearly the same—$b = -0.1023$ and $c = -0.1026$. When σ'_f, ε'_f, b, and c are put into Equation (22), then Equation (22) changes into the following form which predicts the fatigue life of HMAC in the strain-controlled mode:

$$\varepsilon_a = \frac{0.13295}{E} \times \left(2N_f\right)^{-0.1023} + 0.00288 \times \left(2N_f\right)^{-0.1026} \tag{23}$$

7. Fatigue Life Prediction Based on Local Stress-Strain Method

The fatigue life of HMAC was predicted based on the modified Neuber equation. First, the load-time process was transformed into a nominal-time history. Second, local stress and strain of the notched specimen were transformed to nominal stress and strain using the cyclic $\sigma - \varepsilon$ curve and the modified Neuber equation. Fatigue damage was calculated according to the strain fatigue life curve, and the fatigue life was then obtained based on cumulative damage theory. Specific steps for fatigue life prediction based on the modified Neuber equation and local stress-strain method are as follows:

(1) Local strain and local stress in the plane strain state were solved using the results of the load strain and temperature strain, cyclic $\sigma - \varepsilon$ curve and Neuber hyperbolic equations.

$$\varepsilon = \frac{\sigma}{E} + \left(\frac{\sigma}{26.847}\right)^{1.3793} \tag{24}$$

$$\varepsilon\sigma = 0.92 \times 1.976^2 se \tag{25}$$

(2) The fatigue life N_f was determined on the basis of the $\varepsilon_a - N_f$ curve equation.

$$\varepsilon_a = \frac{0.13295}{E} \times \left(2N_f\right)^{-0.1023} + 0.00288 \times \left(2N_f\right)^{-0.1026} \tag{26}$$

8. Case Study

8.1. Structure and Parameters of Pavement

Based on the proposed fatigue life prediction method, a typical HMAC overlay pavement situated in a long steep road in Shandong province of China was selected and its fatigue life was predicted below.

The pavement structure and material parameters are listed in Table 11. Five axle loads (80 kN, 100 kN, 120 kN, 150 kN, and 180 kN) and four varied temperature range (5, 10, 15, 20 °C) were considered in the analysis of nominal stress and strain. For each grade, fatigue life was calculated.

Table 11. Structure and parameters of pavement.

Structure Layer	Material Type	Thickness/cm	Modulus/MPa	Poisson's Ratio
Surface	High Modulus Asphalt Concrete	4	1200	0.25
	High Modulus Asphalt Concrete	6	1200	0.25
Base course	Lean concrete	16~28	10,000~40,000	0.15
Bed course	Cement Stabilized Crushed Stone	20	1700	0.2
Subbase	Graded broken stone	16	250	0.35
Subgrade	Soil	——	40	0.35

8.2. Results

8.2.1. Nominal Strain

According to the traffic axle load and varied temperature range, nominal strains at the bottom of HMAC were calculated using the finite element software, see Tables 12 and 13.

Table 12. Load strain at the bottom of HMAC.

Axle Load/kN	80	100	120	150	180
Load nominal strain	0.2319	0.2634	0.3048	0.3527	0.4156
Load local strain	0.3492	0.3751	0.4163	0.4806	0.5614

Table 13. Temperature strain at the bottom of HMAC.

Varied Temperature Range (°C)	5	10	15	20
Temperature nominal strain	0.1934	0.2486	0.3261	0.4835
Temperature local strain	0.2762	0.3294	0.4351	0.6214

8.2.2. Local Strain

Based on the nominal strain shown in Tables 12 and 13, local strain was calculated using the Equations (24) and (25) (see Tables 12 and 13).

8.2.3. The Analysis of Fatigue Damage

Based on linear cumulative damage hypothesis, damage caused by load and the maximal varied temperature range (20 °C) was calculated, as shown in Table 14. In this table, statistics of traffic loading times were collected from a traffic survey, loading times were calculated using Equation (26), and fatigue damage was the ratio of statistics of traffic loading times and calculated loading times.

Table 14. The analysis of fatigue damage.

Axle Load/kN	80	100	120	150	180
Local strain	1.2892	1.4009	1.4432	1.5057	1.5965
Calculated loading times	79,547	55,698	15,634	9846	5765
Statistics of traffic loading times	627	2568	1931	122	70
Fatigue damage	0.0079	0.0461	0.1235	0.0124	0.0121

8.2.4. Fatigue Life Prediction

As shown in Table 14, the cumulative fatigue damage could be calculated as 0.202, then the pavement fatigue life which is the reciprocal of cumulative fatigue damage was calculated as 4.95, meaning the HMAC overlay might generate reflective cracking in nearly 5 years. Moreover, our in situ survey also showed that a part of the pavement surface cracked almost 4 years after overlaying. This indicates that the proposed method to predict service life of pavement agreed with practical conditions. Furthermore, the current research suggested that a stress absorbing layer should be set before HMAC overlaid between HMAC and concrete pavement.

9. Conclusions

This study provides an experimental investigation into the fatigue performance of high modulus asphalt concretes, including the cyclic $\sigma - \varepsilon$ test and direct tensile fatigue test. A method to predict the fatigue crack initiation life of HMAC was proposed based on the modified Neuber equation and local $\sigma - \varepsilon$ method. The following conclusions can be drawn based on this work:

(1) The cyclic $\sigma - \varepsilon$ curve and modified $\sigma - \varepsilon$ curve of HMAC in plane strain state were established.
(2) Conversion from nominal stress and strain to local strain-stress is proposed based on a modified Neuber equation and the local stress-strain method.
(3) The strain fatigue equation for HMAC is presented based on fatigue tests.
(4) The fatigue life model is able to produce satisfactory crack initiation life predictions using the local stress-strain method based on fatigue test analysis of the specimens and the modified Neuber equation.

Acknowledgments: The research was sponsored by the National Natural Science Foundation (Grant No. 51008033), the Fundamental Research Funds for the Central Universities (Grant No. 310821163502), the Transportation Department of Shandong Province (Grant No. 2008Y007), the Transportation Department of Hainan Province (Grant No. 201000005), and the Transportation Department of Hubei Province of China (Grant No. Ejiaokejiao [2013] 731).

Author Contributions: Professor Mulian Zheng was in charge of the whole research plan and paper submission matters. The first author extends gratitude to Peng Li, Jiangang Yang and Zhengliang Zhang for performing the experiment and analyzing test data, to Hongyin Li and Yangyang Qiu for offering help in doing the experiment.

List of Symbols

ε_a = strain, $\Delta\varepsilon$ = cyclic strain amplitude, $\Delta\varepsilon_e$ = circular elastic, $\Delta\varepsilon_p$ = plastic strain amplitude, σ'_f = fatigue strength coefficient, ε'_f = ductility coefficient, N_f = fatigue life, b, c = regression constants, ε_{ea} = elastic strain, ε_{pa} = plastic strain, K_t = theoretical stress concentration factor of notch, K_ε = strain concentration factors, K_σ = factor of stress concentration, e = nominal strain, S = nominal stress, K_f = fatigue notch factor, M = plastic penetrable degree, g = coefficient of equivalent stress and strain, μ' = elastic-plastic Poisson's ration, considering the compressibility of material, μ = Poisson's ration, E = elasticity modulus, E_s = secant modulus, determined by the monotone tensile curve.

References

1. Forough, S.A.; Nejad, F.M.; Khodaii, A. Energy-based approach to predict thermal fatigue life of asphalt mixes using modified uniaxial test setup. *J. Mater. Civ. Eng.* **2016**, *28*. [CrossRef]

2. Zheng, M.L. Simulation of Permanent Deformation in High-Modulus Asphalt Pavement Using the Bailey-Norton Creep Law. *J. Mater. Civ. Eng.* **2016**. [CrossRef]
3. Monismith, C.L.; Epps, J.A.; Kasianchuk, D.A.; McLean, D.B. *Asphalt Mixture Behavior in Repeated Flexure*; Report TE 70-5; University of California: Berkeley, CA, USA, 1971.
4. Roque, R.; Ruth, B.E.; Dickison, S.W.; Reid, B. *Evaluation of SHRP Indirect Tension Tester to Mitigate Cracking in Asphalt Concrete Pavements and Overlays*; Final Report FDOT B-9885; University of Florida: Gainesville, FL, USA, 1997.
5. Yoo, P.J.; Al-Qadi, I.L. A strain-controlled hot-mix asphalt fatigue model considering low and high cycles. *Int. J. Pavement Eng.* **2010**, *11*, 565–574. [CrossRef]
6. Ameri, M.; Nowbakht, S.; Molayem, M. A study on fatigue modeling of hot mix asphalt mixtures based on the viscoelastic continuum damage properties of asphalt binder. *Constr. Build. Mater.* **2016**, *106*, 243–252. [CrossRef]
7. Li, Y.Q.; Metcalf, J.B. Crack Initiation model from asphalt slab tests. *J. Mater. Civ. Eng.* **2002**, *14*, 303–310. [CrossRef]
8. Seung, K.J.; Seung, K.C. Development of a predictive system for estimating fatigue life of asphalt mixtures using the indirect tensile test. *J. Transp. Eng.* **2012**, *138*, 1530–1540.
9. Fallah, S.; Khodaii, A. Developing a fatigue fracture model for asphalt overlay reinforced with geogrid. *Mater. Struct.* **2015**. [CrossRef]
10. Ker, H.W.; Lee, Y.H.; Wu, P.H. Development of fatigue cracking prediction models using long-term pavement performance database. *J. Transp. Eng.* **2008**, *134*, 477–482. [CrossRef]
11. Mohammad, S.; Kim, Y.R. *Development of a Failure Criterion for Asphalt Mixtures under Different Modes of Fatigue Loading*; Transportation Research Board of the National Academies: Washington, DC, USA, 2014; pp. 117–125.
12. Benedetto, H.D.; de la Roche, C.; Strom, R.L. Fatigue of bituminous mixtures. *Mater. Struct.* **2004**, *37*, 202–216. [CrossRef]
13. Karlaftis, A.G.; Badr, A. Predicting asphalt pavement crack initiation following rehabilitation treatments. *Transp. Res. C* **2015**, *55*, 510–517. [CrossRef]
14. Molenaar, A.A.A. Structure Performance and Design of Flexible Pavements and Asphalt Concrete Overlays. Ph.D. Thesis, Delft University of Technology, Delft, The Netherlands, 1983.
15. AASHTO T 321-07. *Determining the Fatigue Life of Compacted Hot Mix Asphalt (HMA) Subjected to Repeated Flexural Bending*; American Association of State and Highway Transportation Officials: Washington, DC, USA, 2007.
16. Shen, S.; Carpenter, S.H. Development of an asphalt fatigue model based on energy principles. *J. Assoc. Asphalt Paving Technol.* **2007**, *77*, 525–573.
17. Graham, J.A. *Fatigue Design Handbook*; Society of Automotive Engineers: New York, NY, USA, 1968.
18. Edward, T.H. *Performance Prediction Models in the Superpave Mix Design System*; Arizona State University: Tempe, AZ, USA, 2006.
19. Kim, T.W.; Baek, J.; Lee, H.J.; Choi, J.Y. Fatigue performance evaluation of SBS modified mastic asphalt mixtures. *Constr. Build. Mater.* **2013**, *48*, 908–916. [CrossRef]
20. De Jcorreia, A.M.P.; Correia, J.A.F.O. Critical Assessment of a Local Strain-Based Fatigue Crack Growth Model Using Experimental Data Available for the P355NL1 Steel. *J. Pressure Vessel Technol.* **2013**, *135*, 011404.
21. Hu, X.; Hu, S.; Walubita, L.F.; Sun, L. Investigation of fatigue cracking: Bottom-up or top-down. In *Pavement Cracking: Mechanisms, Modeling, Detection, Testing and Case Histories*; CRC Press: Leiden, The Netherlands, 2008; pp. 333–344.
22. Hofko, B. Combining performance based lab tests and finite element modeling to predict life-time of bituminous bound pavements. *Constr. Build. Mater.* **2015**, *89*, 60–66. [CrossRef]
23. Castro, M.; Sanchez, J. Estimation of asphalt concrete fatigue curves—A damage theory approach. *Constr. Build. Mater.* **2008**, *22*, 1232–1238. [CrossRef]
24. Khateeb, G.A.; Shenoy, A. A distinctive fatigue failure criterion. *J. Assoc. Asphalt Paving Technol.* **2004**, *73*, 585–622.
25. Topper, T.H.; Wetzel, R.M.; Morrow, J.D. Neubers rule applied to fatigue of notched specimens. *J. Met.* **1969**, *4*, 22–24.

26. Lee, H.J.; Lee, J.H.; Park, H.M. Performance evaluation of high modulus asphalt mixtures for long life asphalt pavements. *Constr. Build. Mater.* **2007**, *21*, 1079–1087. [CrossRef]
27. Shu, X.; Huang, B.S. Laboratory evaluation of fatigue characteristics of recycled asphalt mixture. *Constr. Build. Mater.* **2008**, *22*, 1323–1330. [CrossRef]
28. Schapery, R.A. A theory of crack initiation and growth in visco-elastic media; I: Theoretical development, II: Approximate methods of analysis, III: Analysis of continuous growth. *Int. J. Fract.* **1975**, *11*, 141–159, 369–388 and 549–562. [CrossRef]
29. Doh, Y.S.; Baek, S.H.; Kim, K.W. Estimation of relative performance of reinforced overlaid asphalt concretes against reflection cracking due to bending more fracture. *J. Constr. Build. Mater.* **2009**, *23*, 1803–1807. [CrossRef]
30. Zappalorto, M.; Lazzarin, P. Some remarks on the Neuber rule applied to a control volume surrounding sharp and blunt notch tips. *Fatigue Fract. Eng. Mater. Struct* **2014**, *37*, 349–358. [CrossRef]
31. Micaeloa, R.; Pereiraa, A.; Quaresman, L.; Cidade, M.T. Fatigue resistance of asphalt binders: Assessment of the analysis methods in strain-controlled tests. *Constr. Build. Mater.* **2015**, *15*, 703–712. [CrossRef]
32. Alireza, K.K.; Mahmoud, A. Laboratory evaluation of strain controlled fatigue criteria in hot mix asphalt. *Constr. Build. Mater.* **2013**, *47*, 1497–1502.
33. Salour, F.; Erlingsson, S. *Pavement Structural Behavior during Spring Thaw*; VTI Report 738 A; Swedish National Road and Transport Research Institute: Linköping, Sweden, 2012.
34. Jaeklin, E.P.; Scherer, J. Asphalt reinforcing using glass fiber grid 'Glasphalt'. In Proceedings of the 3rd International RILEM Conference on Reflective Cracking in Pavement: design and performance of overlay systems, Maastricht, The Netherlands, 2–4 October 1996.

15

Research on Performance of a Dense Graded Ultra-Thin Wearing Course Mixture

Lei Geng [1], Tao Ma [2,*], Junhui Zhang [3], Xiaoming Huang [2] and Pengsen Hu [2]

1 Jiangsu Sinoroad Engineering Research Institute Co. LTD., 19 Lanhua Road, Nanjing 211800, China; gl@sinoroad.com
2 School of Transportation, Southeast University, 2 Sipailou, Nanjing 210096, China; huangxm@seu.edu.cn (X.H.); 220163555@seu.edu.cn (P.H.)
3 State Engineering Laboratory of Highway Maintenance Technology, Changsha University of Science and Technology, Changsha 410114, China; zjhseu@csust.edu.cn
* Correspondence: matao@seu.edu.cn

Abstract: This paper focused on the design and performance characterization of a modified ultra-thin wearing course mixture (M-UWM). A dense graded ultra-thin wearing course mixture with nominal maximum aggregate size of 10 mm was designed and named as UWM10. A multi-chain polyolefin modifier was used to modify the performance of UWM10 to get M-UWM10. Based on different laboratory performance tests including wheel tracking tests, low-temperature bending beam tests, immersion Marshall tests and freeze-thaw splitting tests, the high temperature rutting resistance, low-temperature cracking resistance and moisture resistance of the designed M-UWM10 were evaluated. The texture depth tests and wheel tracking tests were combined to characterize the degradation behaviour of the surface texture depth of M-UWM10. Based on test roads, the bonding conditions between the wearing course layer that consisted of M-UWM10 and its sublayer were evaluated by computed tomography (CT) scanning test and pull out test. Filed texture depth tests were also conducted on the test roads. It is proved that the designed wearing course mixture M-UWM10 shows excellent pavement performance as well as better wearing resistance and interlayer bonding than the traditional wearing course mixture.

Keywords: ultra-thin wearing course; road performance; texture depth; interlayer bonding

1. Introduction

An ultra-thin paving technique, usually with 1.5 to 2.5 centimetres pavement thickness, namely Novachip, originated from France in the late 1970s [1]. It has several advantages like short construction time, good anti-skid performance and good water permeability [1–4]. These distinctive features have been gradually gaining popularity, as ultra-thin wearing layers are being frequently applied in old highway maintenance as well as serving as wearing layers in new constructed highways [2,3,5,6]. Apart from the excellent functional performance, the ultra-thin paving technique is also cost-effective, environmental-friendly and a sustainable construction method [7–9], which is attracting more and more attention from researchers.

Kandhal and Lockett [8] examined the road performance of two Novachip projects in Alabama, USA. After 4.5 years of service, both two Novachip projects showed no significant ravelling, concluding that the cohesion between the ultra-thin wearing layer and underlying asphalt layer is good. However, the ultra-thin wearing layer presents more surface friction than the hot mixture asphalt (HMA) wearing layer.

In order to identify appropriate evaluation indexes for ultra-thin wearing layers, Tan, Yao, Wang, Bian and Yu-Xiang [5] conducted several experiments (thermal stress restrained sample tests, permeability tests, skid resistance tests, and indoor abrasion tests) on asphalt mixture with three different surface layers—stone mastic asphalt mixture with a nominal maximum aggregate size of 10 mm (SMA10), ultra-thin asphalt concrete with a nominal maximum aggregate size of 10 mm (UTAC10) and Novachip Type C. The result shows that frozen broken temperature and frozen broken strength could be regarded as appropriate evaluation indexes for low temperature performance of ultra-thin surface layers. Moreover, sliding attenuation degree and initial British Pendulum Number (BPN) value are proposed to measure the anti-skid performance of the ultra-thin surface layer. Furthermore, the author indicates that the freeze-thaw splitting strength ratio and freeze-thaw splitting strength should be combined to evaluate resistance of water damage performance.

Zeng et al. [10] studied the anti-skid performance, permeability and texture depth of three different asphalt mixtures—dense-graded asphalt mixture with a nominal maximum aggregate size of 13.2 mm (AC13), open graded friction course with a nominal maximum aggregate size of 13.2 mm (OGFC13) and semi-open gradation Novachip Type C. Results show that Novachip Type C as an ultra-thin wearing layer has better overall performance in terms of skid resistance, permeability and texture depth. Furthermore, Novachip Type C was proven to have better shear capability than other types of asphalt mixture, under the same tack coat material and dosage. Moreover, Yang, et al. [11] investigated the moisture stability, high temperature stability and low temperature cracking resistance of the same three types of asphalt mixture (AC-13, OGFC-13 and semi-open gradation Novachip Type C). The result shows that Novachip Type C has better high temperature stability than the other two types of asphalt mixture. However, the low temperature cracking resistance and moisture stability performance of Novachip Type C is worse than the AC-13 asphalt mixture but better than the OGFC-13 asphalt mixture.

Yang, Shen and Gao [4] argue that the current design specification in China [12] does not take temperature into consideration and the interlayer contact is presumed as completely continuous. The actual environmental condition subjected to pavement structure is not sufficiently considered in the specification. Thus, the author conducted an experiment to investigate the shear stress between the ultra-thin wearing layer and the underlying asphalt layer under different temperatures and interlayer contact situations. The result shows that maximum interlayer shear stress is negatively correlated with temperature and interlayer sliding coefficient.

The cohesiveness between the thin surface layer and underlying asphalt layer is proven to be an important factor that influences the performance and durability of the thin surface layer [13]. Thus, Wu [14] proposed a modified pull test method to test cohesiveness between the ultra-thin wearing layer and underlying asphalt layer. Three types of ultra-thin wearing layers (Epoxy modified bitumen, Resin and Cement) were selected. The experiment was conducted under three different temperatures (0, 10 and 20 °C) with a loading speed of 0.025 MPa/s. The result demonstrates the feasibility of the proposed modified test method, provided that the interface of the two layers is the weakest area. Moreover, the author also demonstrated that the Epoxy modified bitumen-based surface layer has better cohesiveness than the other two types of layers.

Pavement performance decreases over time, and thus needs to be rehabilitated once the existing pavement is unable to satisfy the traffic demand. Several researchers also focused on ultra-thin wearing layer recycling since high quality materials are used in the ultra-thin wearing layer. Both environment protection and cost reduction could be achieved by recycling high-quality used materials. Rahaman et al. [15] conducted research to investigate if the reclaimed ultra-thin wearing layer material could be used in the chip seal or Superpave mixtures. Sweep test based on the American Society for Testing and Materials (ASTM) D7000-04 was conducted to observe the chip retention of ultra-thin wearing layer millings. Different percentages (0%, 10%, and 20%) of ultra-thin wearing layer millings were added to the Superpave mixture. Rutting, stripping and moisture sensitivity tests were conducted to evaluate the performance of the Superpave mixture. The sweep test indicates that reclaimed ultra-thin wearing layer millings do not contribute to chip retention. However, an ultra-thin wearing layer milling addition was proven to have a positive effect on the performance of Superpave mixture.

The ultra-thin paving technique has been extensively used in China [16], since its first adoption in the 1990s [17]. However, several drawbacks of the ultra-thin paving technique occurred in engineering practices such as low degree of compaction, moisture damage and rapid attenuation of sliding resistance at a later stage. As to open-graded ultra-thin paving material, the situation could be even worse. Because of the high void ratio of open-graded ultra-thin paving material, water could easily permeate into the pavement structure. It would lower the bonding strength between pavement layers, causing raveling and reducing the durability of ultra-thin paving materials.

In view of the existing problems of the current ultra-thin technique, a dense-graded ultra-thin wearing mixture was designed and then further modified by a multi-chain polyolefin modifier to get the modified ultra-thin wearing mixture in this study. Both laboratory and field tests were conducted to fully evaluate the road performances, wearing resistance and bonding conditions with sub-layers of the modified ultra-thin wearing mixture.

2. Experimental

2.1. Mix Design for Different Asphalt Mixtures

Styrene-butadiene-styrene (SBS) modified asphalt with penetration grade of PG-70, basalt aggregate and limestone powders were used for all of the asphalt mixtures prepared in this study. According to the recommendation by Chinese specification and previous studies, the polyester fibre was used for SMA mixtures. To improve the rutting resistance and the anti-wearing performance of ultra-thin wearing course mixture (UWM), a multi-chain polyolefin modifier was added to the SBS modified asphalt to prepare modified ultrathin wearing course mixtures (M-UWM).

Five different asphalt mixtures were studied including an SMA mixture with nominal maximum aggregate size of 13.2 mm (SMA13), dense-graded asphalt mixture with nominal maximum aggregate size of 13.2 mm (AC13), SMA mixture with nominal maximum aggregate size of 9.5 mm (SMA10), ultrathin wearing course mixture with nominal maximum aggregate size of 9.5 mm (UWM10), and a modified ultrathin wearing course mixture with a nominal maximum aggregate size of 9.5 mm (M-UWM10). The SMA13 and AC13 are commonly used in the pavement surface layer while SMA10, UWM10 and M-UWM10 are mainly used in the wearing course above the pavement surface layer [18]. While the SMA10 is a traditional wearing course mixture, the UWM10 and M-UWM10 are newly developed wearing course mixtures. Based on Marshall mix design, the gradations for different asphalt mixtures are shown in Figure 1 and the Marshall design parameters for different asphalt mixtures are listed in Table 1.

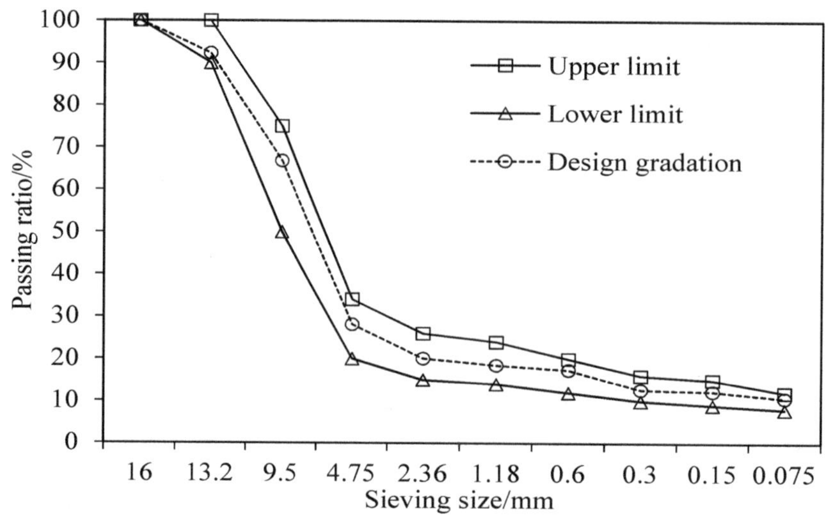

(a)

Figure 1. *Cont.*

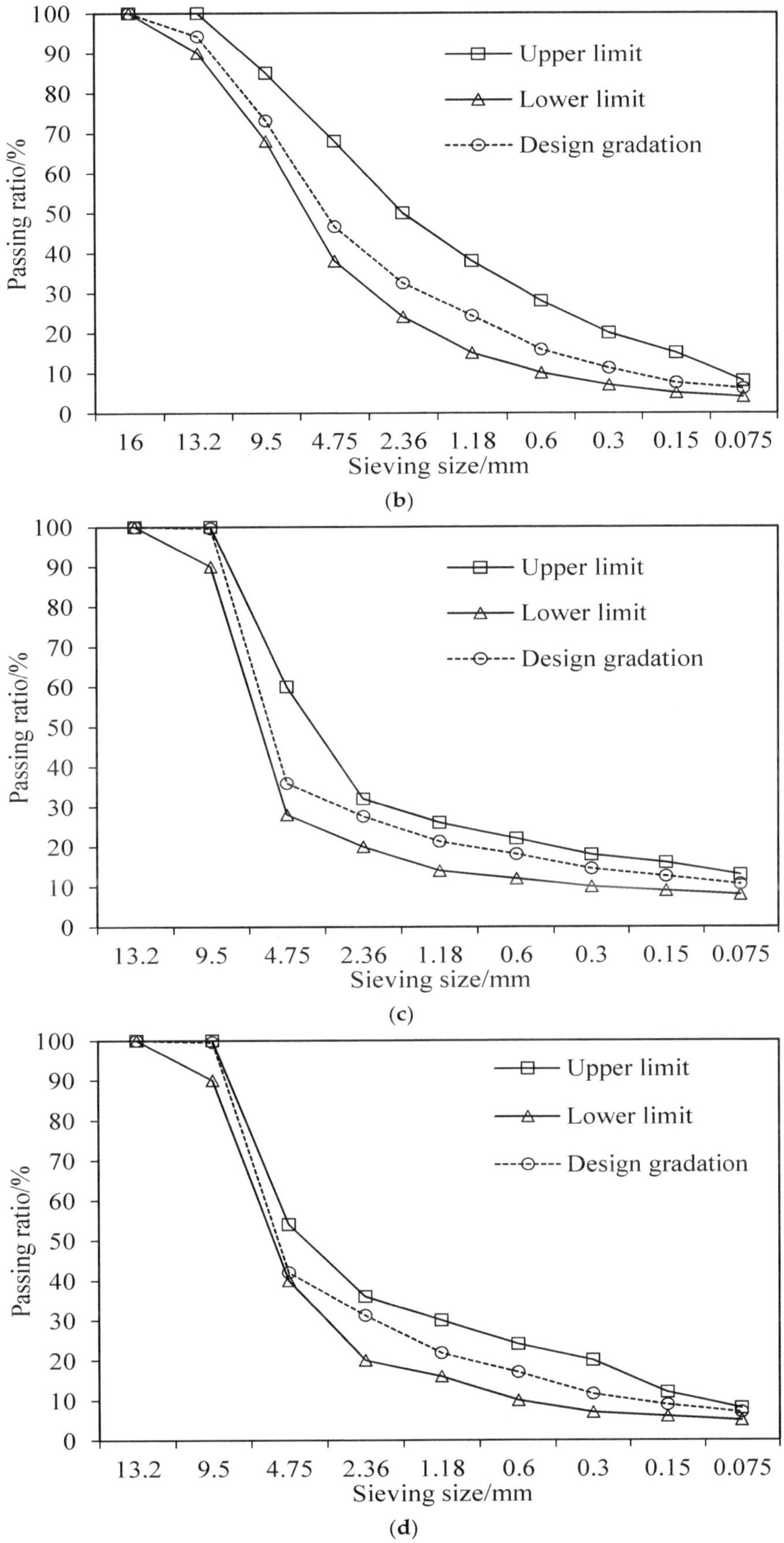

Figure 1. The designed gradation for different asphalt mixtures: (**a**) SMA13; (**b**) AC13; (**c**) SMA10; (**d**) UWM10 and M-UWM10.

Table 1. Marshall design parameters for different asphalt mixtures.

Mixture Type	Asphalt Content (%)	VV (%)	VMA (%)	VFA (%)	Marshall Stability (kN)	Flow Value (0.1 mm)
SMA13	6.0	4.3	18.28	76.7	8.9	26.8
AC13	5.1	4.8	19.90	75.9	8.52	27.4
SMA10	6.3	4.2	18.53	77.5	10.54	28.9
UWM10	5.2	5.1	16.0	68.0	9.83	26
M-UWM10	5.2	4.7	15.6	70.1	12.00	24

VV: percent air voids in bituminous mixtures; VMA: percent voids in mineral aggregate in bituminous mixtures; VFA: percent voids in mineral aggregate that are filled with asphalt in bituminous mixtures.

Based on the designed asphalt mixtures, two test roads were paved. One was paved with AC13 used in the surface layer and M-UWM10 used in the wearing course above the AC13 layer. The other was paved with SMA13 used in the surface layer and SMA10 used in the wearing course above the SMA13 layer.

2.2. Test Procedures

2.2.1. Road Performance Evaluation

Since the designed ultra-thin wearing course mixtures are mainly used in the top surface layer or the wearing course layer of pavement to improve the surface functions instead of structure capacity of asphalt pavement, they should have good road performance, especially excellent rutting resistance and water stability to bear the environment conditioning and wheel loading. Thus, based on the standard test specifications of China, the road performances including permeability, high-temperature rutting resistance, low-temperature cracking resistance, moisture resistance were evaluated by permeability tests, wheel tracking tests, low-temperature bending beam tests, immersion Marshall tests and freeze-thaw splitting tests [19,20]. The wheel tracking test was conducted at 60 °C to get the dynamic stability to describe and compare the high-temperature rutting resistance of different asphalt mixtures. The low-temperature bending beam test was conducted at −10 °C to get the failure strain to describe and compare the low-temperature cracking resistance of different asphalt mixtures. Both immersion Marshall tests and freeze-thaw splitting tests were conducted to get the Marshall strength ratio and tensile strength ratio, separately, in order to describe and compare the moisture resistance of different asphalt mixtures.

2.2.2. Texture Depth Evaluation

To guarantee the driving safety, high-quality skidding resistance is an important characteristic for asphalt mixtures used in the top surface layer or the wearing course layer of pavement. Thus, the surface texture depth (TD) of asphalt mixture and pavement is usually used to describe the skidding resistance. However, the texture depth (TD) test that followed the standard protocol of Chinese specification cannot reveal the wearing resistance of asphalt mixtures, which is another important characteristic for asphalt mixtures used in the top surface layer or the wearing course layer of pavement. Thus, to reveal the degradation behaviour of surface texture of different asphalt mixtures that can represent their wearing resistance, the texture depth test was combined with the wheel tracking test in this study. The designed test processes were summarized as follows:

1. After the rectangular specimen with dimensions of 300 × 300 × 50 mm was prepared for the wheel tracking test, its original texture depth (TD) named K_0 was measured, as shown in Figure 2a.
2. After the silica sands on the specimen surface were washed away, the specimen was submitted to the wheel tracking test for 180 minutes.
3. After the wheel tracking test, the rutting area right underneath the wheel loading positions was cut from the tested specimen and then weighed to get its mass m_1, as shown in Figure 2b.

4. According to the texture depth test, sands were laid on the surface of the cut strip specimen, as shown in Figure 2c, and then the total mass m^2 for the sands and strip specimen was weighed, as shown in Figure 2d.
5. The texture depth K_1 for the cut strip specimen, which represents the rutting area under the wheel loading positions after the wheel tracking test, was determined by Equation (1):

$$K_1 = \frac{(m_2 - m_1)}{\rho_G \cdot s}, \quad (1)$$

where s is the top surface area of the cut strip specimen, and ρ_G is the density of the silica sands.
6. By comparing K_0 and K_1, the degradation degree of the surface texture depth caused by wheel loading can be obtained.

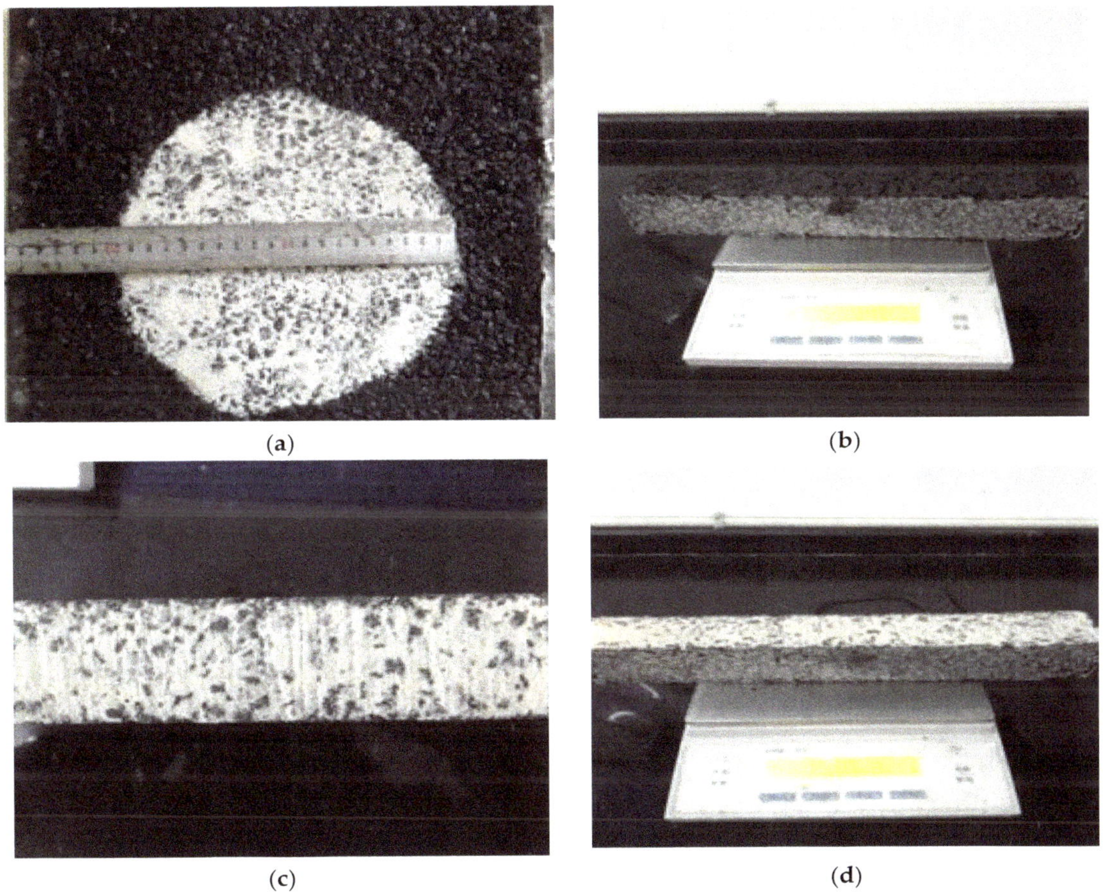

Figure 2. Test processes for combined texture depth test and wheel loading test: (**a**) measuring K_0 for the specimen of wheel loading test; (**b**) measuring m_1 for the cut strip specimen; (**c**) laying sands on the top surface of the strip specimen; (**d**) measuring m_2 for the strip specimen with sands.

2.2.3. Interlay Bonding Evaluation

The interlayer bonding between the wearing course layer and its sublayer has important influences on its long-term durability. Thus, both the CT scanning test and the pull-out test were used to evaluate the interlayer bonding conditions.

The CT scanning test was conducted to evaluate the air voids within the wearing course mixture and between the wearing course layer and its sublayer. The basic principle for the 3D CT scanning technique was explained in Figure 3a and the basic parameters were shown in Table 2. Firstly, the specimen was scanned by transmitted X-rays from multiple directions, and then attenuated X-rays after transmission were gathered by the detector to rebuild 2D or 3D grey level images. The scanned specimen was a standard cylindrical sample cored from pavement and was cut flat at each side by the cutting machine. The test conditions were: voltage 200 kV, current 0.43 mA, and integration time of 300 ms.

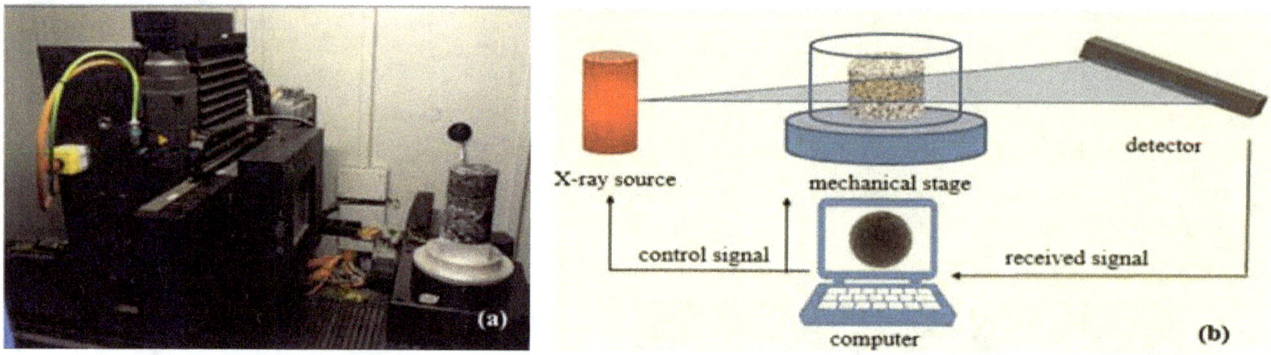

Figure 3. The basic principle and equipment for a computed tomography (CT) scanning test.

Table 2. Basic parameters of the computed tomography (CT) technique.

Type	x-axis (mm)	y-axis (mm)	z-axis (mm)	3D-xy-Pixel Size, 3D-z-Pixel Size (mm)
Specification	200	420	630	0.113

As shown in Figure 4, the pull-out tests were conducted on the field pavements with M-UWM10 wearing course and SMA10 wearing course. The bonding strength between the wearing course layer and its sublayer can be measured.

(a) (b)

Figure 4. Field pull out test.

3. Result and Discussions

3.1. Road Performance Evaluation

Road performance evaluation were conducted for all of the five designed asphalt mixtures including M-UWM10, UWM10, SMA10, SMA13 and AC13. The test results of dynamic stability from the wheel tracking test, failure strain from the low-temperature bending beam test, the Marshall stability ratio from the immersion Marshall test and the tensile strength ratio from a freeze-thaw splitting test were shown in Figure 5, respectively. The requirements for a different road performance index from the Chinese specifications [21,22] were shown in Table 3.

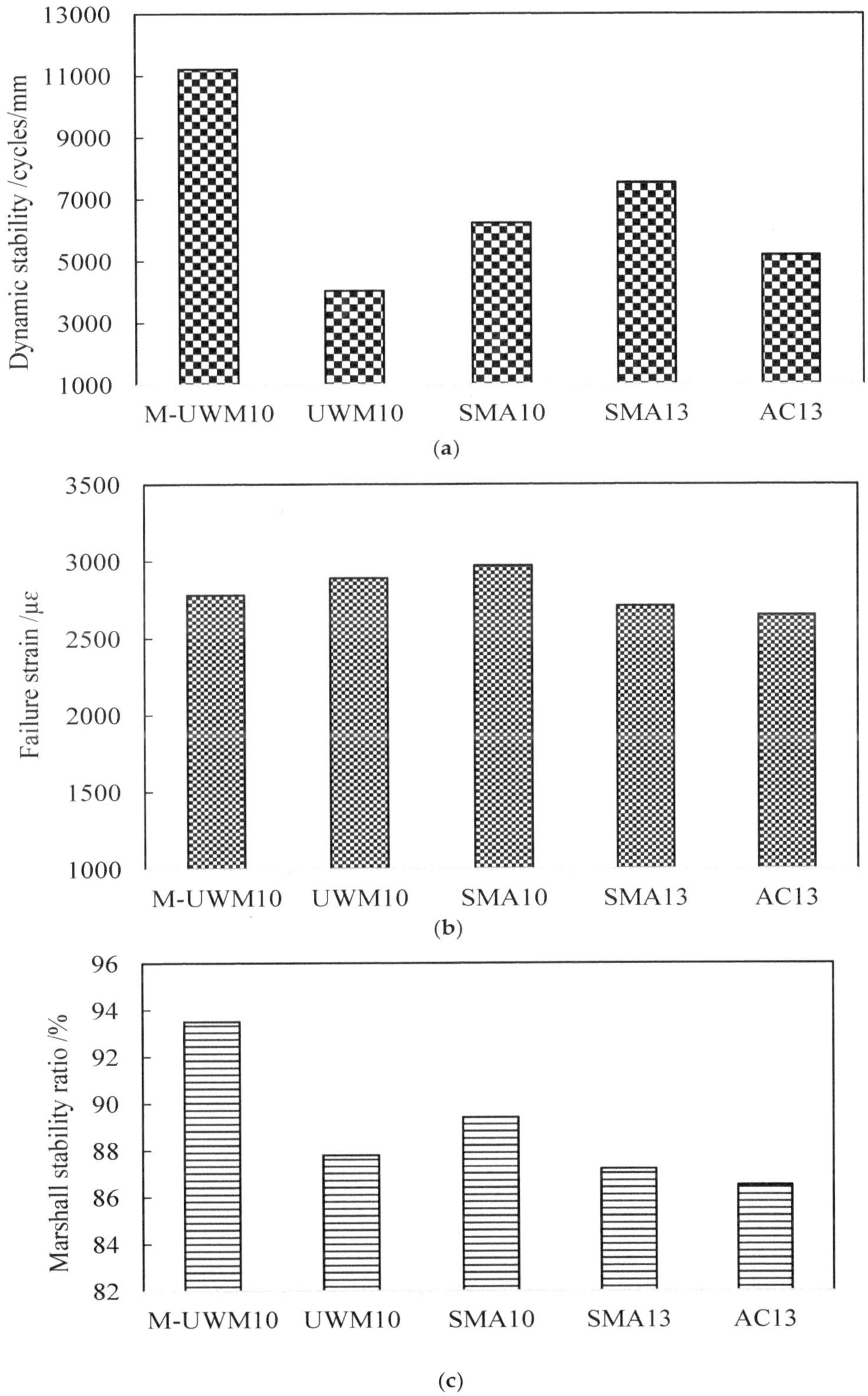

Figure 5. *Cont.*

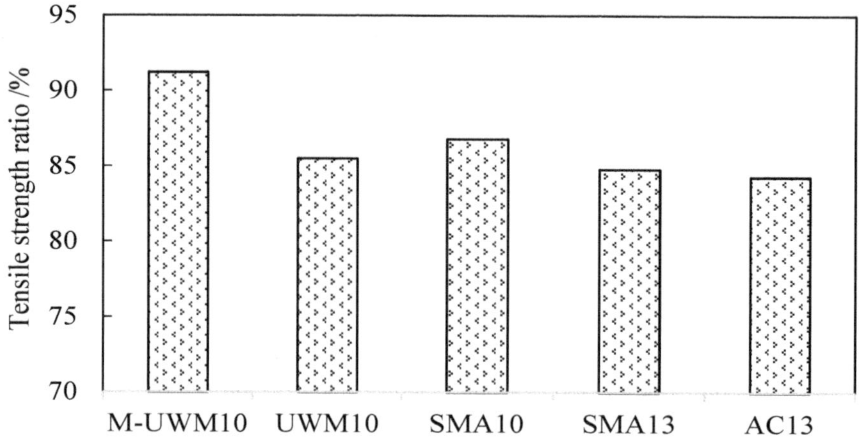

(d)

Figure 5. Test results for road performance evaluation: (**a**) dynamic stability; (**b**) failure strain; (**c**) Marshall stability ratio; (**d**) tensile strength ratio.

Table 3. Pavement performance verification for the M-UWM10 mixture

Performance Index	Dynamic Stability (cycles/mm)	Failure Strain (με)	Marshall Stability Ratio (%)	Tensile Strength Ratio (%)
Requirements	≥3000	≥2500	≥85	≥80
Test method	T0719-2011	T0715-2011	T0709-2011	T0729-2000

First of all, from Figure 5 and Table 3, it can be seen that all of the designed asphalt mixtures can well meet the basic specification requirements. Based on Figure 5a, it can be seen that the M-UWM10 has much higher dynamic stability than the other asphalt mixtures. It proves that the multi-chain polyolefin modifier can well improve the high-temperature rutting resistance of the wearing course. From Figure 5b, it can be seen that the M-UWM10 has a similar failure strain to the other asphalt mixtures. It indicates that the multi-chain polyolefin modifier shows few influences on the low-temperature cracking resistance of the wearing course. From Figure 5c,d, it is clearly seen that the M-UWM10 has the highest Marshall stability ratio and tensile strength ratio. It means that the multi-chain polyolefin modifier can well improve the moisture resistance of the wearing course. Thus, the M-UWM10 has a better high-temperature rutting resistance and moisture resistance than other asphalt mixtures, which are beneficial for the wearing course to bear heavy traffic loading.

3.2. Texture Depth Evaluation

According to the previous test plan, all of the designed asphalt mixtures including M-UWM10, UWM10, SMA10, SMA13, and AC13 were submitted to the combination of the texture depth test and the wheel tracking test. The measured texture depth for different asphalt mixtures before and after the wheel tracking test are shown in Figure 6a, and the loss ratios defined as the texture depth after the wheel tracking test compared to the texture depth before the wheel tracking test are shown in Figure 6b. The rutting depth, for all of the five asphalt mixtures after the wheel tracking test are also shown in Figure 6b, and the correlation between the reduced ratio of texture depth and the rutting depth after the wheel tracking test is shown in Figure 6c. As shown in Figure 7, field texture depth tests were also conducted for the two test roads. Table 4 shows the field measured values of the original texture depth and the texture depth after one year of traffic loading for the two test roads with the M-UWM10 wearing course and the SMA10 wearing course, respectively.

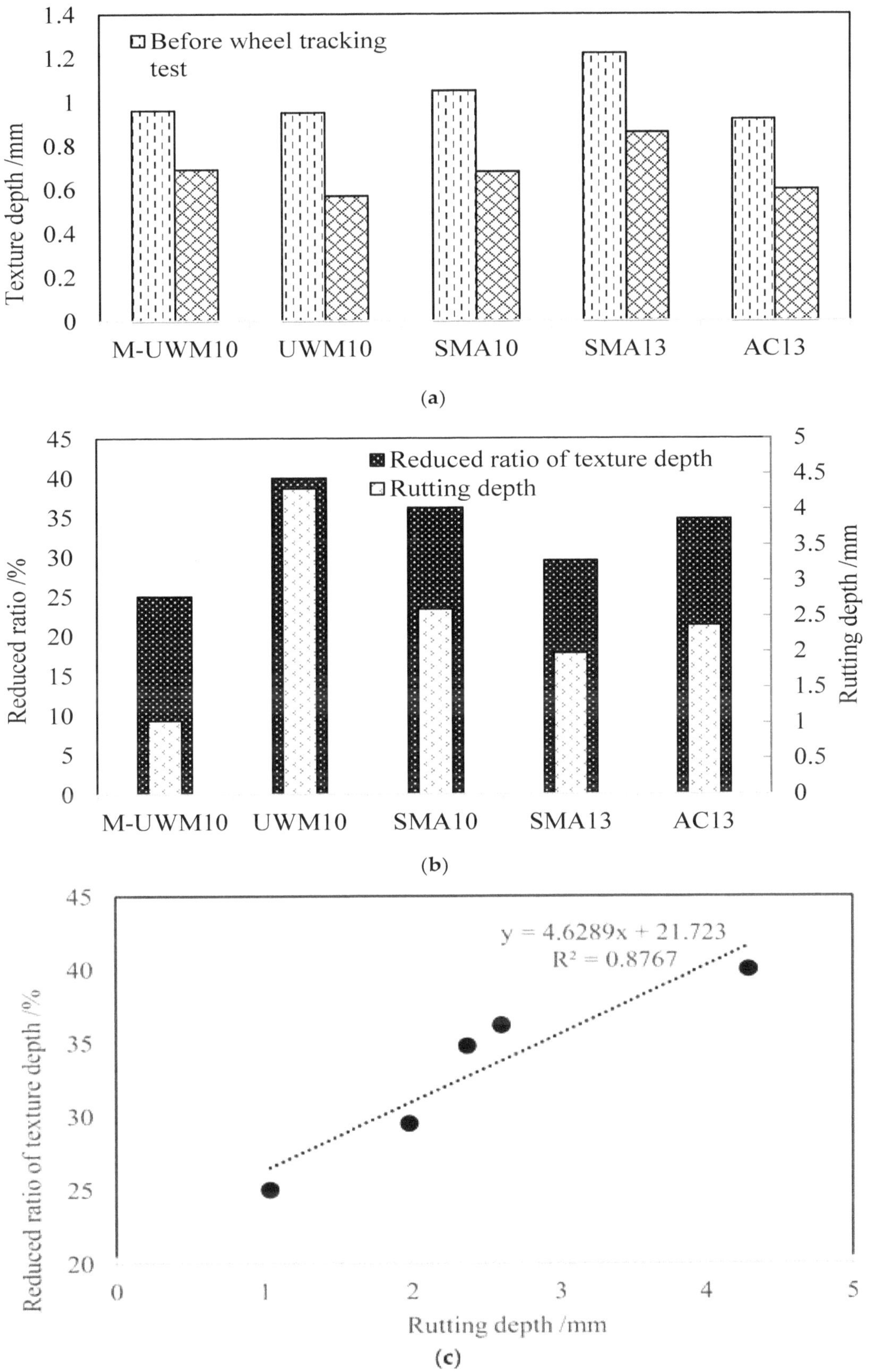

Figure 6. Test results from the combination of the texture depth test and the wheel tracking test: (**a**) texture depth before and after wheel tracking test; (**b**) loss ratio of texture depth after wheel loading; (**c**) correlation between the reduced ratio of texture depth and the rutting depth after the wheel tracking test.

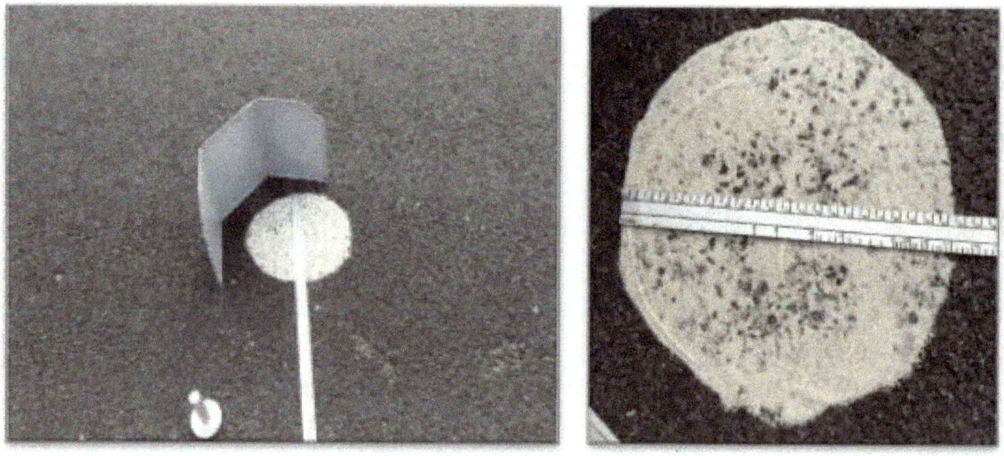

Figure 7. Field texture depth test.

Table 4. Field test results of texture depth test on different test roads.

Type of Wearing Course Mixtures	Original Texture Depth (mm)	Texture Depth after One Year (mm)	Reduced Ratio (%)
M-UWM10	0.85	0.78	8.2
SMA10	0.92	0.8	13.0

From Figure 6a, it can be seen that the SMA mixtures have higher texture depth than the other three asphalt mixtures, which have similar texture depth. It is mainly because the SMA mixtures use gap gradation while the other three asphalt mixtures use dense gradation. From the data after the wheel tracking test shown in Figure 6a, it can be seen that, due to its original high texture depth, SMA13 still have the highest texture depth, the M-UWM10 and SMA10 have similar texture depth, and the UWM10 and AC13 show a similar texture depth, which are the lowest. From Figure 6b, it can be further seen that the UWM10 has the biggest reduced ratio of texture depth, SMA10 and AC13 show a similar reduced ratio, which are lower than that of UMW10, while M-UWM10 and SMA13 have a similar reduced ratio, which are lower than the other three mixtures. Furthermore, the M-UWM10 shows a lower reduced ratio than SMA13.

From Figure 6c, it could be observed that the reduced ratio of texture depth increases with the growing of rutting depth. It indicates that the degradation behaviour of texture depth has a good correlation with the rutting resistance of asphalt mixture, and higher rutting resistance leads to less degradation of texture depth during the wheel tracking test.

The field test results confirm well with the laboratory test analysis. From Table 4, it can be seen that, although the SMA10 has a bigger texture depth than the M-UWM10 before traffic loading, after one year of traffic loading, the SMA10 shows a higher reduced ratio of texture depth than the M-UWM10, and the texture depth of the two mixtures becomes similar to each other.

Thus, the M- UWM10 has similar skidding resistance with other dense graded asphalt mixtures such as AC13 but better wearing resistance than other asphalt mixtures including both the dense graded asphalt mixture and the gap graded asphalt mixture.

3.3. Interlay Bonding Evaluation

Samples consisted of the wearing course mixture and the top surface layer mixture was cored from the two test roads and submitted to CT scanning tests. Examples of the cored samples and the 3D rebuilt images for the two wearing course mixtures including M-UWM10 and SMA10 based on the CT scanning were shown in Figure 8. The CT scanning results for the M-UWM10 sample and the SMA10 sample are shown in Figures 9 and 10, respectively.

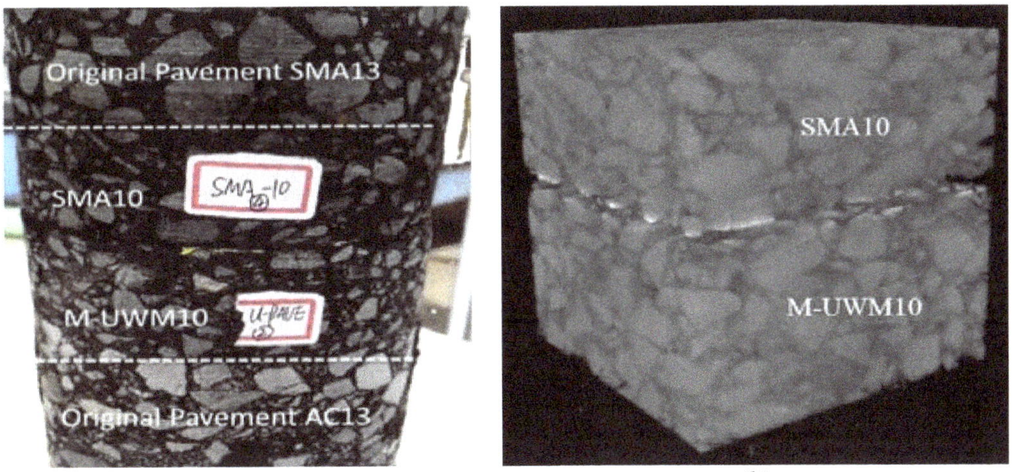

(a) (b)

Figure 8. Examples of the cored samples from test roads and the restructuring graphs of wearing course mixtures: (**a**) cored samples; (**b**) restructuring graphs for M-UWM10 and SMA10.

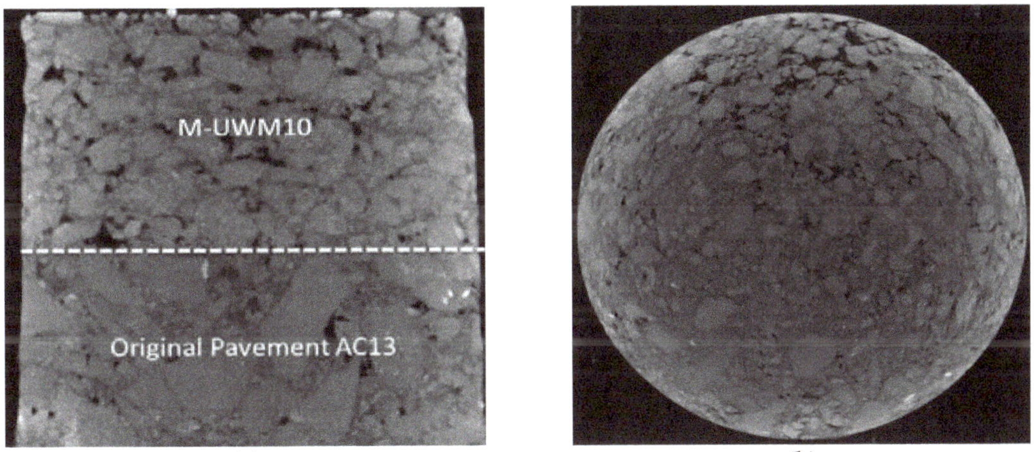

(a) (b)

Figure 9. The scanning images for cored sample from M-UWM10 pavement: (**a**) front view for cored sample; (**b**) the interface between wearing course with M-UWM10 and sub-layer with AC13.

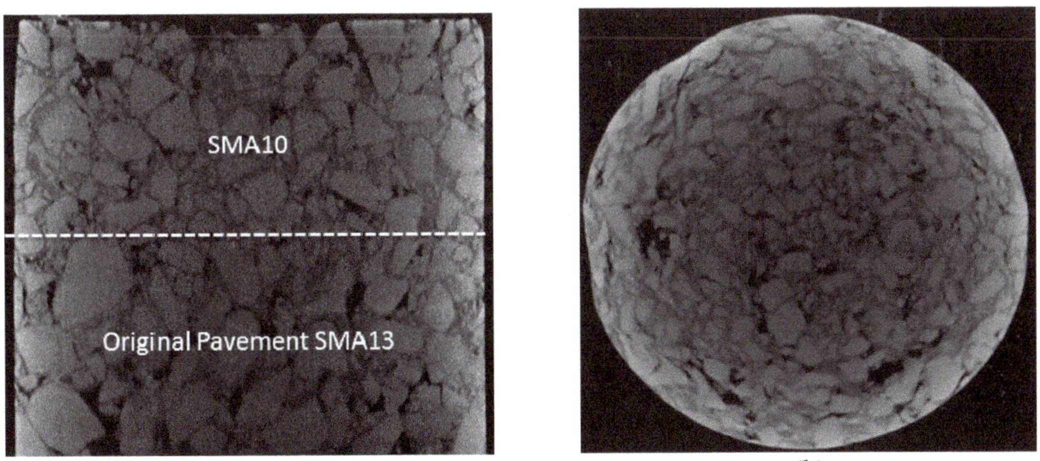

(a) (b)

Figure 10. The scanning images for the cored sample from SMA10 pavement: (**a**) front view for cored sample; (**b**) the interface between wearing course with SMA10 and sub-layer with SMA13.

Based on the CT scanning images of the cored samples, grey processing of the images was conducted and Origin software (version, Manufacturer, City, US State abbrev. if applicable, Country) was used to calculate the air void ratios. The grey level versus probability density distribution curve was depicted in Figure 11. Since the grey level value for air voids was usually less than 30, the air void ratio was calculated by integrating from 0 to 30. The distribution of the air void ratios along the scanning height of the cored samples for the two different wearing course mixtures are shown in Figure 12. The average air void ratio within the wearing course mixture and the average air void ration on the interface between the wearing course mixture and the sub-layer mixture were also calculated. The calculate results for the two different cored samples were shown in Figure 13.

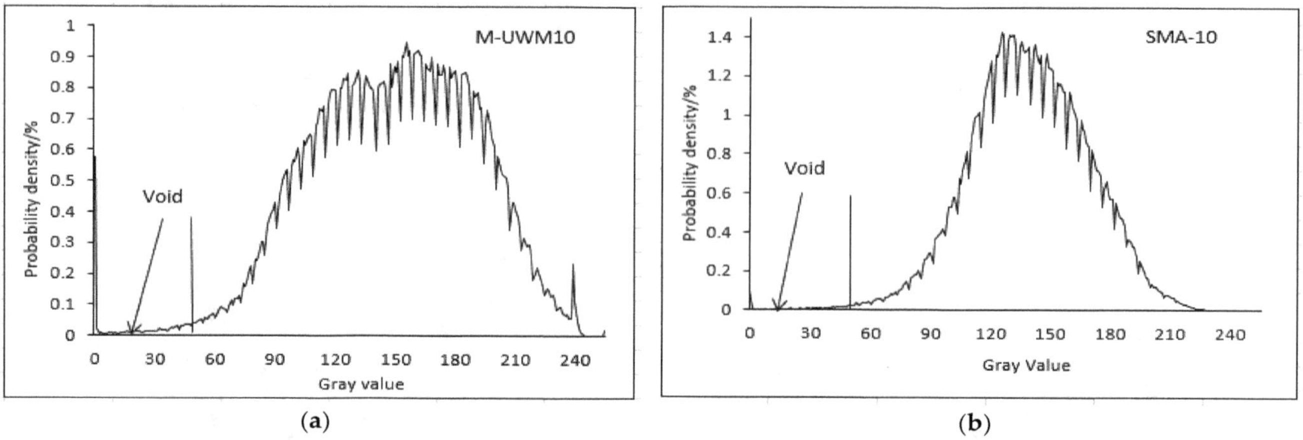

Figure 11. Grey level- probability density distribution curves: (**a**) M-UWM10; (**b**) SMA10.

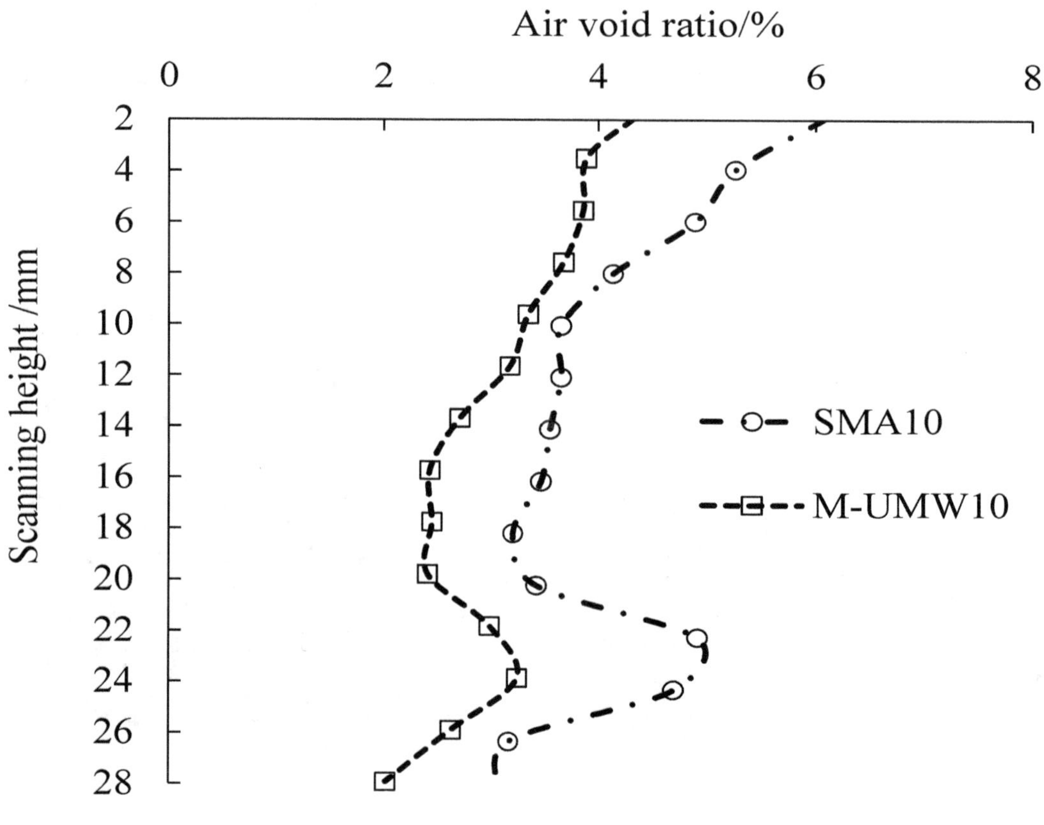

Figure 12. Distribution of air void ratios along the scanning height of the cored samples.

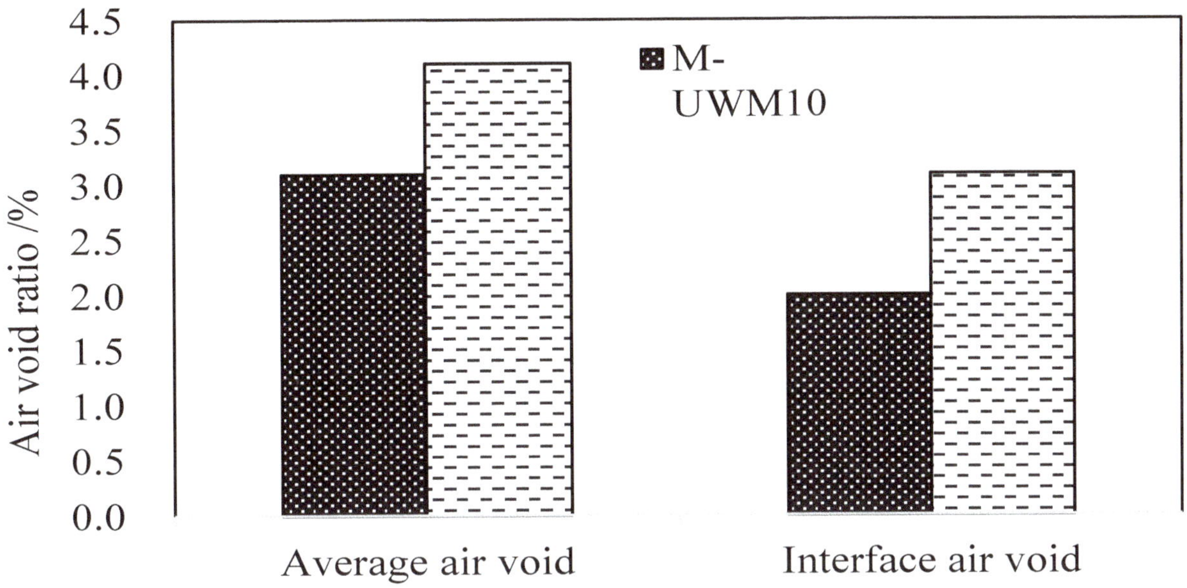

Figure 13. Average air void ratios and interface air void ratios for different cored asphalt mixtures.

From Figures 12 and 13, it can be seen that, no matter for the air void at the same height or for the average air void, the air void ratios of the M-UWM10 are lower than that of the SMA10. It could be attributed to better construction workability of M-UWM10 than SMA10. Meanwhile, it can be seen from Figure 13 that the interface air void ratio of M-UWM10 pavement is lower than that of the SMA10 pavement. Both the smaller average air void ratio and interface air void ratio are helpful to improve the bonding effect between the wearing course and the sub-layer. It is well confirmed by the test results from the field pull out test. Figure 14 shows the pull-out samples from field pavements for M-UWM10 and SMA10. The pull-out force and pull-out stress for the two wearing course mixtures are shown in Figure 15. It can be seen, that the bonding strength between M-UWM10 and its sublayer is much better than that between SMA10 and its sublayer. Thus, the wearing course consisted of M-UWM10 shows better structural integrity with the underneath pavement surface than the wearing course consisted of SMA10.

(**a**) (**b**)

Figure 14. Field pull out test samples: (**a**) M-UWM10; (**b**) SMA10.

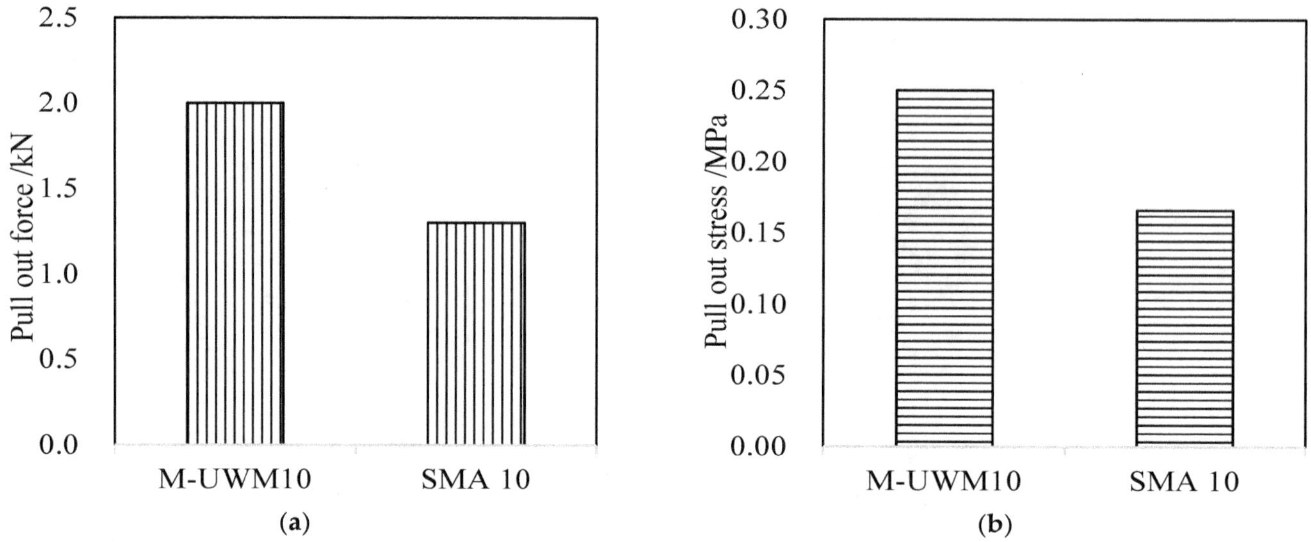

Figure 15. Pull out test results: (**a**) pull out force; (**b**) pull out stress.

4. Conclusions

Based on the laboratory and field tests, the following conclusions can be drawn:

1. A dense-graded ultra-thin wearing course mixture with multi-chain polyolefin modifier and SBS modified asphalt was prepared in this paper. It is proved that the designed asphalt mixture has satisfied high-temperature rutting resistance, low-temperature cracking resistance and moisture resistance to bear the traffic loading and environment effects.
2. The test results from combined tests of the wheel tracking test and the texture depth test proved that the designed wearing course mixture shows satisfied skidding resistance and wearing resistance. Field texture depth test results confirmed that the designed wearing course mixture is promising to keep long-term skidding resistance during traffic loading.
3. Laboratory and field tests based on the test road indicated that the wearing course paved with the designed ultra-thin wearing course mixture can provide satisfied water-proof and interlayer bonding effects, which are helpful to improve the pavement durability.
4. Future work will focus on the fatigue resistance of the ultra-thin wearing course mixture.

Acknowledgments: This work is financially supported by the Natural Science Foundation of China (No. 51378006 and No. 51378123), the Natural Science Foundation of Jiangsu (BK20161421), the Fundamental Research Funds for the Central Universities (No. 2242015R30027), and the Open Fund of the State Engineering Laboratory of Highway Maintenance Technology (Changsha University of Science and Technology, kfj160104). The author would like to thank his collaborators and support from the Jiangsu Sinoroad Engineering Research Institute Co., Ltd and Southeast University.

Author Contributions: Tao Ma conceived and designed the experiments; Pengsen Hu performed the experiments; Junhui Zhang analyzed the data; Xiaoming Huang contributed materials and apparatus; Lei Geng wrote the paper.

References

1. Luo, Y.L. Brief introduction to application of maintenance technology on ultra-thin wearing course for asphalt pavement of expressway. *Guangdong Highway Commun.* **2011**. [CrossRef]
2. Pei, P.; Peng, G.J. Application of ultra-thin wear layer technology in expressway maintenance. *W. China Commun. Sci. Technol.* **2015**, *12*, 23–26. [CrossRef]
3. Qiao, X.L. Application of novachip ultra-thin wearing layer in highway maintenance. *Transport. Standard.* **2013**, *18*. Available online: http://xueshu.baidu.com/s?wd=paperuri%3A%2832b4e61b2b25b0508cfc06e363ef4551%29&filter=sc_long_sign&tn=SE_xueshusource_2kduw22v&sc_vurl=http%3A%2F%2Fen.cnki.com.cn%2FArticle_en%2FCJFDTOTAL-JTBH201318005.htm&ie=utf-8&sc_us=14205075173698836325 (accessed on 6 August 2017).

4. Yang, Y.H.; Shen, Y.; Gao, X.X. Analysis interlayer shear stress of ultra-thin wearing course considering temperature and different interlayer contact conditions. *Appl. Mech. Mater.* **2014**, *505–506*, 102–105. [CrossRef]
5. Tan, Y.Q.; Yao, L.; Wang, H.P.; Bian, X.; Yu-Xiang, Q.I. Performance evaluation indicator of ultra-thin wearing course asphalt mixture. *J. Harbin Inst. Tech.* **2012**, *44*, 73–77.
6. Zhi-Tao, H.U.; Niu, X.X. Application of novachip ultra-thin wearing layer of asphalt pavement on the preventive maintenance engineering of expressways. *Guangdong Highway Commun.* **2009**, *3*. Available online: http://xueshu.baidu.com/s?wd=paperuri%3A%2817fe1a1403e497837a8ed55c1959682d%29&filter=sc_long_sign&tn=SE_xueshusource_2kduw22v&sc_vurl=http%3A%2F%2Fen.cnki.com.cn%2FArticle_en%2FCJFDTOTAL-GDGT200903003.htm&ie=utf-8&sc_us=6153204051113523490 (accessed on 6 August 2017).
7. Chan, S.; Lane, B.; Kazmierowski, T. Pavement preservation—a solution for sustainability. *Transport. Res. Rec.* **2011**, *47*, 36–42. [CrossRef]
8. Kandhal, P.S.; Lockett, L. *Construction and performance of ultrathin asphalt friction course*; NCAT Report No. 97-5; National Center for Asphalt Technology of Auburn University: Auburn, AL, USA, September 1997.
9. Pretorius, F.J.; Wise, J.C.; Henderson, M. Development of application differentiated ultra-thin asphalt friction courses for southern african application. In Proceedings of the 8th Conference on Asphalt Pavements for Southern Africa, Sun City, South Africa, 12–16 September 2004.
10. Zeng, M.L.; Peng, L.Q.; Chao-Fan, W.U.; Tan, B.Y. Experimental study of the performance of ultrathin asphalt friction course. *J. Wuhan Univ. Technol.* **2012**, *4*. U414.
11. Yang, X.; Ling, J.; Chaofan, W.U.; Zeng, Z. Evaluation of performance of novachip ultrathin asphalt friction course. *Highway Eng.* **2013**, *1*. Available online: http://en.cnki.com.cn/Article_en/CJFDTOTAL-ZNGL201301004.htm (accessed on 6 August 2017).
12. Ministry of Transport of the People's Republic of China. *Specifications for design of highway asphalt*; China Communications Press: Beijing, China, 2006.
13. Xiao, F.; Amirkhanian, S.N. Effects of liquid antistrip additives on rheology and moisture susceptibility of water bearing warm mixtures. *Constr. Build. Mater.* **2010**, *24*, 1649–1655. [CrossRef]
14. Wu, S. Assessment of bonding behaviours between ultrathin surface layer and asphalt mixture layer using modified pull test. *J. Adhes. Sci. Technol.* **2015**, *29*, 1508–1521.
15. Rahaman, F.; Musty, H.Y.; Hossain, M. Evaluation of recycled asphalt pavement materials from ultra-thin bonded bituminous surface. In Proceedings of the GeoCongress 2012, Oakland, CA, USA, 25–29 March 2012. [CrossRef]
16. Yang, Y.H.; Liu, Z.; Gao, J.X.; Zhu, G.S. Bonding characteristics analysis of ultra-thin overlay of asphalt pavement layer. *Appl. Mech. Mater.* **2012**, *178–181*, 1245–1249. [CrossRef]
17. Weidong, C.; Jianrong, S.; Hengchun, H. Introduction of Technology of Ultra-thin Asphalt Friction Course. *Petrol. Asphalt* **2005**, *19*, 56–59.
18. Pasetto, M.; Baldo, N. Influence of the aggregate skeleton design method on the permanent deformation resistance of stone mastic asphalt. *Mater. Res. Innovations* **2014**, *18*, S3-96–S3-101. [CrossRef]
19. Ministry of Transport of the People's Republic of China. *Standard test methods of bitumen and bituminous mixtures for highway engineering*; China Communications Press: Beijing, China, 2011.
20. Ma, T.; Geng, L.; Ding, X.; Zhang, D.; Huang, X. Experimental study of deicing asphalt mixture with anti-icing additives. *Constr. Build. Mater.* **2016**, *127*, 653–662. [CrossRef]
21. Ministry of Transport of the People's Republic of China. *Standard specification for construction of highway asphalt pavements*; China Communications Press: Beijing, China, 2004.
22. Ma, T.; Wang, H.; He, L.; Zhao, Y.; Huang, X.; Chen, J. Property characterization of asphalt binders and mixtures modified by different crumb rubbers. *J. Mater. Civil Eng.* **2017**, *29*. [CrossRef]

Permissions

All chapters in this book were first published by MDPI; hereby published with permission under the Creative Commons Attribution License or equivalent. Every chapter published in this book has been scrutinized by our experts. Their significance has been extensively debated. The topics covered herein carry significant findings which will fuel the growth of the discipline. They may even be implemented as practical applications or may be referred to as a beginning point for another development.

The contributors of this book come from diverse backgrounds, making this book a truly international effort. This book will bring forth new frontiers with its revolutionizing research information and detailed analysis of the nascent developments around the world.

We would like to thank all the contributing authors for lending their expertise to make the book truly unique. They have played a crucial role in the development of this book. Without their invaluable contributions this book wouldn't have been possible. They have made vital efforts to compile up to date information on the varied aspects of this subject to make this book a valuable addition to the collection of many professionals and students.

This book was conceptualized with the vision of imparting up-to-date information and advanced data in this field. To ensure the same, a matchless editorial board was set up. Every individual on the board went through rigorous rounds of assessment to prove their worth. After which they invested a large part of their time researching and compiling the most relevant data for our readers.

The editorial board has been involved in producing this book since its inception. They have spent rigorous hours researching and exploring the diverse topics which have resulted in the successful publishing of this book. They have passed on their knowledge of decades through this book. To expedite this challenging task, the publisher supported the team at every step. A small team of assistant editors was also appointed to further simplify the editing procedure and attain best results for the readers.

Apart from the editorial board, the designing team has also invested a significant amount of their time in understanding the subject and creating the most relevant covers. They scrutinized every image to scout for the most suitable representation of the subject and create an appropriate cover for the book.

The publishing team has been an ardent support to the editorial, designing and production team. Their endless efforts to recruit the best for this project, has resulted in the accomplishment of this book. They are a veteran in the field of academics and their pool of knowledge is as vast as their experience in printing. Their expertise and guidance has proved useful at every step. Their uncompromising quality standards have made this book an exceptional effort. Their encouragement from time to time has been an inspiration for everyone.

The publisher and the editorial board hope that this book will prove to be a valuable piece of knowledge for researchers, students, practitioners and scholars across the globe.

List of Contributors

Jaroslav Pokorný and Jiří Šál
Department of Civil Engineering, Faculty of Technology, Institute of Technology and Business, Okružní 517/10, 370 01 České Budějovice, Czech Republic

Radek Ševčík
Department of Civil Engineering, Faculty of Technology, Institute of Technology and Business, Okružní 517/10, 370 01 České Budějovice, Czech Republic
Institute of Theoretical and Applied Mechanics of the Czech Academy of Sciences, Prosecká 809/76, 190 00 Praha, Czech Republic

Katarzyna Konieczna, Piotr Pokorski, Piotr Radziszewski and Jan Bolesław Król
Faculty of Civil Engineering, Warsaw University of Technology, Al. Armii Ludowej 16, 00-637 Warsaw, Poland

Wojciech Sorociak
Faculty of Civil Engineering, Silesian University of Technology, ul. Akademicka 5, 44-100 Gliwice, Poland
Eurovia Polska S.A., Bielany Wrocławskie, ul. Szwedzka 5, 55-040 Kobierzyce, Poland

Dawid Żymełka
Eurovia Polska S.A., Bielany Wrocławskie, ul. Szwedzka 5, 55-040 Kobierzyce, Poland

Farzaneh Tahmoorian and Bijan Samali
Centre for Infrastructure Engineering, Western Sydney University, Kingswood, NSW 2751, Australia

Vivian W.Y. Tam
School of Computing, Engineering, and Mathematics, Western Sydney University, Kingswood, NSW 2751, Australia
College of Civil Engineering, Shenzhen University, Shenzhen 518060, China

John Yeaman
Faculty of Science, Health, Education and Engineering, University of Sunshine Coast, Sippy Downs, Queensland 4556, Australia

Pengfei Liu and Markus Oeser
Institute of Highway Engineering, RWTH Aachen University, Mies-van-der-Rohe-Street 1, D52074 Aachen, Germany

Qinyan Xing, Yiyi Dong and Si Yuan
Department of Civil Engineering, Tsinghua University, 100084 Beijing, China

Dawei Wang
Institute of Highway Engineering, RWTH Aachen University, Mies-van-der-Rohe-Street 1, D52074 Aachen, Germany
Institute of Highway Engineering, Paul-Bonatz-Street 9–11, University of Siegen, D57076 Siegen, Germany

Shuaicheng Guo, Qingli Dai, Xiao Sun and Zhen Liu
Department of Civil and Environmental Engineering, Michigan Technological University, Houghton, MI 49931-1295, USA

Ye Sun
Department of Mechanical Engineering and Engineering Mechanics, Michigan Technological University, 1400 Townsend Drive, Houghton, MI 49931-1295, USA

Henrikas Sivilevičius, Justas Bražiūnas and Olegas Prentkovskis
Department of Mobile Machinery and Railway Transport, Faculty of Transport Engineering, Vilnius Gediminas Technical University, Plytinės g. 27, LT-10105 Vilnius, Lithuania

Cheng Li and Jeramy Ashlock
Department of Civil Construction and Environmental Engineering, Iowa State University, Ames, IA 50011, USA

David White and Pavana Vennapusa
Ingios Geotechnics, Inc., Northfield, MN 55057, USA

Yanping Sheng, Huaxin Chen and Rui Xiong
School of Materials Science and Engineering, Chang'an University, Xi'an 710064, China

Haibin Li and Guijuan Zhao
School of Architecture and Civil Engineering, Xi'an University of Science and Technology, Xi'an 710054, China

Ping Guo
Xi'an Highway Research Institute, Xi'an 710054, Shaanxi, China

Aboelkasim Diab
Department of Civil Engineering, Aswan University, Aswan 81542, Egypt

Zhanping You
Department of Civil and Environmental Engineering, Michigan Technological University, Houghton, MI 49931, USA

Xu Yang
Department of Civil Engineering, Monash University, Clayton, VIC 3800, Australia

Mohd Rosli Mohd Hasan
School of Civil Engineering, Universiti Sains Malaysia, Engineering Campus, Nibong Tebal Penang 14300, Malaysia

Yue Xiao, Chao Li, Miao Wan and Shaopeng Wu
State Key Laboratory of Silicate Materials for Architectures, Wuhan University of Technology, Wuhan 430070, China

Xinxing Zhou
Key Laboratory of Highway Construction and Maintenance in Loess Region, Shanxi Transportation Research Institute, Taiyuan 030006, China

Yefei Wang
Jinan Urban Construction Group Co., Ltd., Jinan 250031, China

Liang Li and Xiuli Du
The Key Laboratory of Urban Security and Disaster Engineering, Ministry of Education, Beijing University of Technology, Beijing 100124, China

Jun Wu
The Key Laboratory of Urban Security and Disaster Engineering, Ministry of Education, Beijing University of Technology, Beijing 100124, China
School of Urban Railway Transportation, Shanghai University of Engineering Science, Shanghai 201620, China

Xuemei Liu
School of Civil Engineering and Built Environment, Queensland University of Technology, Brisbane 4001, Australia

Federico Leonelli, Paola Di Mascio and Laura Moretti
Department of Civil, Constructional and Environmental Engineering, Sapienza University of Rome, Via Eudossiana 18, 00184 Rome, Italy

Antonello Germinario
Laboratorio Principale Prove e Sperimentazioni (ITAF Infrastructure Department), Viale di Marino snc, 00043 Ciampino, Italy

Francesco Picarella and Mauro Cassata
2 Reparto Genio A.M. (ITAF Infrastructure Department), Viale di Marino snc, 00043 Ciampino, Italy

Alberto De Rubeis
Servizio Infrastrutture A.M. (Chief of ITAF Infrastructure Department), Viale dell'Università 4, 00185 Rome, Italy

Jie Wang, Wei Zeng, Yongchun Qin, Songchang Huang and Jian Xu
Research Institute of Highway Ministry of Transport, Beijing 100088, China

Mulian Zheng and Peng Li
Key Laboratory for Special Area Highway Engineering of Ministry of Education, Chang'an University, South Erhuan Middle Section, Xi'an 710064, Shaanxi, China

Jiangang Yang
Civil Engineering Materials Laboratory, East China Jiaotong University, Shuang Gang East Street, Nanchang 330000, Jiangxi, China

Hongyin Li
Shandong Highway Administration Bureau, Shungeng Road of Jinan City, Jinan 250000, Shandong, China

Yangyang Qiu
Jinhua Traffic Planning and Design Institute, Songlian Road of Jin Dong District, Jinhua 321015, Zhejiang, China

Zhengliang Zhang
Anhui Highway Administration Center, Da Bie Shan Road of Hefei City, Hefei 230088, Anhui, China

Lei Geng
Jiangsu Sinoroad Engineering Research Institute Co. LTD., 19 Lanhua Road, Nanjing 211800, China

Tao Ma, Xiaoming Huang and Pengsen Hu
School of Transportation, Southeast University, 2 Sipailou, Nanjing 210096, China

Junhui Zhang
State Engineering Laboratory of Highway Maintenance Technology, Changsha University of Science and Technology, Changsha 410114, China

Index

www.ingramcontent.com/pod-product-compliance
Lightning Source LLC
LaVergne TN
LVHW080952210926
848485LV00004B/822

* 9 7 8 1 6 4 7 2 8 5 0 8 1 *